Mitochondrial Disorders in Neurology

Mitochondrial Disorders in Neurology

Edited by

Anthony H.V. Schapira
Department of Neurosciences
Royal Free Hospital School of Medicine
London; and Institute of Neurology
Queen Square, London, UK

and

S. DiMauro
Department of Neurology
College of Physicians and Surgeons of Columbia
New York, USA

Butterworth–Heinemann Ltd
Linacre House, Jordan Hill, Oxford OX2 8DP

℞ A member of the Reed Elsevier group

OXFORD   LONDON   BOSTON
MUNICH   NEW DELHI   SINGAPORE   SYDNEY
TOKYO   TORONTO   WELLINGTON

First published 1994

British Library Cataloguing in Publication Data
Mitochondrial Disorders in Neurology.—
  (Butterworth International Medical
  Reviews: Neurology; Vol. 14)
  I. Schapira, A. H. V. II. DiMauro, S.
  III. Series
  616.8

ISBN 0 7506 0585 5

Library of Congress Cataloguing in Publication Data
Mitochondrial disorders in neurology/edited by A.H.V. Schapira and S. DiMauro
  p.   cm. — (Butterworth-Heinemann international medical reviews. Neurology : 14)
  Includes index.
  ISBN 0 7506 0585 5
  1. Mitochondrial pathology. 2. Nervous system—Diseases. 3. Neuromuscular
  diseases. I. Schapira, A. H. V. (Anthony Henry Vernon) II. DiMauro, S. III. Series.
  [DNLM: 1. Mitochondria—pathology. 2. Nervous System Diseases—pathology. 3.
  Nervous System Diseases—complications. WL 100 M6835 1994]
  RC347.M53   1994
  DNLM/DC
  for Library of Congress   93–41988
                                                                                 CIP

Typeset by Scribe Design, Gillingham, Kent
Printed in Great Britain at the University Press, Cambridge

Contents

Contributors

E. Bonilla MD
Professor of Clinical Neurology, Columbia-Presbyterian Medical Center, New York, USA

J.B. Clark PhD, DSc
Miriam Marks Department of Neurochemistry, Institute of Neurology, London, UK

J.M. Cooper PhD
Department of Neurosciences, Royal Free Hospital School of Medicine, London, UK

D.C. De Vivo MD
Departments of Neurology and Pediatrics, College of Physicians and Surgeons, Columbia University, New York, USA

S. DiDonato MD
Division of Biochemistry and Genetics, Istituto Nazionale Neurologico C. Besta, Milan, Italy

S. DiMauro MD
Lucy G. Moses Professor of Neurology, Columbia-Presbyterian Medical Center, New York, USA

S.R. Hammans MA, MRCP
University Department of Clinical Neurology, Institute of Neurology, London, UK

A.E. Harding MD, FRCP
University Department of Clinical Neurology, Institute of Neurology, London, UK

viii   *Contributors*

M. Hirano MD
Assistant Professor of Neurology, Columbia-Presbyterian Medical Center, New York, USA

C.T. Moraes PhD
Assistant Professor of Neurology, University of Miami School of Medicine, Miami, USA

J.A. Morgan-Hughes MD, FRCP
The National Hospital, London, UK

B.H. Robinson PhD
Departments of Paediatrics and Biochemistry, University of Toronto; and the Research Institute, The Hospital for Sick Children, Toronto, Canada

L.P. Rowland MD
Neurological Institute, H. Houston Merritt Clinical Research Center, Columbia-Presbyterian Medical Center, New York, USA

A.H.V. Schapira BSc, MD, FRCP
Department of Neurosciences, Royal Free Hospital School of Medicine, London; and Institute of Neurology, Queen Square, London, UK

J.M. Shoffner MD
Departments of Genetics and Molecular Medicine, and Neurology, Emory University School of Medicine, Atlanta, USA

E.A. Schon PhD
Associate Professor of Genetics and Development (in Neurology), Columbia-Presbyterian Medical Center, New York, USA

M.G. Sweeney BSc
University Department of Clinical Neurology, Institute of Neurology, London, UK

D.C. Wallace PhD
Departments of Genetics and Molecular Medicine, and Neurology, Emory University School of Medicine, Atlanta, USA

Preface

There can be few areas of medicine where the practice of phenomenology and biochemistry have had such rapid and complementary growth as in mitochondrial diseases. The description by Luft and colleagues in 1962 of a 35 year old woman with euthyroid hypermetabolism was the first report of a disease primarily associated with defective mitochondrial function. The next 20 years saw a number of papers documenting abnormal muscle and mitochondrial morphology in a variety of clinical presentations from isolated ophthalmoplegia to severe encephalopathy. During this period, analysis of mitochondrial function revealed specific metabolic defects including, most importantly, abnormal oxidative phosphorylation. Such a variety of clinical phenotypes and biochemical defects presented a suitable challenge to those who sought to classify these disorders into distinct groups.

The mid-1980s saw the first attempts to define the molecular basis of the biochemical deficiencies exhibited by these patients. In 1988, Holt and colleagues reported deletions of mitochondrial DNA (mtDNA) in patients with mitochondrial myopathies. This landmark paper was the starting point for a torrent of reports describing mtDNA mutations of various types in several of the established mitochondrial myopathy phenotypes as well as in disorders such as Pearson's syndrome, Leber's hereditary optic neuropathy and diabetes mellitus. The early euphoria of apparent molecular and clinical linkage has given way to a healthy scepticism of the precise role mtDNA mutations play in the pathogenesis of these diseases.

The reports of mitochondrial respiratory chain defects in neurodegenerative diseases as well as senescent tissue have opened up new and challenging vistas for mitochondrial researchers. The eagerness with which these challenges have been undertaken must be tempered by the lessons only recently learned from our research into the mitochondrial myopathies.

In this book we have sought to provide an overview of the mitochondrial research field as it existed at the time of writing. Tremendous advances have been made in the last six years and we believe that this book is published at a time when we can sensibly reflect on these advances and their implications for our understanding of mitochondrial diseases. An understanding of the effects of mitochondrial dysfunction must be based on a sound knowledge of the relevant biochemistry. Thus, several chapters have integrated sections on the basic biochemistry and molecular genetics relevant to their respective subjects. The clinical presentation of mitochondrial disorders is detailed, with that of the respiratory chain defects given most prominence. Chapters on Leber's hereditary optic neuropathy and neurodegenerative disorders complete the spectrum of this work.

As editors we have not sought to provide a 'party line'. Those with a special interest in mitochondrial diseases will know that this would be impossible (bearing in mind the characters that would have had to tow the line!) as well as being premature and misleading. Authors have therefore been free to expound their own hypotheses and the book accommodates a range of controversies from across the mitochondrial divide. In this respect, Professor Rowland's chapter on lumping, splitting and melding has sculpted a fine reflection of some of the more important questions relating to this area.

Finally, we should like to thank the authors for their hard work and dedication in helping to produce this book.

Anthony H.V. Schapira
Salvatore DiMauro

1
The structural organization of the mitochondrial respiratory chain

J.M. Cooper and J.B. Clark

INTRODUCTION

The mitochondrion is an intracellular organelle found in virtually all eukaryotic cells, where it plays a major role in cellular ATP production. It consists of four compartments (Figure 1.1); the inner and outer membranes and two soluble fractions, the matrix and the intermembrane space. The inner membrane and matrix are associated with most of the functional activities of the mitochondria, including those involved with the tricarboxylic acid (TCA) cycle, fatty acid oxidation and ATP generation. The inner membrane is folded forming cristae giving it a much larger surface area than the outer membrane.

The main function of the mitochondrion is ATP synthesis. The enzymes for the TCA cycle and β-oxidation of fatty acids are situated in the mitochondrial matrix where dicarboxylic, tricarboxylic and fatty acids are oxidized generating NADH and $FADH_2$. The overall function of the respiratory chain is the oxidation of NADH and $FADH_2$ and the transport of reducing equivalents along a series of carriers to the terminal electron acceptor oxygen. The consequent decrease in electropotential is conserved in the vectorial movement of protons (H^+) across the inner membrane from the matrix, generating a proton motive force (PMF). The PMF is used for a variety of functions including the transport of charged molecules (proteins, carboxylic acids and ions) and the generation of ATP by ATP synthase. The importance of mitochondrial ATP generation is apparent from the more efficient ATP generation from oxidative phosphorylation (36 mol ATP/mol glucose) as compared with anaerobic glycolysis (2 mol ATP/mol glucose) and the fact that all the ATP generation from fatty acids is by the mitochondrial respiratory chain and oxidative phosphorylating system.

The respiratory chain comprises four multipolypeptide enzyme complexes: Complex I (NADH–ubiquinone reductase); Complex II (succinate–ubiquinone reductase); Complex III (ubiquinol–cytochrome *c* reductase); and Complex IV (cytochrome *c* oxidase); and two mobile electron carriers: ubiquinone and cytochrome *c*. Together with a fifth complex (ATP synthase) they comprise the oxidative phosphorylation system (Figure 1.2). All five complexes are situated in the inner mitochondrial membrane, accessible to their substrates from either within the membrane or the matrix side of the membrane. The pathway of electron

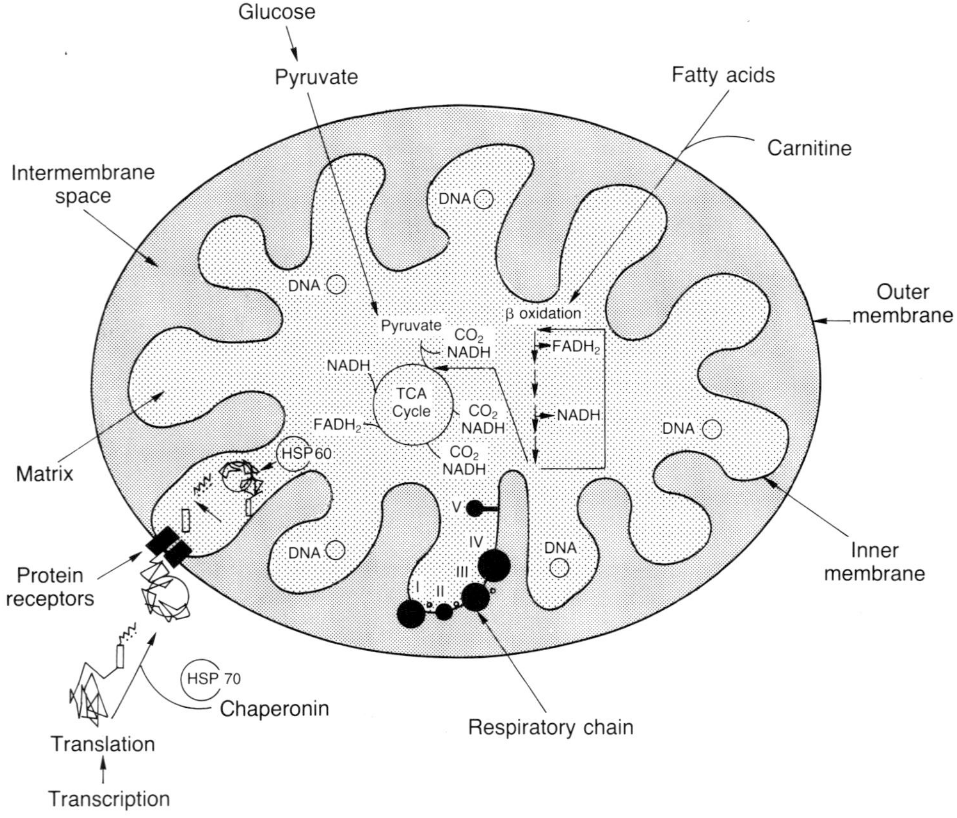

Figure 1.1 The mitochondrion and its associated metabolism

transport and the mechanism and stoichiometry of proton translocation are not completely understood and remain a matter of much debate.

The majority of mitochondrial proteins are encoded by the nucleus, synthesized on cytoribosomes and imported into the mitochondrion. However, mitochondria contain their own DNA (mtDNA) which is circular and double stranded. In mammals it codes for 13 proteins, seven constituting part of Complex I (NADH dehydrogenase ND1, 2, 3, 4, 4L, 5, 6), one in Complex III (cytochrome *b*), three in Complex IV (COI, II, III) and two in Complex V (A6, A6L). It also codes for most of the machinery needed for its translation (see Chapter 2).

COMPLEX I

The mammalian NADH dehydrogenase (Complex 1) is the largest of the respiratory complexes, but probably the least well understood. It is characterized by the catalysis of the rotenone-sensitive oxidation of NADH and the reduction of ubiquinone. In addition to NADH, NADPH can be utilized as a substrate, albeit

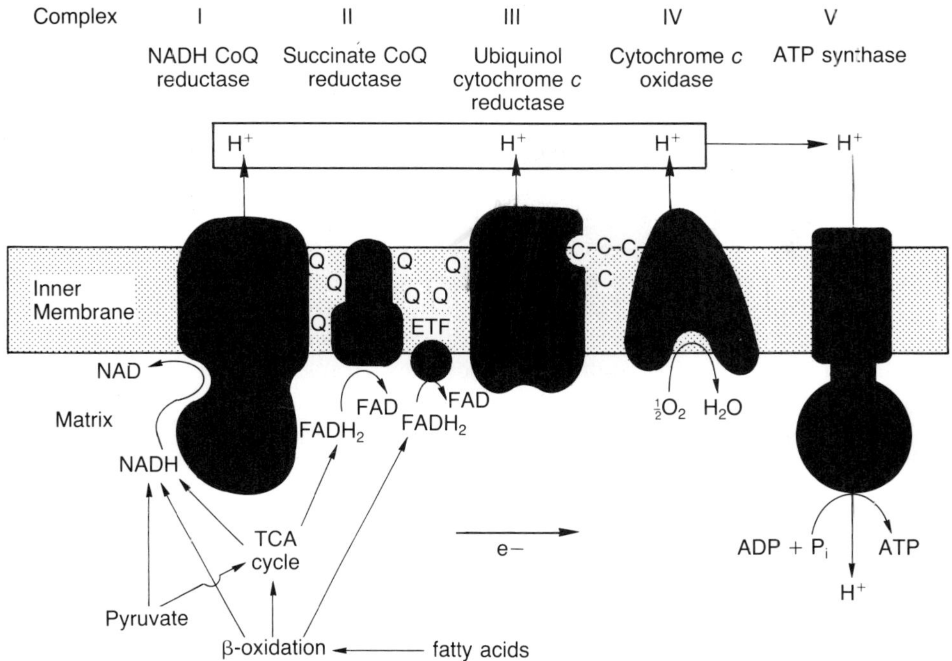

Figure 1.2 The mitochondrial respiratory chain. Q, ubiquinone; C, cytochrome *c*; ETF, electron-transferring flavoprotein

at a much lower rate. It has been proposed that the electron pathway from NADPH is different, not involving all the iron-sulphur centres [1]. Two transhydrogenase activities have also been associated with Complex I preparations, one NADPH–NAD$^+$ (T-D type) and one NADH–NAD$^+$ (D-D type) transhydrogenase. The T-D type is associated with a 110 kDa (or 130 kDa) contaminant of Complex I, while the D-D type has been associated with the 42 kDa subunit [2], although sequence analysis of this subunit has failed to identify a typical NAD$^+$-binding domain [3].

Bovine Complex I contains a number of redox centres including flavin mononucleotide (FMN) and 22–24 atoms of iron and acid-labile sulphur in the form of eight to nine iron-sulphur clusters (five or six binuclear, 2Fe-2S, and three tetranuclear, 4Fe-4S). The precise number and nomenclature of the iron-sulphur clusters in Complex I is the subject of much debate and the confusion is compounded by many of the centres being invisible to electron paramagnetic resonance (EPR) spectroscopy. This area has been reviewed extensively with some attempts to clarify the situation [4]. Electron transport in Complex I proceeds from a two-electron carrier (NADH) to one-electron carriers (iron-sulphur clusters and ubiquinone). A good candidate for this conversion and the primary oxidant of NADH is FMN. Various pathways of electron transport have been proposed for Complex I but with little evidence to support any particular one [5–7]. Conservation of oxidative energy is achieved with outward proton translocation but the exact stoichiometry and mechanism is disputed, varying from $3H^+/2e^-$ to $5H^+/2e^-$ [8,9]. The pathways of electron transport and proton translocation in Complex I are poorly understood and have been reviewed previously [10,11].

Fractionation of the enzyme

Mammalian Complex I has only been satisfactorily purified using the method for the bovine heart enzyme described by Hatefi and co-workers [12]. Other attempts to purify the complex have resulted in either no activity or no rotenone sensitivity [13–15]. This has created a major problem in determining which subunits of Complex I are integral components and which merely copurify with the complex, resulting in the number of subunits in Complex I ranging from 26 to 41, with a total molecular mass of approximately 800 kDa.

Identification of components of Complex I has been aided by its subfractionation using chaotropic agents into three fractions: two soluble fractions, the flavoprotein (FP) fraction and the iron-protein (IP) fraction, and one insoluble hydrophobic (HP) fraction. The FP fraction (comprising three subunits: 51, 24 and 10 kDa) and the IP fraction (comprising six subunits: 75, 49, 30, 18, 15 and 13 kDa) have received the most attention for two main reasons: first, they contain fairly well-defined components, and secondly their constituents are more antigenic and can be readily detected with antibody probes. The HP fraction is ill-defined, containing the remainder of the complex (>15 subunits) and possibly containing contaminants. Known contaminants of Complex I have included the transhydrogenase (110 kDa), Complex III subunits and ATPase α and β subunits. The ability of antisera that cross-react with only four or five Complex I subunits to immunoprecipitate virtually all the detectable subunits of Complex I [16,17] suggests that they are all closely associated with each other. However, the 42 kDa subunit from the HP fraction may only be loosely associated with the complex as it failed to immunoprecipitate with the complex [17]. The cDNAs for many of the subunits in Complex I have been sequenced and the position of cysteine residues and hydrophobic regions in the proteins have helped to identify iron-sulphur proteins and assess the functions of others (Table 1.1).

The flavoprotein fraction

This fraction catalyses the oxidation of NADH by a variety of artificial electron acceptors (e.g. ferricyanide). It is associated with the flavoprotein (FMN), and EPR studies have suggested the presence of a binuclear (N-1b) and a tetranuclear (N-3) centre, which are believed to reside on the 24 and 51 kDa subunits respectively [18]. Photoactivatable derivatives of NAD^+ label four polypeptides of molecular mass 130, 75, 51 and 42 kDa [19]. The 130 kDa polypeptide is the contaminating NADPH-NAD (T-D) transhydrogenase (often running at 110 kDa), and the 42 kDa subunit has been associated with a D-D transhydrogenase [2]. The 51 kDa subunit is the largest subunit in the FP fraction and, because this fraction exhibits the NADH ferricyanide reductase activity, it suggests that the 51 kDa subunit is involved with the NAD^+ binding for NADH ubiquinone reductase activity. The 75 kDa subunit is the largest subunit in the IP fraction, and may also be involved in the NAD^+-binding site.

Table 1.1 Summary of bovine Complex I subunits

Molecular mass (kDa) (from gels)	Fraction	Nuclear mtDNA	EPR data	Molecular mass (Da) (from sequence)	No. of Cys residues	Known Fe-S sequence motifs	Other features/homology	N-terminal sequence
75	IP	N	$2 \times$ 2Fe-2S	75 961	17	4Fe-4S/2Fe-2S	NAD binding, HOX U	TATAAS
51	FP	N	4Fe-4S (N3)	48 416	12	4Fe-4S	FMN/NAD binding, HOX F	SGDTTA
51	HP	M (ND5)		68 323	4	?		MNMFSS
49	IP	N	4Fe-4S/2Fe-2S?	49 175	5	No	Chloroplast DNA (ndhH)	ARQWQP
42	HP	N		36 693	5	No	D-D transhydrogenase	LQYGPL
39	HP	N		39 115	1	No		LHHAVI
39	HP	M (ND4)		52 109	2	No		MLKYII
33	HP	M (ND2)		39 260	0	No		MNPIIF
30	IP	N	4Fe-4S/2Fe-2S?	26 432	2	No	Chloroplast DNA (ndhJ)	ESSAAD
30	HP	M (ND1)		35 675	0	No	DCCD/Rotenone binding	MFMINI
24	FP	N	2Fe-2S (N1b)	23 830	5	2Fe-2S?	HOX F	GAGGAL
23	HP	N		20 196	8	$2 \times$ 4Fe-4S (N2)	Chloroplast DNA (ndhI)	TYKYVN
19	HP	M (ND6)		19 080	1	No		MMLYIV
19	HP	N		19 960	8	No		PGIVEL
18	IP	N		–	–	–		
15	IP	N		–	–	–	Q binding	
13.6	HP	M (ND3)		13 056	1	No		MNLMLA
13	IP	N	2Fe-2S/4Fe-4S?	–	–	–		
10.5	HP	M (ND4L)		10 799	3	No		MSMVYM
10	FP	N		8 438	0	No		SAESGK

IP, iron protein; FP, flavoprotein; HP, hydrophobic fractions. N-Terminal sequences determined from cDNA sequences. HOX F, HOX U, genes coding for the NAD$^+$-reducing hydrogenase from *Alcaligenes eutrophus*.

The amino acid sequence for the 24 kDa subunit [20] and the cDNA sequences for the 51 kDa [21], 24 kDa [22] and 10 kDa [23] subunits have been published. Analysis of the deduced amino acid sequences revealed several stretches of similarity between the 51 and 24 kDa sequences and the α subunit of a soluble NAD$^+$-reducing hydrogenase from *Alcaligenes eutrophus* H16 [24]. This, in conjunction with the results of a comparison with other NAD- or FMN-binding proteins, suggests that the NAD$^+$-binding site within the 51 kDa subunit resides between residues 62 and 99, and the FMN-binding site to be between residues 180 and 234. The arrangement of cysteine residues in the 51 kDa subunit suggests that it contains a tetranuclear iron-sulphur centre [21]. Although the 24 kDa subunit contains five conserved cysteine residues, they are not arranged in any pattern indicative of a known iron-sulphur cluster, suggesting that the 24 kDa subunit does not contribute all the ligands for the EPR-predicted binuclear centre. It is possible that this subunit supplies some of the ligands for the binuclear centre and may share the centre with the 51 kDa subunits, as the 10 kDa subunit has no cysteine residues and therefore is unlikely to be an iron-sulphur protein.

The iron protein fraction

This fraction, similar to the FP fraction, is water-soluble and contains six major polypeptides of molecular mass 75, 49, 30, 18, 15 and 13 kDa; however it contains no enzyme activity. The 15 kDa subunit is loosely associated with the other five and may not be a constituent part of the subcomplex [25]; however, it has been reported to be involved with Q-binding [26]. EPR studies suggest the presence of four iron-sulphur clusters in the IP fraction. Two binuclear clusters (possibly including the N-1a centre) are associated with the 75 kDa subunit, while a binuclear cluster and a tetranuclear (possibly N-4) cluster are associated with the 49, 30 and 13 kDa subunits [18].

The sequences of the cDNAs for the 75 [27], 49 [28] and 30 kDa [29] subunits have been published. The deduced amino acid sequence for the 75 kDa subunit is compatible with it containing one or more iron-sulphur clusters. The cysteine residues in the 49 kDa and 30 kDa subunits are not grouped in a pattern of known iron-sulphur clusters. However, the ligands for the iron-sulphur cluster could be donated from several subunits, in agreement with the loss of the EPR signal upon separation of the subunits [18].

The hydrophobic protein fraction

There are at least 15 subunits in this fraction and EPR studies suggest the presence of a tetranuclear cluster (possibly N-2) and one or two binuclear clusters. The N-2 cluster has a high redox potential relative to the other iron-sulphur clusters; this in conjunction with its presence in a hydrophobic environment makes it the likely electron donor to ubiquinone. Three Complex I polypeptides encoded by the mitochondrial DNA, ND1, ND3 and ND4L, have been identified in the HP fraction [4]. Although the other four mitochondrially encoded ND subunits are also believed to reside in this fraction, positive localization has yet to be made.

In addition to the ND subunits, the cDNAs for the following subunits have been sequenced: 42 kDa, 39 kDa [3], 23 kDa [30] and 19 kDa [31] subunits. Neither the

42 nor the 39 kDa subunits have amino acid sequences indicative of iron-sulphur proteins. The 23 kDa and 19 kDa subunits possess a number of cysteine residues. In the 23 kDa subunit they are arranged in a manner suggestive of two tetra-nuclear iron-sulphur clusters, one of which could be the N-2 cluster; however, in the 19 kDa subunit they do not resemble the motifs of known iron-sulphur proteins.

Mitochondrial DNA

The seven mtDNA-encoded subunits of Complex I (ND1, 2, 3, 4, 4L, 5 and 6) are believed to be located in the HP fraction. The ND subunits have hydrophobic sequences compatible with their presence in the hydrophobic fraction and suggesting they are transmembranous. Of the mtDNA products, only ND5 possesses four conserved residues, making it a possible candidate for an iron-sulphur protein. The photoactivatable Complex I inhibitor dihydrorotenone [32] or an analogue [33] labelled the ND1 subunit (33 kDa) suggesting that ND1 is involved with ubiquinone reduction and possibly its binding. The sites of inhibition of other Complex I inhibitors have also been suggested to involve the ND subunits (see later). This involvement of ND1 in the binding site of various inhibitors suggests that it has an important functional role, although it does not contain conserved cysteine residues (Table 1.1) and is therefore unlikely to be an iron-sulphur protein. It may, however, have an important structural role, or be involved in protein translocation as suggested by its binding of *NN'*-dicyclohexylcarbodiimide (DCCD). Using [^{14}C]DCCD, Yagi [34] found two subunits, of molecular mass 49 and 29 kDa, to be involved in DCCD binding in Complex I. The labelling of the 29 kDa subunit followed the profile of Complex I inhibition and has therefore been suggested to be the subunit involved in proton translocation and possibly electron transfer. This subunit has been identified as ND1 [35].

Topology

The complex is believed to exist as a dimer, although the evidence for this form in the membrane is not conclusive [36]. How the three subfractions of Complex I (IP, FP, HP) relate to each other and whether they have any structural significance is uncertain. However, studies on mitochondria and inverted submitochondrial particles using cross-linking probes, hydrophilic and hydrophobic probes and antisera to specific subunits have shed some light on the membrane orientation and interrelationship with other constituent Complex I polypeptides.

The use of hydrophilic probes and specific antisera suggests that all three subunits in the FP fraction (51, 24, 10 kDa) are exposed on the matrix side of the membrane [10,37,38]. All six IP subunits were accessible from the matrix side of the membrane, with the 75, 49, 30 and 15 kDa subunits spanning the membrane [16,37,39]. Of the subunits in the HP fraction, only the 33 kDa subunit could be identified, and this was found to be transmembranous [37]. Only subunits from the HP fraction were labelled by hydrophobic probes [10], including the 42, 39 and 33 kDa subunits. The lack of any labelling of either IP or FP subunits, even the transmembranous ones, is consistent with the lack of hydrophobic stretches in their sequences and supports the hypothesis that the IP and FP domains are surrounded by the HP domain in the membrane [10,40] (Figure 1.3).

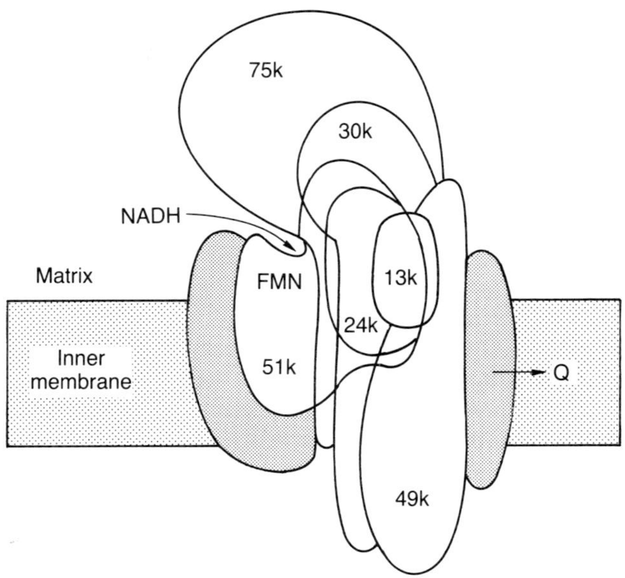

Figure 1.3 Schematic representation of bovine Complex I

The spatial arrangement of Complex I subunits was determined using a variety of cross-linking reagents with Complex I. The 51 kDa subunit was found to cross-link to the 75 kDa, 24 kDa and a number of smaller hydrophobic subunits. In addition, many of the IP subunits were cross-linked to each other and also to several smaller hydrophobic subunits (Figure 1.3) [41]. It is clear that the IP and FP subunits are located peripherally and predominantly on the matrix side of the membrane. However, the data suggesting that several subunits are transmembranous (75, 49, 30 and 15 kDa) are in contrast with the sequence data which suggest that the 75, 49 and 30 kDa subunits could not form membrane-spanning regions. This dilemma can be accommodated by the model depicted in Figure 1.3 [10,40] which suggests that the IP and FP subunits are surrounded and protected from the lipid bilayer by the hydrophobic (HP) subunits. It is difficult to accommodate the transmembranous orientation of the 75, 49 and 30 kDa subunits without such a model. To date the ND products, particularly the larger ones, are the only Complex I subunits sufficiently hydrophobic to interact significantly with the lipid bilayer. This may mean that either they are the only proteins in direct contact with the lipid environment, or other integral membrane proteins have yet to be sequenced.

Isoforms

There is some evidence from studies on immunoprecipitated rat mitochondria that the composition of Complex I differs between tissues. Complex I from liver and kidney appeared to lack an 18.5 kDa subunit present in heart, skeletal muscle and brain Complex I, but contain a 17 kDa subunit, lacking in heart, skeletal muscle and brain [42]. Analysis of the Complex I subunits that have been sequenced to date did not indicate the presence of multiple gene copies, suggesting that those subunits at least do not exist in tissue-specific forms.

Inhibitors of Complex I

Complex I activity is sensitive to a number of inhibitors, including rotenone [43], piericidin A [44], 1-methyl-4-phenylpyridinium (MPP$^+$) [45], diphenyleneiodonium (DPI$^+$) [46] and amytal [47]. The enzymic definition of Complex I is the rotenone-inhibitable NADH CoQ$_1$ reductase activity. Rotenone inhibits the reduction of ubiquinone but not ferricyanide, with its site of inhibition being between the highest potential iron-sulphur cluster and ubiquinone. A 33 kDa subunit was the main photolabelled product of Complex I using tritiated dihydrorotenone, with some labelling of a 51 kDa, 42 kDa and lower molecular mass subunits [32]. Under anaerobic conditions, the 51 kDa subunit was labelled more extensively, suggesting that it may play a role in rotenone inhibition. The 33 kDa subunit was positively identified as ND1, and it was assumed that the 51 kDa subunit was the large subunit in the FP fraction. However, it is equally likely to be ND5 which has been shown to co-migrate with this subunit under the conditions used [48]. The smaller subunits could be any, but the possibility that they were all ND subunits cannot be ruled out. Piericidin, like rotenone, inhibits Complex I between the high-potential iron-sulphur centre and ubiquinone, inhibiting the reduction of ubiquinone but not that of ferricyanide.

MPP$^+$ is a less potent inhibitor of Complex I, causing maximal inhibition at concentrations of 10 mM and its inhibition is readily reversed by simple dilution. The presence of MPP$^+$ impedes the binding of rotenone [49] and piericidin [50] to Complex I, suggesting that rotenone, piericidin and MPP$^+$ inhibit at either the same or similar sites in Complex I. Using a photoaffinity derivative of pethidine, an analogue of MPP$^+$, a number of subunits in Complex I were labelled which included ND1 and possibly other mtDNA products [51]. DPI$^+$ covalently labels a subunit in the HP fraction with a molecular mass of 23.5 kDa [52]. Inhibition by DPI$^+$ is enhanced by the presence of rotenone, suggesting that their sites of inhibition are different, but rotenone binding may induce a conformational change in Complex I, enhancing the binding of DPI$^+$. The barbiturate, amytal, has also been shown to inhibit Complex I activity. Although it is believed to inhibit at a similar site to rotenone [53], it is not as potent an inhibitor as rotenone [54].

COMPLEX II

Complex II (succinate dehydrogenase, succinate–ubiquinone reductase) is the smallest of the respiratory chain complexes, and is composed of only four subunits. The two larger subunits (70 kDa and 27.5 kDa) constitute the hydrophilic succinate dehydrogenase portion of the complex. The two smaller hydrophobic subunits (15.5 kDa and 13.5 kDa) provide the link between succinate dehydrogenase and ubiquinone. Reconstitution of succinate dehydrogenase with the two smaller subunits restores the succinate–ubiquinone reductase activity and the thenoyl-trifluoroacetone (TTFA) sensitivity.

The 70 kDa flavoprotein (Fp) subunit contains the FAD covalently bound [55], while the three iron-sulphur centres, which are believed to include a binuclear (2Fe-2S), a trinuclear (3Fe-xS) and a tetranuclear (4Fe-4S) cluster, reside on the 27.5 kDa subunit (Ip subunit) [56]. The amino acid sequence of the iron-protein (Ip) subunit has been determined [57] with a calculated molecular mass of 28 655 Da. It contains three conserved clusters of cysteine residues believed to contribute the ligands for the binuclear, trinuclear and tetranuclear iron-sulphur clusters.

There is evidence that a cytochrome b haem is associated with this complex but its function is not apparent. It is not reducible by succinate, but there is a correlation between the cytochrome b content of the complex and succinate–ubiquinone reductase activity [58]. The b haem is distinct from the cytochrome b in Complex III [59], and is associated with the two smaller subunits.

Complex II is the only complex in the respiratory chain and oxidative phosphorylation system (Complexes I–V) that does not contain a subunit encoded by the mtDNA. It is also the only complex that does not involve proton translocation. An obvious inference from this is that mtDNA products are somehow involved with proton translocation in their respective complexes.

COMPLEX III

Complex III (ubiquinol–cytochrome c reductase) is situated in the middle of the respiratory chain accepting electrons from Complexes I, II, electron-transferring flavoprotein (ETF, covered in Chapter 8) and other flavoproteins via ubiquinol. In bovine heart, Complex III is composed of 11 subunits which include the following redox groups: low-potential cytochrome $b566$, high-potential cytochrome $b562$, cytochrome c_1 and a high-potential binuclear iron-sulphur centre (the Rieske iron-sulphur centre).

The proton motive Q cycle of electron transfer and proton translocation [60,61] has been proposed for the mechanism in Complex III. This involves QH_2 oxidation (centre o) and Q reduction (centre i). Two protons are released at centre o; one electron passes to the iron-sulphur centre and on to cytochromes c_1 and c, and the other electron proceeds via the low- and high-potential b-reducing coenzyme Q to the semiquinone at centre i. The semiquinone is reduced to QH_2 by a second electron and two associated protons.

Structure and composition

Ubiquinol–cytochrome c reductase preparations have been made from mitochondria from a number of sources including bovine heart [62,63], rat liver [64] and *Neurospora crassa* [65]. Complex III has been resolved into 11 different subunits which have been labelled according to their mobility on the gel system of Schägger *et al.* [66] using the Roman numerals I–XI (Table 1.2).

Subunits I and II (47 kDa and 45 kDa) are referred to as the core proteins and they are thought to be part of the QH_2 reduction centre (centre i). The cDNA sequence of core I [67] and core II [68] suggests that they are both largely hydrophilic with no regions that could either form membrane-spanning helices or anchor them to the membrane. The core proteins are not associated with any of the redox centres in Complex III. However, in *N. crassa*, removal of the core

Table 1.2 Summary of bovine Complex III subunits

Subunit	Molecular mass (kDa) (from gels)	Molecular mass (Da) (from sequence)	Features	N-Terminal sequence
I	47	35 833	Core (PEP)	TATYA
II	45	44 620	Core (MPP)	SLKVA
III	35	43 000	(mtDNA) b-566 b-562	MTNIR
IV	31	–	C_1, Cyt.c binding	SDLEL
V	25	21 536	2Fe-2S	SHTDI
VI	13.4	13 389	–	AGRPA
VII	9.5	–	Quinone binding (QP-C)	GRQFG
VIII	9.2	–	'Hinge'	GDPKE
IX	8	–	DCCD binding	MLSVA
X	7.2	–	–	VAPTL
XI	6.4	6 363	–	MLTRF

PEP, processing enhancing protein; MPP, mitochondrial processing peptidase.

proteins abolishes the electron flow through the complex, which can be restored on reconstitution of the cores, suggesting that the core subunits influence the function of the complex [69]. In *Saccharomyces cerevisiae*, core I mutants do not insert the haem into apocytochrome b and lack a functional complex, while deficiencies in core II lead to a decrease in cytochrome b and a decreased activity. This suggests that the core proteins are essential for assembly of the complex [70]. Their role in protein import and assembly is highlighted in *N. crassa* where core I is identical with the mitochondrial processing enhancing protein (PEP) [71], and has a significant sequence homology with the mitochondrial processing peptidase (MPP) (see section on biosynthesis) [72]. In addition, the PEP, MPP, core I and core II of *S. cerevisiae* have significant sequence homologies, and, together with the equivalent proteins from *N. crassa*, constitute a protein family [73–76]. This homology is extended to bovine core I and II. This is most prominent between the N-terminus of bovine core I and the PEP of *S. cerevisiae*, and the C-terminus of bovine core I and core I of *S. cerevisiae* [67]. This suggests that bovine core I may be bifunctional, enabling electron transfer and also processing of proteins during import, the C-terminus having a role in Complex III and the N-terminus having a PEP function. There appears to be only a single gene, and not two similar genes, suggesting that core I and PEP in bovine mitochondria may be the same protein as in *N. crassa*. Subunit III has an apparent molecular mass of 35 kDa. It is encoded by the mtDNA [77,78] and is associated with both the low- (566) and high- (562) potential b haems. It is a hydrophobic subunit and has eight or nine membrane-spanning regions [79] and therefore lies predominantly within the membrane. There are four histidine residues, conserved in mitochondria from various sources, occurring in two distinct regions, which are thought to contribute the ligands for the two b haems [79]. Subunit IV (31 kDa) is associated with the c_1 haem. The amino acid sequence of the bovine heart subunit has been determined [80], and the cDNA for the human subunit has been sequenced [81], and the structural organization of the gene has been investigated [82] and localized to chromosome 8 [83]. The haem which is covalently bound to two cysteine residues lies on the intermembrane side of the membrane where it binds cytochrome c [84].

Subunit V (25 kDa) contains a binuclear iron-sulphur centre (the Rieske iron protein) with a high redox potential (+280 mV). It is the primary acceptor of one electron from the hydroquinone. The amino acid sequence [85] and cDNA sequence [86] of this subunit have been determined. There are four conserved cysteine residues near the C-terminus which is consistent with the redox centre being located on the intermembrane side of the membrane where it is a constituent part of the QH_2 oxidation centre. The five larger subunits are fairly well conserved across species; it is the six smaller subunits that are not always present in Complex III preparations from other species. Subunit VI (13.4 kDa) is often wrongly referred to as the quinone-binding protein (QP-C) (see below). The cDNAs for the bovine and human subunits have been sequenced [87,88] and the human gene has been sequenced [89] and localized to chromosome 8 [90].

A photoaffinity azido-Q derivative of ubiquinone was found to label two Complex III subunits, designated subunits III and VII [91] on the gel system of Schägger *et al.* [92]. Subunit III is undoubtedly cytochrome b and subunit VII is the 9.5 kDa subunit sequenced by Borchart *et al.* [93], and not subunit VI as sequenced by Wakabayashi *et al.* [87]. The confusion appears to have risen as a result of the different mobilities of these subunits on different gel systems. Subunit VII (11 kDa) is therefore believed to be a ubiquinone-binding protein (QP-C). Its amino acid sequence has been determined [93] and has a calculated molecular mass of 9.5 kDa; it is consequently often referred to as the 9.5 kDa Q-binding subunit (QP-C). The amino acid sequences of subunits VIII (9.2 kDa), IX (8 kDa), X (7.2 kDa) and XI (6.4 kDa) have been determined [66,94–96]. Subunits VIII and X are believed to be associated with subunit IV and therefore on the intermembrane space side of the membrane. Subunit VIII has been referred to as the 'hinge protein' because it is thought to be involved with electron transfer between cytochromes c_1 and c [97]. Subunit IX is the DCCD-binding subunit of Complex III and may participate in protein translocation.

Topology

González-Halphen *et al.* [98] have reassessed the arrangement of Complex III subunits in the light of the number and nomenclature of subunits identifiable using the separation system described by Schägger *et al.* [92]. Using N-terminal sequence analysis, subunit VII was identified to run as two bands and not subunit VIII as proposed by Schägger *et al.* [92]. Using protease digestion, bilayer-intercalated probes and cross-linking reagents, González-Halphen *et al.* [98] have proposed the subunits in Complex III to be arranged in a manner depicted in Figure 1.4. Protease digestion of mitochondria and submitochondrial particles suggests that subunits II and VI of beef heart mitochondria are sited on the matrix side of the membrane, while subunits V and XI are on the cytoplasmic side. Membrane-intercalated probes suggest that cytochrome b is the main intramembrane subunit.

Inhibitors

There are a number of inhibitors of ubiquinol-cytochrome c reductase activity, including antimycin A, myxathiozol and stigmatellin. The binding of antimycin is in the vicinity of the high-potential haem b and is affected by mutations of cytochrome b [99]. However, photoaffinity derivatives of antimycin A label a

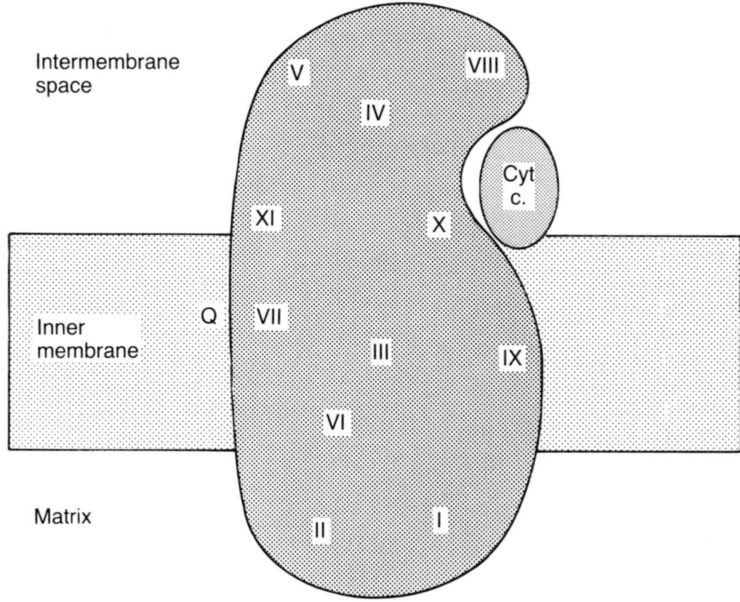

Figure 1.4 Schematic representation of bovine Complex III. Roman numerals depict the relative positions of the subunits. Q, ubiquinone; Cyt c., cytochrome *c*

11.5 kDa subunit of Complex III [100] which is believed to be in the vicinity of the high-potential haem *b*. Myxathiazol on the other hand binds to or near the low-potential haem *b*, as suggested from spectroscopic data and cytochrome *b* mutants resistant to myxathiazol [101].

COMPLEX IV

Complex IV (cytochrome *c* oxidase) is the terminal complex of the respiratory chain and catalyses the oxidation of cytochrome *c*, the reduction of oxygen and also the translocation of protons. Many reviews have been written on various aspects of this enzyme including its kinetics [102], structure and assembly [103] and structure and function [104].

In eukaryotes, the four prosthetic groups, haems *a* and a_3 and two copper atoms (Cu_a and Cu_b), are associated with Complex IV. Haem *a* and Cu_a are at the low-potential centre of Complex IV accepting a pair of electrons from cytochrome *c*, whereas haem a_3 and Cu_b are the high-potential centre forming the oxygen-binding site.

Fractionation

The complexity of the terminal oxidase varies from three subunits in *Paracoccus denitrificans* to 13 in mammals. Cytochrome *c* oxidase has been purified from a

Table 1.3 Summary of bovine Complex IV subunits

Subunit	Nuclear/mtDNA	Molecular mass (Da) (from sequence)	Features	N-Terminal sequence
I	M	56 993	Cu_a, Cu_b a, a_3	MFI
II	M	26 049	Cyt c binding	MAY
III	M	29 918	DCCD binding Cyt. c binding	MTH
IV	N	17 153	–	AHG
Va	N	12 436	–	SHG
Vb	N	10 670	–	ASG
VIa	N	9 419	H + L isoforms	ASA (SSG)
VIb	N	10 068	–	AED
VIc	N	8 480	–	STA
VIIa	N	6 244	H + L isoforms	FEN
VIIb	N	5 900	–	SGY
VIIc	N	5 541	–	SHY
VIII	N	4 962	H + L isoforms	ITA (IHS)

H, heart isoform; L, liver isoform. The N-terminal sequences for the liver isoforms are in parentheses.

variety of mammalian species (rat, pig, bovine and human) and tissues (liver, heart, kidney and skeletal muscle), and the separation of these on sodium dodecyl sulphate/polyacrylamide gel electrophoresis (SDS/PAGE) revealed 13 subunits [105]. The subunit numbering systems in eukaryotes vary between that of Capaldi *et al.* [106] and Kadenbach and Merle [105]. However, the latter nomenclature seems to be the most widely used and will therefore be applied here (Table 1.3).

Structure and function

Subunits I, II and III of bovine Complex IV (CO I, II, III) correspond to the mtDNA products and are homologous to the three subunits in *P. denitrificans* and the three large subunits in lower eukaryotes. In mammals the nuclear-encoded subunits are labelled IV, Va, Vb, VIa, VIb, VIc, VIIa, VIIb, VIIc and VIII according to Kadenbach's nomenclature. All eukaryotes share subunits IV, Va, Vb and VIc, while yeast and higher eukaryotes also share subunits VIIa and VIIb; however, subunits VIa, VIb, VIIb and VIII are restricted to the mammalian enzyme, and whether these are authentic components or fortuitous contaminants has been the subject of much debate.

Mitochondrial DNA subunits

The mtDNA products contain the catalytic core of the complex, containing all the prosthetic groups involved in electron transfer and proton pumping. The prosthetic groups have been localized to subunits I and II. Evidence of this lies with the comparison with the complex from *P. denitrificans* which contains the prosthetic groups but is composed of only subunits I, II and III. Removal of subunit III from either *P. denitrificans* or the eukaryotic complex does not affect the haem or copper content.

The DNA sequences of the human mitochondrial DNA products have been determined [77]. The two conserved cysteine residues and two conserved histidine residues in subunit II are believed to form the ligands for Cu_a, while the seven conserved histidine residues in subunit I are thought to contribute the ligands for Cu_b and haem a and a_3 [107]. DCCD inhibits the Complex IV proton-pumping function, but only partially inhibits electron transfer. DCCD has been demonstrated to bind to subunit III [108], in agreement with the suggestion that this subunit is involved in proton pumping.

There are two binding sites for cytochrome c, a high-affinity site and a low-affinity site. Using photoaffinity derivatives of cytochrome c, the high-affinity site has been localized to subunit II, possibly near the Cu_a site [109,110]. Using a different derivatization of cytochrome c, subunit III was labelled [111], and it has been suggested that both subunits II and III are involved with the binding of cytochrome c, possibly from different monomers of the enzyme.

Nuclear subunits

The sequences of all the bovine heart Complex IV subunits have been determined [112–121]. However, the function of the nuclear-encoded products is less well defined. In yeast, null mutants in subunit VIIc can still assemble the complex, but the activity is altered. However, null mutants in subunits IV, Va, Vb, VIc or VIIa fail to assemble the complex, implying that these subunits are either needed structurally or involved in the importation/assembly process. This could explain why these subunits are absent from prokaryotes where proteins are not imported.

Cytochrome oxidase catalyses the step of the respiratory chain with the greatest free energy fall, and could act as a regulator of respiratory chain activity. This could explain the requirement for isoforms of some subunits to meet the metabolic requirements of different tissues.

Topography

Electron microscopic analysis of two-dimensional crystals of Complex IV suggests that it is a Y shape, with one domain protruding from the membrane into the intermembrane space and two domains protruding into the matrix [122,123]. From the reported sequences, subunits I, II and III possess 12, 2 and 7 putative transmembranous sequences respectively. Subunits IV, VIa, VIc, VIIa, VIIb, VIIc and VIII all have a hydrophobic region in the middle of their sequence with hydrophilic regions at the N- and C-termini. All these subunits are labelled with one or more bilayer-intercalating agents, suggesting that they project into the lipid bilayer and are probably membrane spanning.

Subunits Va, Vb and VIb do not have hydrophobic sequences capable of spanning the membrane, although subunits Va and Vb have been reported to do so [124,125]. Protease digestion studies suggest that subunits IV, VIc and VIIb have their N-termini and subunit VII has its C-terminus on the matrix side of the membrane, while VIb is situated on the C side [126]. Cross-linking analysis suggests that all the nuclear-encoded subunits are arranged around subunits I and/or II.

Isoforms

In mammalian cytochrome oxidase, a number of nuclear-encoded subunits exist as isoforms. In beef, subunits VIa, VIIa and VIII exist as two forms: a heart (H) form, expressed predominantly in the heart and skeletal muscle, and a liver (L) form, expressed predominantly in the liver, brain and kidney. The homologies between the two subunits are: 55% (VIa), 70% (VIIa) and 50% (VIII). The H form is transcribed predominantly in the heart and skeletal muscle while the L form is transcribed in all tissues, although subunit VIII is not expressed in skeletal muscle. This suggests that in skeletal muscle at least, the H form of the VIII subunit is controlled at the level of transcription while the L form is post-transcriptionally regulated [127]. Although subunit VIII exists as the H and L forms in beef and rat, in human tissues there is only evidence of the L form [120]. Subunit VIIc exists as two isoforms with different presequences but identical mature proteins. There is evidence that this subunit has different fetal and adult isoforms [128] which may participate in developmental regulation of this complex.

Many of the subunits of Complex IV appear to have multiple coding regions on Southern blots; however, some of these have been identified as pseudogenes [129,130].

Inhibitors

Inhibitors of Complex IV include carbon monoxide, cyanide and azide. They all react with cytochrome oxidase forming a complex with a_3 which is facilitated when the a_3 is in a reduced state [131,132].

COMPLEX V

Complex V (ATP synthase)

The mitochondrial ATP synthase complex is composed of two main domains, designated F_1 and F_0. The F_1 domain contains the catalytic centre and the water-soluble part of the ATPase which protrudes into the mitochondrial matrix, whereas the F_0 domain is a hydrophobic component embedded in the inner membrane, and is concerned with proton translocation. The intact complex has been isolated from a variety of sources with ATP-P_i exchange which is sensitive to uncouplers and inhibitors (oligomycin) [133,134].

Composition and structure

The eukaryotic ATP synthase is structurally more complicated than the prokaryotic counterpart in that 14 or more polypeptides are found in the eukaryotic complex and only eight in the prokaryotic (*Escherichia coli*). A comparison of the

Table 1.4 Summary of bovine Complex V subunits

		Eukaryote (bovine heart)	Prokaryote (E. coli)
F_1 fraction	Total molecular mass (Da)	371 000	381 000
	Subunit stoichiometry	α_3 β_3 γ δ ϵ	α_3 β_3 γ δ ϵ
		Molecular mass (Da)	
	α	55 000	α
	β	51 600	β
	γ	30 100	γ
	δ	15 000	ϵ
	ϵ	5 600	–
F_0 fraction			
	OSCP	21 000	δ
	F_6 (coupling factor 6)	8 000	–
	a (ATPase 6)*	25 000	a
	b	24 570	b
	c (DCCD-binding protein)	7 400	c
	d	19 000	–
	A6L (ATPase 8)*	7 965	–
	IF_1 (inhibitor protein)	9 600	–
	e	14 000	

The components of the F_1/F_0 fraction relate to the eukaryotic Complex V. The prokaryotic equivalents from *E. coli* are listed for comparison. It should be noted that the δ subunit of the prokaryotic complex is part of the bacterial F_1 fraction although its homology is much closer to the OSCP (oligomycin sensitivity conferring protein) subunit of the F_0 fraction of the mammalian complex. * denotes mitochondrial encoded subunits.

known subunits of the Complex V from bovine heart and *E. coli* is made in Table 1.4. The eukaryotic Complex V also consists of two domains with the F_0 fraction containing nine subunits and the F_1 fraction five subunits. Electron microscopy studies suggest a 'lollipop' type of configuration with certain subunits forming the headpiece (F_1 fraction) which projects into the mitochondrial matrix. This headpiece is joined by a stalk to components integrated with the membrane, these two parts consisting mainly of subunits of the F_0 fraction.

Function of subunits

The F_1 fraction which forms the 'head' of the lollipop structure seen under electron microscopy contains the catalytic centre for ATP synthesis (hydrolysis). The F_1 is bound by the stalk to the membrane sector (F_0), the latter forming a specific proton-conducting pathway which is intimately involved in the production of ATP from ADP + P_i [133].

F_1 subunits

The bovine F_1 binds 6 mol of adenine nucleotide/mol, 3 to each of the α and β subunits [133]. The β subunit appears to contain the catalytic site on the basis of the binding of inhibitory ATP analogues. In addition, the inhibitors, efrapeptin, aurovertin and the IF_1 (inhibitor protein and part of the F_0 fraction), also specifically bind to the β subunit [133].

F_0 subunits

Whilst F_6 has been shown to be required for the binding of F_1 to F_0 giving rise to ATPase activity, this is not oligomycin sensitive until oligomycin sensitive conferring protein (OSCP) is added. It is assumed that these two subunits are involved in proton translocation between the membrane (F_0) and the active site of ATP synthesis (F_1) in some, as yet, unspecified way [133].

SUBUNIT b
Subunit b is a water-soluble protein which appears to be needed for energy transfer to and from the F_1-ATPase and is present with a 1:1 stoichiometry with respect to F_1. It contains an essential dithiol which is implicated in the mechanism of energy transduction and transfer. Considerable controversy still exists regarding its apparent molecular mass but it does seem to be homologous with the prokaryotic b subunit [134,135].

SUBUNIT c (DCCD-BINDING PROTEIN)
This is so called because DCCD reacts covalently with the free carbonyl of a specific glutamate residue (Glu-58 in bovine heart) of this small molecular mass proteolipid and inhibits proton translocation through F_0.

IF_1 (ATPase INHIBITOR PROTEIN)
This water-soluble protein binds to the β subunit of the F_1 fraction inhibiting its function. It binds with a 1:1 stoichiometry and requires both ATP and Mg^{2+} for effective inhibition. There has been some suggestion of a regulatory role for this subunit [134].

The other subunits identified (see Table 1.4) do not as yet have any clearly defined role in the mechanism of ATP synthesis.

Topography

Studies on the topography of the ATP synthase have been made by a variety of techniques including reconstitution from isolated subunits [136], chemical cross-linking investigations [137] and immunochemical approaches using antibodies raised to specific subunits [138]. A basic consensus has been obtained which fits reasonably with the structure seen under the electron microscope with a head (F_1) projecting into the mitochondrial matrix which is linked by a stalk to a membrane section (F_0) embedded in the inner membrane.

The headpiece (F_1) contains three copies of each of the α and β subunits which are arrayed in a hexagonal orientation (Figure 1.5), together with one copy of the γ subunit which connects the α and β subunits and one copy of each of the δ and ε subunits. The latter act as part of the connection between the headpiece containing the catalytic and regulatory centres (β and α subunits respectively) and the membrane section. Recent studies [137] have suggested that the α and β subunits may be ellipsoidal in shape and extend down towards the membrane surface. In

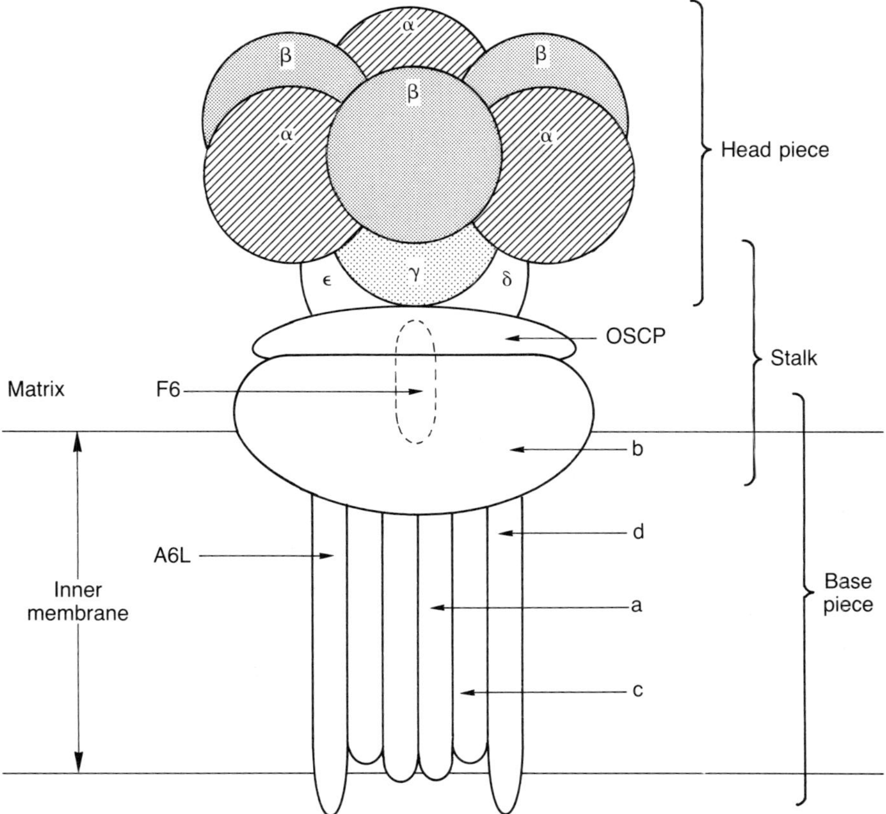

Figure 1.5 Schematic representation of bovine Complex V

doing so, they encompass part of the stalk and certain of its subunits. This contrasts with the more globular arrangement proposed previously [136].

The stalk or the region which connects the headpiece with the membrane may be considered as being made up of subunits deriving from the two fractions – those from the F_1 fraction (one copy of the α, δ and ε subunits) and the more hydrophilic subunits of the F_0 fraction (OSCP, b) and two copies of F_6. Subunit b is placed closer to the membrane than the OSCP with the two copies of F_6 either side of the stalk [137,138]. The base or membrane component contains one copy of subunits a, A6L and subunit d and six copies of subunit c (DCCD-binding protein) formed into a ring. Other components have yet to be positioned [137]. An outline of this model is shown in Figure 1.5.

Genetics

As previously indicated, the eukaryotic F_0–F_1-ATP synthase (Complex V) contains two subunits coded for by mtDNA, whereas the other 12 are all nuclear encoded,

Table 1.5 Composition of respiratory chain and phosphorylation system

Respiratory chain complex	Total no. of subunits	No. encoded by MtDNA
Complex I (NADH-CoQ reductase)	25–41	7 – ND1, 2, 3, 4, 4L, 5, 6
Complex II (Succinate-CoQ reductase)	4	0
Complex III (CoQH$_2$-cytochrome *c* reductase)	11	1 – Cytochrome *b*
Complex IV (Cytochrome oxidase)	13	3 – COX. subunits I, II, III
Complex V (ATP synthase)	14	2 – ATPase subunits 6 + A6L

cytoplasmic synthesized and imported into the mitochondria. Both of the mtDNA-encoded subunits [subunit a (ATPase 6) and A6L (ATPase 8] belong to the membrane (F$_0$) fraction. Ten of the 12 nuclear-encoded subunits, the exceptions being the F$_1$ δ subunit and the F$_0$ subunit e [135], have been cloned and sequenced.

Isoforms

More interesting from the point of view of the clinical expression of defects in Complex V is the observation that some of the nuclear-encoded subunits are encoded by more than one nuclear gene. The bovine F$_1$ α subunit has been identified as being expressed in heart and liver by two distinct genes [139]. The DCCD-binding factor (subunit c) is also coded for by two different genes which, however, give rise to the same mature protein. In humans, the genes are located on different chromosomes, 17 and 12, and are expressed in different ratios in distinct tissues. Thus the ratio of the gene expressed on chromosome 17 to that on chromosome 12 is 1:3 in liver but 1:1 in heart. As can be appreciated, this may have important implications in the tissue expression of clinical defects.

Biosynthesis of respiratory chain complexes

The organization, structure, transcription and translation of mtDNA has been extensively reviewed [140–142]. It has some unique characteristics and has been characterized in a number of species. In humans, it is a circular double-stranded molecule 16.5 kb in length which codes for 13 polypeptides, two ribosomal RNAs and 22 tRNAs, sufficient to translate the mtDNA [77]. The enzyme complexes of the mammalian mitochondrial oxidative phosphorylation system are unique in that they are coded for by both mitochondrial and nuclear DNA. Whilst the majority (>80%) are coded for by nuclear DNA and are cytoplasmically synthesized, all 13 mtDNA-encoded proteins are known components of the respiratory chain; seven subunits in Complex I, one subunit in Complex III, three subunits in Complex IV and two subunits in Complex V (Table 1.5) [140]. In view of the mitochondrial

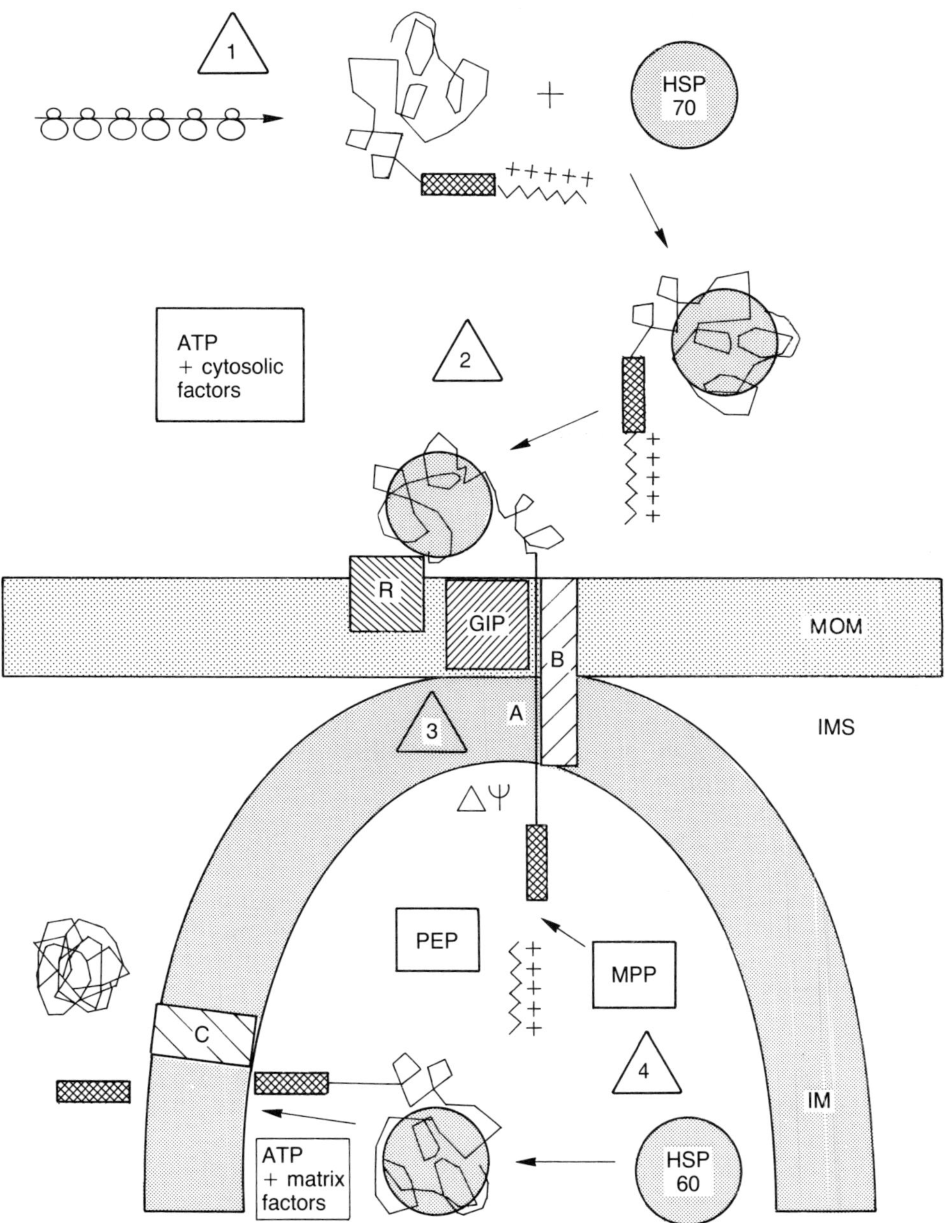

Figure 1.6 Scheme for importation of proteins into mitochondria. 1, Precursor synthesis (HSP 70, 70 kDa heat-shock protein); 2, Membrane binding and insertion [R, receptor – MOM 72 and 19; GIP, general insertion proteins – MOM 7, 8, 30 and 38 (MOM, mitochondrial outer membrane)], 3, Membrane translocation (A, membrane-contact sites; B, possible translocating protein; ΔΨ, membrane potential); 4, Intramitochondrial processing, sorting and complex assembly (C, putative inner membrane receptor/peptidase). MPP, mitochondrial processing peptidase; PEP, processing enhancing protein; HSP 60, 60 kDa heat shock protein; IM, mitochondrial inner membrane

localization of the oxidative phosphorylation system, this means that the majority of the proteins concerned must be imported across the mitochondrial membranes and assembled into the various complexes inside the mitochondria. Much of our understanding of the mechanisms involved has been derived from work carried out on simpler eukaryotic systems such as *Neurospora* [143], which will be briefly described.

The process whereby a nuclear-encoded and cytoplasmically synthesized polypeptide is transported into mitochondria and assembled into the respiratory chain may be divided into four parts (Figure 1.6).

(1) PRECURSOR SYNTHESIS

The majority of mitochondrial proteins are encoded by the nucleus and are synthesized on cytosolic polysomes, many with N-terminal presequences of between 20 and 80 amino acid residues which act as recognition signals for the importation process [144]. These presequences usually contain positively charged residues the neutralization or deletion of which impairs protein import. The exact importance of these residues is unclear, although it may be associated with responding to the electrical potential across the inner mitochondrial membrane or creating an amphipathic α-helix which aids membrane penetration [144]. In addition, signalling information for the targeting of mitochondrial proteins may be found in the mature protein (i.e. not on the presequence); indeed some cytosolic precursors appear to contain all their signalling information in the mature protein (e.g. cytochrome *c*, ADP/ATP carrier) [143]. Protecting the precursor protein from protease breakdown while at the same time permitting it to assume a transport-competent conformation are two other important parts of the process. Both ATP and a 70 kDa heat-shock protein (HSP) (chaperonins) have been implicated in this process [145,146].

(2) MEMBRANE BINDING AND INSERTION

Recent studies have shown that distinct specific receptor sites exist on the outer membrane to which precursor proteins may bind with high affinity [143]. Precursor proteins are then imported directly from these receptor sites without release from the mitochondrial membranes. Two particular mitochondrial outer membrane (MOM) proteins of molecular mass 19 and 72 kDa (MOM 19 and 72 respectively) appear to be located on the outside of the outer membrane and act as the initial receptor for the polypeptides which are subsequently transferred on to a receptor complex – the general insertion protein (GIP). The latter is formed by at least four distinct proteins, MOM 7, MOM 8, MOM 30 and MOM 38 [147] and involves the insertion of the polypeptide into the outer membrane.

(3) MEMBRANE TRANSLOCATION

Most proteins that are imported into the mitochondrial matrix via the GIP require energization of the inner mitochondrial membrane for their translocation across it. This appears to be achieved at 'contact sites' between the two membranes and may be inhibited by discharge of the electrical potential across the inner membrane by uncouplers [143].

Cytochrome *c* however, is not translocated across the inner membrane but is released into the intermembrane space, after the covalent addition of haem by cytochrome *c* haem lyase to its apoform. However, it would appear that the processing of cytochrome *c* is anomalous in these respects [143].

(4) INTRAMITOCHONDRIAL PROCESSING, SORTING AND ASSEMBLY

Once translocated across the inner membrane into the matrix, the N-terminal presequences are cleaved in a one- or two-step process to yield the mature protein. Two proteins appear to be necessary for this process – a Mn^{2+}- or Zn^{2+}-dependent MPP of molecular mass 57 kDa [148], together with a PEP of 52 kDa. The peptidase alone has very limited activity, suggesting that the PEP in some way enhances the peptidase activity. One possibility is that PEP is involved in the translocation process and facilitates the precursor protein access to the MPP catalytic site, However, processing may take place in more than one step on either side of the inner membrane. There is also evidence that these processing activities may be coordinated with other modifications such as prosthetic group introduction, e.g. haem, iron-sulphur centres or flavin nucleotides. For example, the attachment of haem to the apoform of cytochrome c_1 is a prerequisite for the second proteolytic processing step [143]. Recent reports suggest that the PEP and MPP are either the same as or related to cores I and II of Complex III of the mitochondrial respiratory chain (see earlier section). Why these proteins reside in Complex III or possess a similar sequence is not clear. However, cores I and II have been suggested to be involved with the assembly of Complex III.

The folding of the now mature polypeptide (with prosthetic group if appropriate) into the conformation suitable for incorporation and assembly into the functional complex has been found to be a complicated procedure which is assisted by another member of the chaperonin protein family (HSP 60) and ATP hydrolysis [143]. Indeed, there have been suggestions that HSP 60 may also be involved in the translocation process as well [149]. Sorting of the various nuclear-encoded mitochondrial polypeptide components into both their mitochondrial subcompartments (outer and inner membrane, intermembrane space and matrix) and their functional complexes in the case of the respiratory chain (Complexes I–V), and how these processes are regulated in a coordinated fashion with the production of the intramitochondrially synthesized components, is not known. The mechanisms whereby the final assembly of the oligomeric respiratory chain complexes are organized are poorly understood, and equally poorly defined are the means whereby the rate of importation of nuclear-encoded components is coordinated with respect to the mitochondrially synthesized components. Further research on simple eukaryotic systems is needed to augment our understanding of the processes.

REFERENCES

1. van Belzen, R. and Albracht, S.P.J. (1989) The pathway of electron transfer in NADH:Q oxido-reductase. *Biochimica Biophysica Acta*, **974**, 311–320
2. Chen, S. and Guillory, R.J. (1984) Identification of the NADH–NAD$^+$ transhydrogenase peptide of the mitochondrial NADH–CoQ reductase (complex I). *The Journal of Biological Chemistry*, **259**, 5124–5131
3. Fearnley, I.M., Finel, M., Skehel, J.M. and Walker, J.E. (1991) NADH: ubiquinone oxidoreductase from bovine heart mitochondria cDNA sequences of the import precursors of the nuclear-encoded 39 kDa and 42 kDa subunits. *Biochemical Journal*, **278**, 821–829
4. Ragan, C.I., Ohnishi, T. and Hatefi, Y. (1986) Iron sulphur proteins of mitochondrial NADH–ubiquinone reductase (complex I). In *Iron Sulfur Protein Research* (eds H. Matsubara *et al.*), Japan Science Society Press, Tokyo/Springer-Verlag, Berlin, pp. 220–231
5. King, T.E. and Suzuki, H. (1984) Ubiquinone proteins with emphasis of QP-N. In *Biomedical and Clinical Aspects of Coenzyme Q* (eds K. Folkers and Y. Yamamura), Elsevier, Amsterdam, Vol. 4, pp. 43–55

6. Ragan, C.I. (1985) In *Coenzyme Q. Biochemistry, Bioenergetics and Clinical Application of Ubiquinone* (ed. G. Lenaz), John Wiley, New York, pp. 315–336

7. Krishnamoorthy, G. and Hinkle, P.C. (1988) Studies on the electron transfer pathway, topography of iron-sulfur centres, and site of coupling in NADH-Q oxidoreductase. *The Journal of Biological Chemistry*, **263**, 17566–17575

8. Lawford, H.G. and Garland, P.B. (1972) Proton translocation coupled to quinone reduction by reduced nicotinamide-adenine dinucleotide in rat liver and ox heart mitochondria. *Biochemical Journal*, **130**, 1029–1044

9. Freedman, J.A. and Lemasters, J.J. (1984) Thermodynamics of reverse electron-transfer across site-1-ATP/2E- is greater than one. *Biochemical and Biophysical Research Communications*, **125**, 8–13

10. Ragan, C.I. (1987) Structure of NADH–ubiquinone reductase (complex I). *Current Topics in Bioenergetics*, **15**, 1–36

11. Ragan, C.I. (1990) Structure and function of an archetypal respiratory chain complex: NADH–ubiquinone reductase. *Biochemical Society Transactions*, **18**, 515–516

12. Hatefi, Y., Haavik, A.G. and Griffiths, D.E. (1962) Studies on the electron transfer system. *The Journal of Biological Chemistry*, **237**, 1676–1680

13. Paech, C., Friend, A. and Singer, T.P. (1982) Simplified isolation and molecular composition of NADH dehydrogenase of the respiratory chain. *Biochemical Journal*, **203**, 477–481

14. Ogata, K., Shimomura, Y. and Ozawa, T. (1982) Molecular profile of purified complex I (NADH–CoQ oxidoreductase) of mitochondrial electron-transfer chain. *Biochemistry International*, **4**, 621–627

15. Haines, A.M.R., Cooper, J.M., Morgan-Hughes, J.A., Clark, J.B. and Schapira, A.H.V. (1992) One-step immunoaffinity purification of complex I subunits from beef heart mitochondria. *Protein Expression and Purification*, **3**, 223–227

16. Smith, S. and Ragan, C.I. (1980) The organization of NADH dehydrogenase polypeptides in the inner mitochondrial membrane. *Biochemical Journal*, **185**, 315–326

17. Cleeter, M.W.J. and Ragan, C.I. (1985) The polypeptide composition of the mitochondrial NADH: ubiquinone reductase complex from several mammalian species. *Biochemical Journal*, **230**, 739–746

18. Ohnishi, T., Ragan, C.I. and Hatefi, Y. (1985) EPR studies of iron-sulfur clusters in isolated subunits and subfractions of NADH-ubiquinone oxidoreductase. *The Journal of Biological Chemistry*, **260**, 2782–2788

19. Chen, S. and Guillory, R.J. (1981) Studies on the interaction of arylazido-β-alanyl NAD$^+$ with the mitochondrial NADH dehydrogenase. *The Journal of Biological Chemistry*, **256**, 8318–8332

20. von Bahr-Lindström, H., Galante, Y.M., Persson, M. and Jörnvall, H. (1983) The primary structure of subunit II of NADH dehydrogenase from bovine-heart mitochondrial. *European Journal of Biochemistry*, **134**, 145–150

21. Patel, S.D., Aebersold, R. and Attardi, G. (1991) cDNA-derived amino acid sequence of the NADH-binding 51 kDa subunit of the bovine respiratory NADH dehydrogenase reveals striking similarities to a bacterial NAD$^+$-reducing hydrogenase. *Proceedings of the National Academy of Sciences of the USA*, **88**, 4225–4229

22. Nishikimi, M., Hosokawa, Y., Toda, H., Suzuki, H. and Ozawa, T. (1988) The amino acid sequence of the 24 kDa subunit, and iron-sulfur protein, of rat liver mitochondrial NADH dehydrogenase deduced from cDNA sequence. *Biochemical and Biophysical Research Communications*, **157**, 914–920

23. Skehel, J.M., Pilkington, S.J., Runswick, M.J., Fearnley, I.M. and Walker, J.E. (1991) NADH: ubiquinone oxidoreductase from bovine heart mitochondria. Complementary DNA sequence of the import precursor of the 10 kDa subunit of the flavoprotein fragment. *FEBS Letters*, **282**, 135–138

24. Tran-Betcke, A., Warnecke, U, Böcker, C., Zaborosch, C. and Friedrich, B. (1990) Cloning and nucleotide sequences of the genes for the subunits of NAD-reducing hydrogenase of *Alcaligenes eutrophys* H16. *Journal of Bacteriology*, **172**, 2920–2929

25. Ragan, C.I., Galante, Y.M. and Hatefi, Y. (1982) Purification of three iron-sulfur proteins from the iron-protein fragment of mitochondrial NADH-ubiquinone oxidoreductase. *Biochemistry*, **21**, 2518

26. Suzuki, H. and Ozawa, T. (1986) An ubiquinone-binding protein in mitochondrial NADH–ubiquinone reductase (complex I). *Biochemical and Biophysical Research Communications*, **138**, 1237–1242

27. Runswick, M.J., Gennis, R.B., Fearnley, I.M. and Walker, J.E. (1989) Mitochondrial NADH:ubiquinone reductase: complementary DNA sequence of the import precursor of the bovine 75-kDa subunit. *Biochemistry*, **28**, 9452–9459

28. Fearnley, I.M., Runswick, M.J. and Walker, J.E. (1989) A homologue of the nuclear coded 49 kd

subunit of bovine mitochondrial NADH-ubiquinone reductase is coded in chloroplast DNA. *The EMBO Journal*, **8**, 665–672

29. Pilkington, S.J., Skehel, J.M. and Walker, J.E. (1991) The 30-kilodalton subunit of bovine mitochondrial complex I is homologous to a protein coded in chloroplast DNA. *Biochemistry*, **30**, 1901–1908

30. Dupuis, A., Skehel, J.M. and Walker, J.E. (1991) A homologue of a nuclear-coded iron-sulfur protein subunit of bovine mitochondrial complex I is encoded in chloroplast genomes. *Biochemistry*, **30**, 2954–2960

31. Dupuis, A., Skehel, J.M. and Walker, J.E. (1991) NADH–ubiquinone oxidoreductase from bovine mitochondria. cDNA sequence of 19 kDa cysteine-rich subunit. *Biochemical Journal*, **277**, 11–15

32. Earley, F.G.P., Patel, S.D., Ragan, C.I. and Attardi, G. (1987) Photolabelling of a mitochondrially encoded subunit of NADH dehydrogenase with [^{1}H]dihydrorotenone. *FEBS Letters*, **219**, 108–113

33. Earley, F.G.P. and Ragan, C.I. (1984) Photoaffinity labelling of mitochondrial NADH dehydrogenase with arylazidomorphigenin, an analogue of rotenone. *Biochemical Journal*, **224**, 525–534

34. Yagi, T. (1987) Inhibition of NADH–ubiquinone reductase activity by N,N'-dicyclohexylcarbodiimide and correlation of this inhibition with the occurrence of energy-coupling site 1 in various organisms. *Biochemistry*, **26**, 2822–2828

35. Yagi, T. and Hatefi, Y. (1988) Identification of the dicyclohexylcarbodiimide-binding subunit of NADH–ubiquinone oxidoreductase (complex I). *The Journal of Biological Chemistry*, **263**, 16150–16155

36. van Belzen, R., van Gaalen, M.C.M., Cuypers, P.A. and Albracht, S.P.J. (1990) New evidence for the dimeric nature of NADH:Q oxidoreductase in bovine-heart submitochondrial particles. *Biochimica et Biophysica Acta*, **1017**, 152–159

37. Patel, S.D., Cleeter, M.W.J. and Ragan, C.I. (1988) Transmembrane organization of mitochondrial NADH dehydrogenase as revealed by radiochemical labelling and cross-linking. *Biochemical Journal*, **256**, 529–535

38. Han, A.L., Yagi, T. and Hatefi, Y. (1988) Studies of the structure of NADH: ubiquinone oxidoreductase complex: topography of the subunits of the iron-sulfur flavoprotein component. *Archives of Biochemistry and Biophysics*, **267**, 490–496

39. Han, A.L., Yagi, T. and Hatefi, Y. (1989) Studies of the structure of NADH: ubiquinone oxidoreductase complex: topography of the subunits of the iron-sulfur protein component. *Archives of Biochemistry and Biophysics*, **275**, 166–173

40. Hatefi, Y. and Stempel, K.E. (1969) Isolation and enzymatic properties of DPNH dehydrogenase. *Journal of Biological Chemistry*, **244**, 2350–2357

41. Patel, S.D. and Ragan, C.I. (1988) Structural studies on mitochondrial NADH dehydrogenase using chemical cross-linking. *Biochemical Journal*, **256**, 521–528

42. Clay, V.J. and Ragan, C.I. (1988) Evidence for the existence of tissue specific isoenzymes of mitochondrial NADH dehydrogenase. *Biochemical and Biophysical Research Communications*, **157**, 1423–1428

43. Lindahl, P.E. and Oberg, K.E. (1961) The effect of rotenone on respiration and its point of attack. *Experimental Cell Research*, **23**, 228–237

44. Hall, C., Wu, M., Crane, F.L., Takahashi, N., Tamura, S. and Folkers, K. (1966) Pericidin – a new inhibitor of mitochondrial electron transport. *Fed. Proc. Fed. Am. Soc. Exp. Biol.*, **25**, 530

45. Nicklas, W.J., Vyas, I. and Heikkla, R.E. (1985) Inhibition of NADH-linked oxidation in brain mitochondria by 1-methyl-4-phenylpyridine, a metabolite of the neurotoxin, 1-methyl-4-phenyl-1,2,5,6-tetrahydropyridine. *Life Science*, **36**, 2503–2508

46. Holland, P.C., Clark, M.G., Bloxham, D.P. and Lardy, H.A. (1973) Mechanism of action of the hypoglycemic agent diphenyleneiodonium. *The Journal of Biological Chemistry*, **248**, 6050–6056

47. Jalling, O., Lindberg, O. and Ernster, L. (1955) On the effect of substituted barbiturates on mitochondrial respiration. *Acta Chemica Scandinavica*, **9**, 198–199

48. Gibb, G.M. and Ragan, C.I. (1990) Identification of the subunits of bovine NADH dehydrogenase which are encoded by the mitochondrial genome. *Biochemical Journal*, **265**, 903–906

49. Krueger, M.J., Singer, T.P., Casida, J.E. and Ramsay, R.R. (1990) Evidence that the blockade of mitochondrial respiration by the neurotoxin 1-methyl-4-phenylpyridinium (MPP$^+$) involves binding at the same site as the respiratory inhibitor, rotenone. *Biochemical and Biophysical Research Communications*, **169**, 123–128

50. Ramsay, R.R., Krueger, M.J., Youngster, S.K. and Singer, T.P. (1991) Evidence that the inhibition sites of the neurotoxin amine 1-methyl-4-phenylpyridinium (MPP$^+$) and of the respiratory chain inhibitor piericidin A are the same. *Biochemical Journal*, **273**, 481–484

51. Werner, S. (1989) Photoaffinity labeling of mitochondrial NADH: ubiquinone reductase with pethidine analogues. *Biochemical Pharmacology*, **38**, 1807–1818

52. Ragan, C.I. and Bloxham, D.P. (1977) Specific labelling of a constituent polypeptide of bovine heart mitochondrial reduced nicotinamide-adenine dinucleotide-ubiquinone reductase by the inhibitor diphenyleneiodonium. *Biochemical Journal*, **163**, 605–615

53. Horgan, D.J., Singer, T.P. and Casida, J.E. (1968) Studies on the respiratory chain-linked reduced nicotinamide adenine dinucleotide dehydrogenase. XIII. Binding sites of rotenone, piericidin A, and amytal in the respiratory chain. *The Journal of Biological Chemistry*, **243**, 834–843

54. Ernster, L., Dallner, G. and Azzone, G.F. (1963) Differential effects of rotenone and amytal on mitochondrial electron and energy transfer. *The Journal of Biological Chemistry*, **238**, 1124–1131

55. Walker, W.H. and Singer, T.P. (1970) Identification of the covalently bound flavin of succinate dehydrogenase as 82 (histidyl) flavin adenine dinucleotide. *The Journal of Biological Chemistry*, **245**, 4224–4225

56. Ohnishi, T. (1987) Structure of the succinate–ubiquinone oxidoreductase (Complex II). *Current Topics in Bioenergetics*, **15**, 37–65

57. Yao, T., Wakabayashi, S., Matsuda, S., Matsubara, H., Yu, L. and Yu, C.A. (1986) Amino acid sequence of the iron-sulphur protein subunit of beef heart succinate dehydrogenase. In *Iron Sulfur Protein Research* (ed. S. Matsubara *et al.*), Japan Science Society Press, Tokyo/Springer-Verlag, Berlin, pp. 240–244

58. Ackrell, B.A.C., Ball, M.B. and Kearney, E.B. (1980) Peptides from complex II are active in reconstitution of succinate–ubiquinone reductase. *The Journal of Biological Chemistry*, **255**, 2761–2769

59. Nobrega, F.G. and Tzagoloff, A. (1980) Assembly of the mitochondrial membrane system. *The Journal of Biological Chemistry*, **255**, 9828–9837

60. Mitchell, P. (1976) Possible molecular mechanisms of the proton motive function of cytochrome systems. *Journal of Theoretical Biology*, **62**, 327–367

61. Mitchell, P. and Moyle, J. (1985) In *Coenzyme Q. Biochemistry, Bioenergetics and Clinical Application of Ubiquinone* (ed. G. Lenaz), Wiley, New York, pp. 145–163

62. Rieske, J.S., Zaugg, W.S. and Hansen, R.E. (1964) Studies on the electron transfer system. *The Journal of Biological Chemistry*, **239**, 3023–3030

63. Nalecz, M.J., Bolli, R. and Azzi, A. (1985) Molecular conversion between monomeric and dimeric states of the mitochondrial cytochrome b-c_1 complex: isolation of active monomers. *Archives of Biochemistry and Biophysiology*, **236**, 619–628

64. Gellerfors, P., Johansson, T. and Nelson, B.D. (1981) Isolation of the cytochrome bc_1 complex from rat-liver mitochondria. *European Journal of Biochemistry*, **115**, 275–278

65. Weiss, H. and Kolb, H.J. (1979) Isolation of mitochondrial succinate ubiquinone reductase, cytochrome c reductase and cytochrome c oxidase from *Neurospora crassa* using nonionic detergent. *European Journal of Biochemistry*, **99**, 139–149

66. Schägger, H., Borchart, U., Aquila, H., Link, T.A. and von Jagow, G. (1985) Isolated and amino acid sequence of the smallest subunit of beef heart bc_1 complex. *FEBS Letters*, **190**, 89–94

67. Gencic, S., Schägger, H. and von Jagow, G. (1991) Core I protein of bovine ubiquinone–cytochrome c reductase; an additional member of the mitochondrial-protein-processing family cloning of bovine core I and core II cDNAs and primary structure of the proteins. *European Journal of Biochemistry*, **199**, 123–131

68. Hosokawa, Y., Suzuki, H., Toda, H., Nishikimi, M. and Ozawa, T. (1989) Complementary DNA encoding core protein II of human mitochondrial cytochrome bc_1 complex. *The Journal of Biological Chemistry*, **264**, 13483–13488

69. Linke, P. and Weiss, H. (1986) Reconstitution of ubiquinone cytochrome c reductase from *Neurospora* mitochondria with regard to subunit I and subunit II. *Methods in Enzymology*, **126**, 201–210

70. Crivellone, M.D., Wu, M. and Tzagoloff, A. (1988) Assembly of the mitochondrial membrane system. *The Journal of Biological Chemistry*, **263**, 14323–14333

71. Schulte, U., Arretz, M., Schneider, H., Tropschug, M., Wachter, E., Neupert, W. and Weiss, H. (1989) A family of mitochondrial proteins involved in bioenergetics and biogenesis. *Nature*, **339**, 147–149

72. Schneider, H., Arretz, M., Wachter, E. and Neupert, W. (1990) Matrix processing peptidase of mitochondria. *Journal of Biological Chemistry*, **265**, 9881–9887

73. Jensen, R.E. and Yaffe, M.P. (1988) Import of proteins into yeast mitochondria. The nuclear Mas2 gene encodes a component of the processing protease that is homologous to the Mas1-encoded subunit. *The EMBO Journal*, **7**, 3863–3871

74. Witte, C., Jensen, R.E., Yaffe, M.P. and Scatz, G. (1988) Mas1, a gene essential for yeast mitochondrial assembly, encodes a subunit of the mitochondrial processing protease. *The EMBO Journal*, **7**, 1439–1447

75. Oudshoorn, P., Van Steeg, H., Swinkels, B.W., Schoppink, P. and Grivell, L.A. (1987) Subunit II of yeast QH2 – cytochrome-c oxidoreductase – nucleotide-sequence of the gene and features of the protein. *The EMBO Journal*, **163**, 97–103

76. Tzagoloff, A., Wu, M. and Crivellone, M. (1986) Assembly of the mitochondrial-membrane system – characterization of Cor1, the structural gene for the 44-kilodalton core protein of yeast. *The Journal of Biological Chemistry*, **261**, 17163–17169

77. Anderson, S., Bankier, A.T., Barrel, B.G., de Bruijn, M.H.L., Coulson, A.R., Drouin, J., Eperon, J.C., Nierlich, D.P., Roe, B.A., Sanger, F., Schreier, P.H., Smith, A.J.H., Staden, R. and Young, J.G. (1981) Sequence and organization of the human mitochondrial genome. *Nature*, **290**, 457–465

78. Anderson, S., de Bruijn, M.H.L., Coulson, A.R., Eperon, J.C., Sanger, F. and Young, J.G. (1982) Complete sequence of bovine mitochondrial DNA – conserved features of the mammalian mitochondrial genome. *Journal of Molecular Biology*, **156**, 683–717

79. Saraste, M. (1984) Location of haem binding sites of the mitochondrial cytochrome c_1. *FEBS Letters*, **166**, 367–372

80. Wakabayashi, S., Matsubara, H., Kim, C.H., Kawai, K. and King, T.E. (1980) The complete amino acid sequence of beef heart cytochrome c_1. *Biochemical and Biophysical Research Communications*, **97**, 1548–1554

81. Nishikimi, M., Suzuki, H., Ohta, S., Sakurai, T., Shimomura, Y., Tanaka, M., Kagawa, Y. and Ozawa, T. (1987) Isolation of a cDNA clone for human cytochrome c_1 from a λgt11 expression library. *Biochemical and Biophysical Research Communications*, **145**, 34–39

82. Suzuki, H., Hosokawa, Y., Nishikimi, M. and Ozawa, T. (1989) Structural organization of the human mitochondrial cytochrome c_1 gene. *Journal of Biological Chemistry*, **264**, 1368–1374

83. Nishikimi, M., Suzuki, H., Yamaguchi, H., Matsukage, A., Yoshida, M.C. and Ozawa, T. (1988) Assignment of the human cytochrome c_1 gene to chromosome 8. *Biochem. Int.*, **16**, 655–660

84. Li, Y., Leonard, K. and Weiss, H. (1981) Membrane-bound and water-soluble cytochrome c_1 from *Neurospora* mitochondria. *European Journal of Biochemistry*, **116**, 199–205

85. Schägger, H., Borchart, U., Machleidt, W., Link, T.A. and von Jagow, G. (1987) Isolation and amino acid sequence of the 'Rieske' iron sulfur protein of beef heart ubiquinone: cytochrome *c* reductase. *FEBS Letters*, **219**, 161–168

86. Nishikimi, M., Hosokawa, Y., Toda, H., Suzuki, H. and Ozawa, T. (1990) The primary structure of human rieske iron-sulfur protein of mitochondrial cytochrome bc_1 complex deduced from cDNA analysis. *Biochemistry International*, **20**, 155–160

87. Wakabayashi, S., Takao, T., Shimonishi, Y., Kuramitsu, S., Matsubara, H., Wang, T., Zhang, Z. and King, T.E. (1985) Complete amino acid sequence of the ubiquinone binding protein (QP-C), a protein similar to the 14,000-dalton subunit of the yeast ubiquinal–cytochrome *c* reductase complex. *The Journal of Biological Chemistry*, **260**, 337–343

88. Suzuki, H., Hosokawa, Y., Toda, H., Nishikimi, M. and Ozawa, T. (1988) Cloning and sequencing of a cDNA for human mitochondrial ubiquinone-binding protein of complex III. *Biochemical and Biophysical Research Communications*, **156**, 987–994

89. Suzuki, H., Hosokawa, Y., Toda, H., Nishikimi, M. and Ozawa, T. (1989) Isolation of a single nuclear gene encoding human ubiquinone-binding protein in complex III of mitochondrial respiratory chain. *Biochemical and Biophysical Research Communications*, **161**, 371–378

90. Hosokawa, Y., Suzuki, H., Nishikimi, M., Matsukage, A., Yoshida, M.C. and Ozawa, T. (1990) Chromosomal assignment of the gene for the ubiquinone-binding protein of human mitochondrial cytochrome bc_1 complex. *Biochemistry International*, **21**, 41–44

91. Yu, L., Yang, F.D. and Yu, C.A. (1985) Interaction and identification of ubiquinone binding proteins in ubiquinol cytochrome *c* reductase by azido-ubiquinone derivatives. *The Journal of Biological Chemistry*, **260**, 963–973

92. Schägger, H., Link, T.A., Engel, W.D. and von Jagow, G. (1986) Isolation of the eleven protein subunits of the bc_1 complex from beef heart. *Methods in Enzymology*, **126**, 224–237

93. Borchart, U., Machleidt, W., Schägger, H., Link, T.A. and von Jagow, G. (1986) Isolation and amino acid sequence of the 9.5 kDa protein of beef heart ubiquinol:cytochrome c reductase. *FEBS Letters*, **200**, 81–86

94. Wakabayashi, S., Matsubara, H., Kim, C.H. and King, T.E. (1982) Structural studies of bovine heart cytochrome c_1. *The Journal of Biological Chemistry*, **252**, 9335–9344

95. Borchart, U., Machleidt, W., Schägger, H., Link, T.A. and von Jagow, G. (1985) Isolation and amino acid sequence of the 8 kDa DCCD-binding protein of beef heart ubiquinol:cytochrome c reductase. *FEBS Letters*, **191**, 125–130

96. Schägger, H., von Jagow, G., Borchart, U. and Machleidt, W. (1983) Amino acid sequence of the smallest protein of the cytochrome c_1 subcomplex from beef heart mitochondria. *Hoppe-Seyler's Zeitschrift für Physiologische Chemie*, **364**, 307–311

97. King, T.E. (1983) Cardiac cytochrome c_1. *Advances in Enzymology and Related Areas of Molecular Biology*, **54**, 267–366
98. González-Halphen, D., Lindorfer, M.A. and Capaldi, R.A. (1988) Subunit arrangement in beef heart complex III. *Biochemistry*, **27**, 7021–7031
99. Roberts, H., Smith, S.C., Marzuki, S. and Linnane, A.W. (1980) Evidence that cytochrome *b* is the antimycin binding component of the yeast mitochondrial cytochrome bc_1 complex. *Archives of Biochemistry and Biophysiology*, **200**, 387–395
100. Das Gupta, U. and Rieske, J.S. (1973) Identification of a protein component of the antimycin binding site of the respiratory chain by photoaffinity labelling. *Biochemical and Biophysical Research Communications*, **54**, 1247–1252
101. Thierbach, G. and Michaelis, G. (1982) Mitochondrial and nuclear myxothiazol resistance in *Saccharomyces cerevisiae*. *Molecular and General Genetics*, **186**, 501–506
102. Cooper, C.E. (1990) The steady-state kinetics of cytochrome *c* oxidation by cytochrome oxidase. *Biochimica et Biophysica Acta*, **1017**, 187–203
103. Capadli, R.A. (1990) Structure and function of cytochrome *c* oxidase. *Annual Reviews in Biochemistry*, **59**, 569–596
104. Capaldi, R.A. (1990) Structure and assembly of cytochrome *c* oxidase. *Archives of Biochemistry and Biophysics*, **280**, 252–262
105. Kadenbach, B. and Merle, P. (1981) On the function of multiple subunits of cytochrome *c* oxidase from higher eukaryotes. *FEBS Letters*, **135**, 1–11
106. Capaldi, R.A., Takamiya, S., Zhang, Y-Z, Gonzalez-Halphen, D. and Yamamura, W. (1986) Structure of cytochrome *c* oxidase. *Current Topics in Bioenergics*, **15**, 91–112
107. Holm, L., Saraste, M. and Wikstrom, M. (1987) Structural models of the redox centres in cytochrome oxidase. *The EMBO Journal*, **6**, 2819–2823
108. Prochaska, L., Bisson, R. and Capaldi, R.A. (1980) Structure of the cytochrome *c* oxidase complex: labelling by hydrophilic and hydrophobic protein modifying reagents. *Biochemistry*, **19**, 3174–3179
109. Bisson, R., Azzi, A., Gutweniger, H., Colonna, R., Montecucco, C. and Zanotti, A. (1978) Interaction of cytochrome *c* with cytochrome *c* oxidase. *The Journal of Biological Chemistry*, **253**, 1874–1880
110. Bisson, R., Steffens, G.C.M., Capaldi, R.A. and Buse, G. (1982) Mapping of the cytochrome *c* binding site on cytochrome *c* oxidase. *FEBS Letters*, **144**, 359–363
111. Malatesta, F. and Capaldi, R.A. (1982) Localisation of $cysteine_{115}$ in subunit III of beef heart cytochrome *c* oxidase to the C side of the mitochondrial inner membrane. *Biochemical and Biophysical Research Communications*, **109**, 1180–1185
112. Zeviani, M., Nakagawa, M., Herbert, J., Lomax, M.I., Grossman, L.I., Shorbany, A.A., Miranda, A.F., DiMauro, S. and Schon, E.A. (1987) Isolation of a cDNA clone encoding subunit IV of human cytochrome c oxidase. *Gene*, **55**, 205–217
113. Taanman, J.W., Schrage, C., Ponne, N., Bolhuis, P., de Vries, H. and Agsteribbe, E. (1989) Nucleotide sequence of cDNA-encoding subunit VIb of human cytochrome *c* oxidase. *Nucleic Acids Research*, **17**, 1766
114. Goto, Y., Amuro, N. and Okazaki, T. (1989) Nucleotide sequence of cDNA for rat brain and liver cytochrome *c* oxidase subunit IV. *Nucleic Acids Research*, **17**, 2851
115. Goto, Y., Amuro, N. and Okazaki, T. (1989) Nucleotide sequence of cDNA for rat liver and brain cytochrome *c* oxidase subunit VIa. *Nucleic Acids Research*, **17**, 6388
116. Droste, M., Schon, E.A. and Kadenbach, B. (1989) Nucleotide sequence of cDNA encoding subunit VA from rat heart cytochrome *c* oxidase. *Nucleic Acids Research*, **17**, 4375
117. Gopalan, G., Droste, M. and Kadenbach, B. (1989) Nucleotide sequence of cDNA encoding subunit IV of cytochrome *c* oxidase from fetal rat liver. *Nucleic Acids Research*, **17**, 4376
118. Fabrizi, G.M., Rizzuto, R., Nakase, H., Mita, S., Kadenbach, B. and Schon, E.A. (1989) Sequence of a cDNA specifying subunit VIa of human cytochrome *c* oxidase. *Nucleic Acids Research*, **17**, 6409
119. Fabrizi, G.M., Rizzuto, R., Nakase, H., Mita, S., Kadenbach, B. and Schon, E.A. (1989) Sequence of a cDNA specifying subunit VIIa of human cytochrome *c* oxidase. *Nucleic Acids Research*, **17**, 7107
120. Rizzuto, R., Nakase, H., Darras, B., Francke, U., Fabrizi, G.M., Mengel, T., Walsh, F., Kadenbach, B., DiMauro, S. and Schon, E.A. (1989) A gene specifying subunit VIII of human cytochrome *c* oxidase is localised to chromosome II and is expressed in both muscle and non-muscle tissues. *The Journal of Biological Chemistry*, **264**, 10595–10600
121. Koga, Y., Fabrizi, G.M., Mita, S., Arnaudo, E., Lomax, M.I., Agua, M.S., Grossman, L.I. and Schon, E.A. (1990) Sequence of a cDNA specifying subunit VIIc of human cytochrome c oxidase. *Nucleic Acids Research*, **18**, 684

122. Fuller, S.D., Capaldi, R.A. and Henderson, R. (1979) Structure of cytochrome c oxidase in deoxy-cholate derived 2 dimensional crystals. *Journal of Molecular Biology*, **134**, 305–327
123. Deatherage, J.F., Henderson, R. and Capaldi, R.A. (1982) 2-dimensional structure of cytochrome c oxidase vesicle crystals in negative stain. *Journal of Molecular Biology*, **158**, 487–499
124. Jarausch, J. and Kadenbach, B. (1985) Structure of the cytochrome c oxidase complex of rat-liver. 1. Studies on nearest-neighbor relationship of polypeptides with cross-linking reagents. *European Journal of Biochemistry*, **146**, 211–217
125. Jarausch, J. and Kadenbach, B. (1985) Structure of the cytochrome c oxidase complex of rat liver. 2. Topological orientation of polypeptides in the membrane as studied by proteolytic digestion and immunoblotting. *European Journal of Biochemistry*, **146**, 219–225
126. Zhang, Y-Z., Lindorfer, M.A. and Capaldi, R.A. (1988) Orientation of the cytoplasmically made subunits of beef heart cytochrome c oxidase determined by protease digestion and antibody binding experiments. *Biochemistry*, **27**, 1389–1394
127. Lightowers, R., Ewart, G., Aggeler, R., Zhang, Y-Z., Calavetta, L. and Capaldi, R.A. (1990) Isolation and characterisation of the cDNAs encoding two isoforms of subunit C_{IX} of bovine cytochrome c oxidase. *The Journal of Biological Chemistry*, **265**, 2677–2681
128. Lomax, M.I., Coucouvanis, E., Schon, E.A. and Barald, K.F. (1990) Differential expression of nuclear genes for cytochrome c oxidase during myogenesis. *Muscle & Nerve*, **13**, 330–337
129. Bachman, N.J., Lomax, M.I. and Grossman, L.I. (1987) Two bovine genes for cytochrome c oxidase subunit IV: a processed gene and an expressed gene. *Gene*, **55**, 219–229
130. Cao, X., Hengst, L., Schlerf, A., Droste, M., Mengel, T. and Kadenbach, B. (1989) Complexity of nucleus-encoded genes of mammalian cytochrome c oxidase. *Annals of the New York Academy of Sciences*, **550**, 337–347
131. van Gelder, B.F. and Muijers, A.O. (1966) On cytochrome c oxidase II. The ratio of cytochrome a to cytochrome a_3. *Biochimica et Biophysica Acta*, **118**, 47–57
132. Tzagoloff, A. and Wharton, D.C. (1965) Studies on the electron transfer system. LXII. The reaction of cytochrome oxidase with carbon monoxide. *The Journal of Biological Chemistry*, **240**, 2628–2633
133. Hatefi, Y., Ragan, C.I. and Galante, T. (1985) The enzymes and the enzyme complexes of mitochondrial oxidative phosphorylatic system. In *The Enzymes of Biological Membranes*, 2nd edn (ed. A.M. Martonosi), Plenum, New York and London, pp. 1–56
134. Hatefi, Y. (1985) The mitochondrial electron-transport and oxidative-phosphorylation system. *Annual Reviews of Biochemistry*, **54**, 1015–1069
135. Walker, J.E., Fearnley, I.M., Lutter, R., Todd, R.J. and Runswick, M.J. (1990) Structural aspects of proton-pumping ATPase. *Philosophical Transactions of the Royal Society of London, Series B – Biological Sciences*, **326**, 367–378
136. Homamoto, T. and Kagawa, Y. (1985) H+-ATPase as an energy converting enzyme. In *The Enzymes of Biological Membranes*, 2nd edn (ed. A.M. Martonosi), Plenum, New York and London, pp. 149–176
137. Joshi, S. and Burrows, R. (1990) ATP synthase complex from bovine heart mitochondria – subunit arrangement as revealed by nearest neighbor analysis and susceptibility to trypsin. *The Journal of Biological Chemistry*, **265**, 14518–14525
138. Hekman, C., Tomich, J.M. and Hatefi, Y. (1991) Mitochondrial ATP synthase complex – membrane topography and stoichiometry of the F_0 subunits. *The Journal of Biological Chemistry*, **266**, 13564–13571
139. Gay, N.J. and Walker, J.E. (1985) 2 genes encoding the bovine mitochondrial ATP synthase proteolipid specify precursors with different import sequences and are expressed in a tissue-specific manner. *The EMBO Journal*, **4**, 3519–3524
140. Clayton, D.A. (1984) Transcription of the mammalian mitochondrial genome. *Annual Review of Biochemistry*, **53**, 573–594
141. Tzagoloff, A. and Myers, A.M. (1986) Genetics of mitochondrial biogenesis. *Annual Review of Biochemistry*, **55**, 249–285
142. Attardi, G. and Schatz, G. (1988) Biogenesis of mitochondria. *Annual Review of Cell Biology*, **4**, 289–333
143. Pfanner, N. and Neupert, W. (1990) The mitochondrial protein import apparatus. *Annual Review of Biochemistry*, **59**, 331–353
144. Roise, D. and Schatz, G. (1988) Mitochondrial presequences. *The Journal of Biological Chemistry*, **263**, 4509–4511
145. Murakami, H., Pain, D. and Blobel, G. (1988) 70 kD heat shock related protein is one of at least 2 distinct cytosolic factors stimulating protein import into mitochondria. *Journal of Cell Biology*, **107**, 2051–2057

146. Ostermann, J., Horwich, A.L., Neupert, W. and Hartl, F-U. (1989) Protein folding in mitochondria requires complex-formation with HSP60. *Nature*, **341**, 125–130
147. Sollner, T., Rassow, J., Wiedmann, M., Schlossmann, J., Keil, P., Neupert, W. and Pfanner, N. (1992) Mapping of the protein import machinery in the mitochondrial outer membrane by cross-linking of translocation intermediates. *Nature*, **355**, 84–87
148. Pollock, R.A., Hartl, F-U., Cheng, M.Y., Ostermann, J., Horwich, A. and Neupert, W. (1988) The processing peptidase of yeast mitochondria – the 2 cooperating components MPP and PEP are structurally related. *The EMBO Journal*, **7**, 3493–3500
149. Cheng, M.Y., Hartl, F-U., Martin, J., Pollock, R.A., Kalousek, F., Neupert, W., Hallberg, E.M., Hallberg, R.L. and Horwich, A. (1989) Mitochondrial heat shock protein HSP60 is essential for assembly of proteins imported into yeast mitochondria. *Nature*, **337**, 620–625

2
Mitochondrial DNA and the genetics of mitochondrial disease

Eric A. Schon

INTRODUCTION

The mitochondrial diseases are a heterogeneous group of disorders that are characterized by defects in mitochondrial function. Because the mitochondrion contains its own DNA, heritable errors leading to mitochondrial dysfunction can arise from errors in the nuclear genome or the mitochondrial genome. In addition, some mitochondrial diseases are not inherited, while others are caused by environmental factors. The last few years have seen a veritable explosion in our knowledge of the genetics of mitochondrial diseases. We will summarize here much of what has been learned, with particular focus on the aetiology and pathogenesis of these disorders.

ORGANIZATION AND EXPRESSION OF THE HUMAN MITOCHONDRIAL GENOME

The human mitochondrial genome is a small 16 569 bp circle of double-stranded DNA [1]. It has a highly asymmetric base composition: one strand (the 'light' strand) is pyrimidine-rich (i.e. many Ts and Cs), while the complementary 'heavy' strand is correspondingly purine-rich (many As and Gs). The mitochondrial DNA (mtDNA) contains 37 genes (Figure 2.1): 13 polypeptide-coding genes (mRNAs), 22 transfer RNA genes (tRNAs) and two ribosomal RNA genes (12S and 16S rRNAs). All 13 polypeptides are components of the respiratory chain/oxidative phosphorylation system, which is located in the mitochondrial inner membrane. The latter include genes encoding seven subunits of Complex I or NADH–ubiquinone oxidoreductase (ND1, ND2, ND3, ND4, ND4L, ND5 and ND6), one subunit of Complex III or ubiquinol–cytochrome c oxidoreductase (cyt b), three subunits of Complex IV or cytochrome c oxidase (CO I, CO II and CO III) and two subunits of Complex V or ATP synthetase (ATPase 6 and ATPase 8). Each of these four complexes also contains subunits encoded by nuclear genes, which are imported from the cytoplasm and assembled together with the mtDNA-encoded subunits. All the subunits of Complex II (succinate–ubiquinone oxidoreductase) are nuclear DNA-encoded. While mitochondria have their own transcriptional and translational

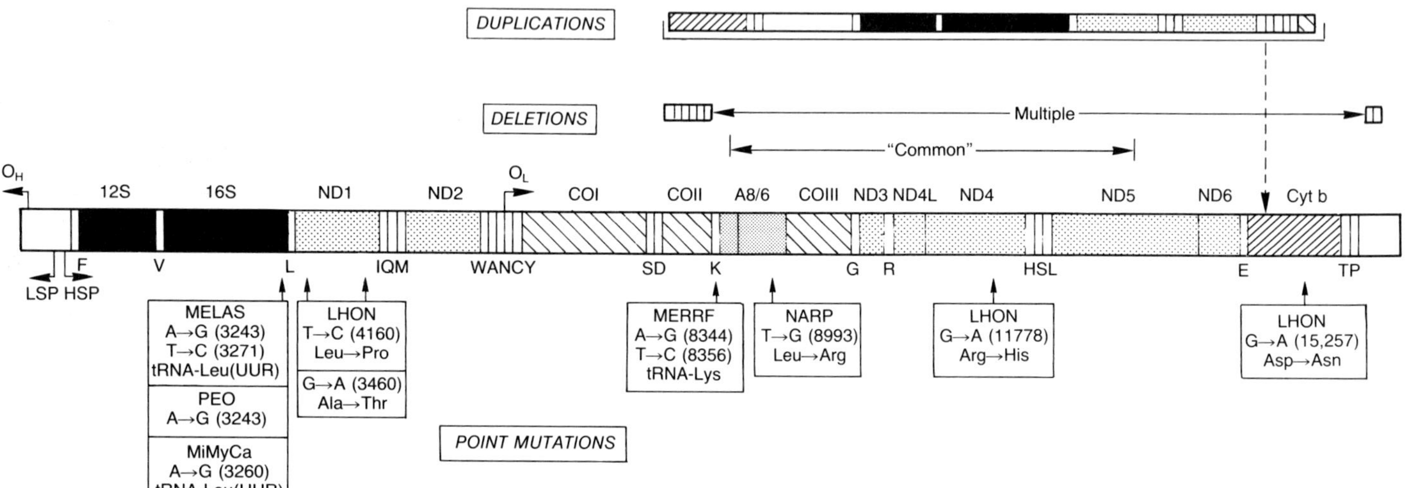

Figure 2.1 Morbidity map of the human mitochondrial genome. Linearized map of the normally circular human mitochondrial genome. The structural genes for the mitochondrial-encoded 12S and 16S ribosomal RNAs, the subunits of NADH–coenzyme Q oxidoreductase (ND), cytochrome *c* oxidase (CO), cytochrome *b* (Cyt b) and ATP synthase (A), and 22 tRNAs (one-letter amino acid nomenclature) are shown. The origins of light-strand (O_L) and heavy-strand (O_H) replication, and of the promoters for initiation of transcription from the light strand (LSP) and heavy strand (HSP), are shown by arrows. Mutations in mtDNA (point mutations, deletions and duplications) associated with mitochondrial diseases (see the text) are shown. MELAS, mitochondrial encephalomyopathy with lactic acidosis and stroke-like episodes; PEO, progressive external ophthalmoplegia; MiMyCa, mitochondrial myopathy and cardiomyopathy; LHON, Leber's hereditary optic neuropathy; MERRF, myoclonic epilepsy with ragged red fibres; NARP, neuropathy, ataxia and retinitis pigmentosa

machinery and a genetic code that differs from the nuclear 'universal' code, most of the other proteins located within mitochondria are encoded by the nuclear genome. These nuclear gene products are synthesized on cytoplasmic ribosomes and are subsequently imported into the mitochondria.

The replication and transcription of human mtDNA is unusual [2]. There are two origins of replication: an origin of heavy-stranded replication (O_H) and an origin of light-strand replication (O_L). The terms 'light' and 'heavy' refer to the differential buoyancy of the separated strands in alkaline caesium chloride gradients. O_H is located in the only region of the genome that contains no known genes. This control region is called the D-loop, because a large segment of the heavy strand is displaced by a piece of DNA that is synthesized in the region. There is only one promoter region for transcription, located, like O_H, in the D-loop region [3,4]. The heavy- and light-strand promoters (HSP and LSP respectively) initiate transcription of their respective strands to produce giant nearly-full-circle-length polycistronic transcripts. The tRNA genes are located at strategic points on the circle, in such a way that they separate all but one of the individual rRNA and mRNA genes (Figure 2.1). These precursor RNAs are then cleaved by endogenous nucleases precisely at the 5' and 3' ends of the tRNAs, thereby releasing the mature tRNAs and the flanking rRNAs and mRNAs.

Translation of the mRNAs is on mitochondrial ribosomes (which contain, of course, the 12S and 16S rRNAs). If all transcription were polycistronic, each rRNA would be synthesized in molar amounts equal to the amount of any individual mRNA. However, the mitochondrion needs to synthesize enough rRNAs to satisfy the translation requirements of 13 mRNAs at once. For this reason, transcription of the HSP (which encodes the rRNAs) is done in two versions: besides synthesis of the full-length polycistronic H-strand transcript, a shorter transcript, extending from HSP to the end of the 16S rRNA gene, is also synthesized. This short transcript is made at about 25 times the rate of the full-length transcript [5]. In this way, sufficient amounts of 12S and 16S rRNAs are made available for protein translation. Importantly, transcription of this shorter polycistronic RNA is terminated at a site within the tRNA$^{Leu(UUR)}$ gene, immediately downstream of the end of the 16S rRNA. A specific 'terminator' protein binds to a DNA region within the tRNA$^{Leu(UUR)}$ gene to block the continued travel of the RNA polymerase on the DNA; the polymerase falls off the DNA, and transcription is terminated [6–8]. When the termination protein is not bound, the polymerase can continue on the DNA and synthesize a full-length transcript. This mechanism of transcription termination has been proposed to play a role in the pathogenesis of MELAS (see below).

MATERNAL INHERITANCE AND POPULATION GENETICS OF HUMAN mtDNA

Most somatic cells contain multiple mtDNAs in each mitochondrion – approximately five mitochondrial genomes per organelle [9]. On the other hand, unfertilized eggs [10–12] and platelets [13] contain an average of only one mtDNA per mitochondrion. Furthermore, mitochondria, and mtDNAs, are inherited only from the mother [14], although there is recent evidence that paternal mitochondria can be inherited at an extremely low level [15]. Genetic diseases arising from errors in the mitochondrial genome are therefore maternally inherited: mothers can transmit the defect to all of their children (both daughters and sons), but only daughters will transmit the defect to their children.

Three other key concepts of mitochondrial genetics, especially as they relate to mitochondrial diseases, are heteroplasmy, mitotic segregation and threshold effects [16]. Heteroplasmy is a condition in which not all the mtDNAs in a specified cell, tissue, organ or individual are identical. Normally, individuals are fundamentally homoplasmic, that is, all mitochondrial genomes are identical. Rarely, a genome may suffer a mutation – usually a point mutation, with transition mutations [a purine (A or G) replaced by a purine, or a pyrimidine (C or T) by a pyrimidine] favoured over transversion (a purine replaced by a pyrimidine or vice versa). If that mutation is fixed – mtDNA has very poor DNA repair systems [17–20] – a new mtDNA genotype will coexist with the unmutated mtDNA population. This condition is known as heteroplasmy. Mitotic segregation refers to the fact that both mtDNA replication and mitochondrial division are stochastic processes unrelated to the cell cycle or to the timing of nuclear DNA replication. Thus, a dividing cell may potentially donate a different complement of organelles and genomes to its progeny. This process becomes clinically important if an individual contains a heteroplasmic population of both wild-type and mutated mtDNAs causing a mitochondrial disease. The phenotypic expression of a mutation may vary in both space (among cells or tissues) and time (during development or during the course of a lifespan), based merely on the random processes of mitotic segregation. Of course, there may also be active selection processes going on as well, in which certain cells may either eliminate or concentrate a population of mutant mtDNAs. These effects will combine to generate, for example, a respiratory chain deficiency in some tissues but not others, but only if the number of mutant mtDNAs exceeds a certain threshold. This threshold varies from tissue to tissue, and is related to the requirements for aerobic respiration and energy production, with brain, retina and muscle (including heart) exhibiting the highest energy requirements.

LARGE-SCALE REARRANGEMENTS OF mtDNA

Four distinct disorders have been associated with giant deletions or insertions of the mitochondrial genome. Surprisingly, most of them are not maternally inherited.

The first disorder is progressive external ophthalmoplegia (PEO). PEO may be manifested either as ocular myopathy (OM) alone, or as part of a more severe disorder called Kearns–Sayre syndrome (KSS). KSS is a multisystem mitochondrial disorder defined by the presence of ophthalmoplegia, onset before age 20, and at least one of the following: high cerebrospinal fluid (CSF) protein content, blockage in heart conduction, or ataxia [21]. Morphologically, KSS patients display ragged red fibres (RRF), which are diagnostic of massive mitochondrial proliferation in muscle. KSS is ultimately fatal. OM shares with KSS ocular myopathy and, often, RRF in muscle, but unlike KSS, there is no systemic involvement, and the disease is rarely fatal. Biochemically, both KSS and OM often show reduced respiratory chain enzyme activity, particularly that of cytochrome *c* oxidase (COX) [22].

Historically, PEO and KSS were difficult diseases to classify, because patients presenting with these disorders were almost all sporadic, with no apparent genetic component. However, Holt *et al.* [23] opened a new field of investigation when they found deletions of mtDNA in nine of 25 patients with 'mitochondrial myopathies', defined by the appearance of morphologically abnormal organelles in the muscle biopsy. Although they did not describe the clinical syndromes of the patients, further investigations [24–28] confirmed that spontaneously occurring giant deletions – up to 9 kb of the mitochondrial genome – were a hallmark of KSS and some cases of PEO. These deletions were clearly observable as a faster-migrating hybridizing fragment in Southern blot hybridizations. All patients with deletions have been heteroplasmic. The size and location of the deletions, and the number of deleted mtDNAs (ΔmtDNAs) relative to the number of normal mitochondrial genomes, differed among patients, and did not appear to be correlated to the presentations or the severity of the disease phenotype [25–31]. However, one particular deletion has been found in about one-third of all patients with deletions, and has therefore been called the 'common deletion' [27,32–38]. It is 4977 bp long, and removes DNA between the ATPase 8 and the ND5 genes. Importantly, the deletion breakpoint is flanked precisely at the edges by a perfect 13 bp direct repeat [33]. In fact, more than half of all sequenced deletion breakpoints are flanked by these direct repeats, called Class I by Mita *et al.* [34], implying that some type of homologous recombination event may be associated with the generation of these deletions. Another mechanism, called 'slipped mispairing' [27,33,37], has also been invoked to explain the generation of ΔmtDNAs.

The common deletion has also been identified in a second ostensibly unrelated disorder. This is Pearson's syndrome, a haematopoietic disease characterized by pancytopenia [36], in which the ΔmtDNA population is most pronounced in blood. Since then, other ΔmtDNAs and partial duplications of mtDNA have also been found in this disease [36].

ΔmtDNAs have also been found in other patients with different combinations of clinical phenotypes. There is one report of ΔmtDNAs in a patient with KSS and features of Lowe syndrome, which is normally considered an X-linked disease [39], while ΔmtDNAs in patients with features of KSS and MELAS have also been reported [40].

Because only a single species of ΔmtDNA is usually found in any individual with sporadic PEO, KSS or Pearson's disease, it is likely that the population of deleted molecules is a clonal expansion of an initial mutation event occurring early in oogenesis or embryogenesis; however, it is currently not clear whether a woman with KSS can transmit the disease to her children [41,42]. A number of observations support this idea. First, the ΔmtDNA population in KSS appears to be generalized to all tissues of the body [43,44]. Second, the ΔmtDNA population in muscle in KSS increases and accumulates during the lifespan of the patient [45]. Third, even though ΔmtDNAs are most prevalent in blood in Pearson's syndrome, they can be found at lower levels in other tissues, and in fact, if the patients lives long enough, he or she will eventually begin to show the symptoms of KSS, as the ΔmtDNA population accumulates in muscle [46]. Finally, ΔmtDNAs in ocular myopathy seem confined to muscle, but are rarely found (by Southern blot analysis [26]) in other tissues, implying that the deletion event occurred somewhat later in embryogenesis, with the ΔmtDNA population confined to mesoderm [47].

Besides deletions, duplications of mtDNA have also been identified in KSS [48,49] and in a kidney disorder [50]. While not as frequent, it is believed that the consequences of a duplication are similar to that of a deletion. In fact, the rare duplications may actually be recombination intermediates on their way to resolution into deletions, but which got 'frozen' and expanded in the patient's mtDNA.

It had long been thought that deletions could remove any region of the mtDNA except for O_H or O_L, because deletion of the replication origins would render the genome 'sterile': a ΔmtDNA of this type would not be able to replicate, and would rapidly be lost in the population. In fact, the rare ΔmtDNAs that were identified in the D-loop region tended to confirm this idea, as they removed HSP, but not LSP (which is required for synthesis of a primer RNA involved in replication), O_H, or O_L [51,52]. However, truly gigantic deletions have recently been described that challenge these assumptions, as two groups have identified ΔmtDNAs greater than 10 kb in size, and which encompass O_L [53,54]. Furthermore, the ΔmtDNA population described by Ballinger *et al.* [54] is maternally inherited, and was found in a family presenting with hereditary diabetes. It is unknown at present how a ΔmtDNA lacking O_L replicates; these findings will require further confirmation and analysis.

ΔmtDNAs have been studied extensively by a variety of methods. It appears that they are transcribed, but are apparently not translated [55]; even genes not encompassed by the deletion do not seem to be translated [55–57]. The lack of translation may be due to the fact that the deletions remove essential tRNAs that are required for protein synthesis. However, this explanation implies that mitochondria, which contain multiple mtDNAs, must be homoplasmic for the deletion, and that they do not exchange genomes (by organelle fusion) with mitochondria containing wild-type genomes which ought to complement the mutation. On the other hand, the deletion may be a dominant mutation [58]; even if the mtDNA population within the mitochondrion is heteroplasmic, the ΔmtDNA subpopulation may prevent overall translation, perhaps by altering the overall tRNA pool, or by titrating out scarce tRNAs (e.g. on fusion mRNAs transcribed across the deletion breakpoint).

The issue of complementation and dominance has become a major focus of investigation, as it is key to further work aimed towards treatment and therapy. It now appears that complementation may occur in some situations. Using a tissue-culture system, it has been observed that there is a threshold number of ΔmtDNAs (approximately 70%) required before translation is inhibited [59]. The situation may be even more complicated, as the degree of respiratory chain impairment may depend on the specific mtDNA genes which are deleted, as well as the number of tRNA genes deleted [60,61].

Besides sporadic ΔmtDNAs in KSS and PEO, familial inheritance of mtDNA deletions has also been demonstrated in a mitochondrial myopathy with PEO [62–64]. This disease is not sporadic, but is inherited in an autosomal dominant manner. The deletions differed among family members, and different deletions coexisted within the same muscle biopsies of different affected family members. Thus, as opposed to the 'clonal' deletions found in sporadic PEO, familial PEO causes multiple deletions that are apparently generated in 'real time' over the life-span of the individual. Clinically, the net effects in the two disorders are similar, even though the aetiological causes are probably different, as the fundamental genetic defect in autosomal dominant PEO most likely resides in a gene product affecting the proclivity of the mtDNA to suffer deletions (presumably during mtDNA replication).

Besides autosomal-dominant PEO, multiple deletions have also been associated with patients presenting with multiple symmetrical lipomas [65], recurrent myoglobinuria [66], progressive mitochondrial encephalomyopathy [67] and myo-neuro-gastrointestinal encephalopathy (MNGIE) (E. Ciafaloni, personal communication).

DELETIONS OF mtDNA IN NORMAL AGING

Following up on the findings of 'unique' ΔmtDNAs in spontaneous KSS/PEO and of multiple ΔmtDNAs in familial PEO, Ikebe *et al.* [68] searched for the presence of ΔmtDNAs in brain tissue of normal subjects and of patients with Parkinson's disease (PD). While they could not observe any ΔmtDNAs by Southern blot hybridization analysis, they did find them using the more sensitive polymerase chain reaction (PCR). They focused on the 'common deletion', and found that this ΔmtDNA was present in brain from all five patients with PD analysed. (Whether or not the amount of ΔmtDNA is elevated in brain from PD patients is currently controversial [69,70].) The same ΔmtDNA was also present in brain from normal age-matched controls but not in younger subjects. Using semi-quantitative methods [69], this group estimated that the common deletion was present at a level of about 0.3% in control striatum and at about 5% in PD striatum. The common deletion was also observed by PCR in liver from all 34 subjects examined over the age of 50 [71]. Cortopassi and Arnheim [72] found that heart and brain tissue from normal adults contained the common deletion, but the deletion was not present in fetal heart or brain. They estimated that there was one ΔmtDNA per 1000 normal mtDNAs (i.e. 0.1%) in the heart muscle of middle-aged adults.

ΔmtDNAs were also observed in patients with cardiac disease. In one study, the amount of common deletion ΔmtDNA was estimated to reach high levels in patients with ischaemic heart disease (up to 0.85% ΔmtDNA), as well as in other cardiac pathologies (up to 0.16%), as compared with controls [73]. In a second study, a different, 7.4 kb ΔmtDNA was found in 15 of 29 patients aged 31–70 and in all 14 patients above age 70 [74]. Finally, ΔmtDNAs were found by PCR in cardiomyocytes of patients with hypertrophic or dilated cardiomyopathy [75].

Using a new quantitative PCR method to measure the amount of ΔmtDNA [76], it has been estimated that there is a 10 000-fold increase in the amount of common deletion in total muscle DNA during the lifespan of normal individuals; this specific ΔmtDNA represented about 0.1% of total mtDNA in the oldest samples studied. This relationship appeared to be exponential, with a 'break' in the curve around age 40. Analysis of the DNA derived from autopsied somatic tissues showed that slowly dividing tissues, such as muscle and brain, had the greatest amount of ΔmtDNA, while more rapidly dividing tissue, such as liver, had much lower amounts of ΔmtDNA [72,76,77].

PCR fragments corresponding to other unique deletions found in KSS patients have also been found in aged normal individuals [78]. This implies that the common deletion, while frequently found in KSS patients, is certainly not the only ΔmtDNA present in aged muscle, and that it is quite likely that aged muscle harbours numerous (hundreds or even thousands) of species of ΔmtDNAs, and that the total amount of ΔmtDNAs may reach levels that could be physiologically significant in terms of the decline in oxidative metabolism in aging. Clearly, this is an area of investigation that will receive a lot of attention in the near future (see Chapter 12).

POINT MUTATIONS IN mtDNA

In the past four years, seven maternally inherited mitochondrial diseases with distinct clinical phenotypes have been associated with point mutations in mtDNA, all of which result in neurological or neuromuscular disorders.

Leber's hereditary optic neuropathy (LHON)

LHON was the first mitochondrial disease to be defined at the molecular level (see Chapter 10). Wallace and co-workers found maternal inheritance in a number of LHON pedigrees of a G→A transition at mtDNA position 11 778, in codon 340 of the ND4 gene of Complex I [79,80]. This mutation was non-conservative, and converted the codon from Arg to His. The mutation is homoplasmic in some families but heteroplasmic in others [81].

Not all patients with LHON carry the nucleotide (nt) 11 778 mutation. Other mutations have now been found that are also associated with LHON. These include mutations at nt 3460 in the ND1 gene and at nt 4160, also in the ND1 gene [82–84]. A new mutation has been described at position 15 257 in the cytochrome *b* gene [85,86], and a mutation at nt 13 708 in the ND5 gene may act in concert with the cytochrome *b* mutation [85]. It now appears that specific mutations may act together, perhaps synergistically, to produce the LHON phenotype. Moreover, there may be a locus on the X-chromosome that interacts with the mutated LHON mtDNA genotype [87,88]. This may have relevance to the known increased penetrance of the disease in males.

Myoclonic epilepsy with ragged red fibres (MERRF)

MERRF is maternally inherited [89] and is characterized by myoclonus, ataxia, weakness, generalized seizures and hearing loss. Muscle biopsies show RRF [90]. MERRF has been found to be associated with an A→G transition at nt 8344 in the tRNA^{Lys} gene [91,92]. The consequences of this mutation have been analysed in an *in vitro* system [93], and appear to result in decreased protein synthesis of mtDNA-encoded polypeptides, as well as synthesis of a few aberrant protein species of currently unknown origin. The nt 8344 mutation is not the exclusive cause of MERRF, as a number of clinically well-defined MERRF patients do not harbour this mutation [94–96]. Moreover, a second mtDNA mutation has also been found in a MERRF patient, and interestingly, it too is in the tRNA^{Lys} gene, at position 8356 (G. Silvestri, personal communication).

Mitochondrial encephalomyopathy, lactic acidosis and stroke-like episodes (MELAS)

The third disorder (MELAS) is characterized by seizures, migraine-like headaches, lactic acidosis, episodic vomiting, short stature and recurrent cerebral stroke-like episodes, causing hemiparesis, hemianopia or cortical blindness [97]. MELAS is associated with two different point mutations, and like the two MERRF mutations, both are the same gene. The first mutation, found in about 80% of cases, is an A→G transition at nt 3243 in the tRNA$^{Leu(UUR)}$ gene [98,99]; the second mutation, found in about 10% of cases, in an A→G transition at nt 3271, also in the tRNA$^{Leu(UUR)}$ gene [100]. The mutations are heteroplasmic, with the proportion of mutated mtDNAs usually exceeding 80% in muscle [101,102]. The MELAS-3243 mutation is unusual in its location, as it lies precisely within the region of the tRNA$^{Leu(UUR)}$ gene which binds the factor for termination of transcription for synthesis of the 'short form' of the H-strand transcripts. In fact, it has been determined that, *in vitro*, this mutation decreases the binding affinity of purified termination factor on a mutated DNA template [103,104]. However, other studies, both in tissue culture [105,106] and *in vivo* (using *in situ* hybridization methods [107]), showed that transcription termination is probably not a major factor in the pathogenesis of MELAS. Rather, a severe decrease in overall protein synthesis has been observed, which is associated with reduced oxygen consumption of cells harbouring exclusively mutant genomes [104,106]. However, in a heteroplasmic situation *in vitro*, fewer than 10% wild-type genomes are sufficient to maintain respiratory competency [104]. Interestingly, the studies of King *et al.* [106] have also implicated a novel transcript, termed by them RNA 19, in the pathogenesis of the disease. This transcript is actually a composite precursor RNA derived from the 16S rRNA, tRNA[Leu(UUR)], and ND1 gene regions (which are contiguous in the mtDNA). Because RNA 19 contains 16S rRNA, it might be incorporated into ribosomes; a model for pathogenesis, based on these observations, has been proposed [106,108].

Maternally inherited PEO

The tRNA[Leu(UUR)] mutation at position 3243 is not confined exclusively to MELAS. As noted above, about half of patients with PEO did not have observable mtDNA deletions. Within this group of 'deletion-minus' PEO patients, approximately one-third had a positive family history; most of these patients have now been found to harbour the MELAS-3243 mutation [109]. Significantly, the percentage of the mutation in muscle in these patients is approximately 50–70, as opposed to 'authentic' MELAS patients, who usually harbour 80–90% mutated mtDNAs, implying that the difference between the two clinical presentations may be one of threshold effects. However, there are qualitative differences between the two disorders. Specifically, patients with PEO (both with ΔmtDNA and with the 3243 mutation) have a higher percentage of ragged red fibres that are COX-deficient, whereas MELAS patients with the 3243 mutation usually have COX-positive RRF. Moreover, MELAS-3243 patients almost never have ptosis or PEO, whereas conversely PEO-3243 patients almost never have stroke-like episodes. These differences suggest that other mutations (either nuclear or mitochondrial) could be present that might exacerbate (or ameliorate) the effects of the 3243 mutation.

Mitochondrial myopathy and cardiomyopathy (MiMyCa)

A maternally inherited adult-onset mitochondrial myopathy and cardiomyopathy (coined MiMyCa by Zeviani) is associated with a point mutation in yet another position in the tRNA[Leu(UUR)] gene, at nt 3260 [110].

Neuropathy, ataxia and retinitis pigmentosa (NARP)

NARP is maternally inherited, and is characterized by developmental delay, retinitis pigmentosa, dementia, seizures, ataxia, proximal neurogenic muscle weakness and sensory neuropathy. It is associated with a T→G transversion at mtDNA position 8893 at codon 155 of the ATPase 6 gene (i.e. subunit 6 of Complex V), which converts the codon from Leu to Arg [111]. The presence of the mutation is concordant with the presence of the pathology, indicating that the mutation is truly aetiological, and is not merely a neutral polymorphism. The mutation has always been found to be heteroplasmic, and the clinical severity appears to be correlated to the amount of mutant mitochondrial genomes in heteroplasmic family members.

Leigh syndrome

A remarkable aspect of the NARP mutation is that, when the proportion of the mutation is high (>90%), it produces a clinically different phenotype, Leigh syndrome [112]. Leigh syndrome is a heterogeneous fatal disorder. In many patients the main biochemical defect is COX deficiency, while in others there is pyruvate dehydrogenase or pyruvate decarboxylase deficiency. These forms of Leigh syndrome are inherited as a mendelian trait, and the COX-deficient form has been demonstrated to be autosomal recessive [113]. However, the infants harbouring the NARP nt 8993 mutation show the clinical features of Leigh syndrome, but none of the aforementioned biochemical defects. Because the mutation is in the ATPase 6 gene, it has been postulated that the fundamental defect lies in the proton channel portion of Complex V (the F_0 portion), with the positively charged arginine residue interfering with the mobility of protons as they traverse the channel in the F_0 'stalk' of the ATPase complex (B. Robinson, personal communication).

DEPLETIONS OF MITOCHONDRIAL DNA

In the last two years, a new class of mitochondrial diseases has emerged. These disorders are due not to qualitative errors in mtDNA, but rather to quantitative depletions of mtDNA in specific tissues [114,115]. Two related mtDNA depletion diseases, which can be distinguished by the age of onset, are apparently inherited as mendelian traits. In the 'early-onset' form, symptoms begin at birth, with death ensuing within a few months. Patients in this category had tissue-specific defects, but the particular tissue affected in any one patient varied (including one infant who died of a pure myopathy, and whose second cousin died of a pure hepatopathy [116]). Patients with the 'later-onset' form (i.e. onset at about 1 year of age) had only myopathies. In all the patients, affected tissues were characterized by respiratory chain defects, mitochondrial proliferation and a severe depletion in the amount of mtDNA. There was no observable biochemical or genetic defect in unaffected tissues.

The presence of mtDNA depletion could be observed by Southern blot hybridization analysis of *Pvu*II-digested total DNA from affected tissues, by *in situ* hybridization of affected muscle sections with radiolabelled mtDNA probes, and by immunohistochemistry of affected muscle sections with anti-DNA antibodies. In this latter analysis, nuclei stained positively in both control and patient muscle sections, but the reticulated staining of the cytoplasm (due to immunoreactivity to mtDNA within mitochondria in the cytoplasm) was absent in fibres from the patients.

The two clinical subtypes of mtDNA depletion differ in a number of ways. The degree of mtDNA depletion in the early-onset form is severe (up to 98% reduction in mtDNA compared with controls), whereas in the late-onset cases it is less pronounced (approximately 70–90% reduction). Second, in the early-onset myopathies, all muscle fibres have little or no immunoreactive mtDNA and are COX-deficient, whereas affected muscle in the late-onset cases shows a segmental pattern of depletion (some fibres appear normal with respect to both COX activity and mtDNA content, while others are deficient in both respects). Third, as opposed to the early-onset form, all the late-onset cases reported to date have been pure myopathies, with no other affected tissues.

In addition to this mendelian-inherited mtDNA depletion disorder, an environmental cause of mtDNA depletion has now also been described, which is associated with azidothymidine (AZT)-induced myopathy in acquired immunodeficency syndrome (AIDS). AIDS patients treated long-term with zidovudine (AZT) develop a myopathy with RRF [117,118]. It has been found that the amount of muscle mtDNA in these patients is reduced significantly [119]. The depletion seems to be due specifically to the AZT treatment, and is not a secondary effect of the human immunodeficiency virus (HIV) infection, because no mtDNA depletion was found in AIDS patients who had not been treated with AZT. The depletion is apparently reversible [119,120], as mtDNA analysis of muscle from a patient in whom AZT treatment had been halted showed that muscle RRF decreased while the mtDNA content increased significantly 4 months after AZT treatment ceased. These findings are consistent with the observation that AZT, a nucleoside analogue, is readily incorporated *in vitro* into mtDNA by the mitochondrial DNA polymerase (polymerase γ), thus inhibiting mtDNA daughter-strand synthesis during replication [121]. They also imply that other nucleoside analogues, such as dideoxycytidine (ddC [122]) and dideoxyinosine (ddI), which are also being used to treat AIDS patients, may also have effects on mtDNA. It is therefore noteworthy that about 5% of HIV-positive children who were treated with ddI developed pigmentary retinal lesions [123], a prominent feature of a number of mitochondrial disorders.

The effects of AZT on mitochondria and on mtDNA are now being studied in greater detail [124–128].

A NOVEL *IN VITRO* SYSTEM TO STUDY DEFECTS IN mtDNA

In 1988, King and Attardi [129] isolated a human cell line without mtDNA (called ρ^0, using the nomenclature established for yeast). Specifically, a ρ^0 line was derived by long-term exposure of a human cell line to ethidium bromide, an inhibitor of mtDNA replication. These ρ^0 cells have a very low level of cellular respiration, with the rate of O_2 consumption being 5% that of the parental line. They are also pyruvate and uridine auxotrophs, and are deficient in thymidine kinase activity as well.

Significantly, these cells can be repopulated with exogenous mitochondria, using specific selection schemes [130]. The ρ^0 cells are repopulated by forming cybrids (i.e. cytoplasmic hybrids) between the ρ^0 cells and cytoplasts (i.e. enucleated cells) from the mtDNA donor cell line. After cell fusion, cells are replated in medium containing bromodeoxyuridine (BrdU) and lacking either pyruvate or uridine. These media permit only the growth of ρ^0 cells which had fused with cytoplasts containing functional mitochondria, because ρ^0 cells are not able to grow in the absence of uridine or pyruvate, and donor cells are not able to grow in the presence of BrdU. Cybrid cell lines can also be isolated by injection of isolated single mitochondria into ρ^0 cells, followed by selection in the absence of uridine [129,130].

Clearly, this tissue culture system is a giant step forward in the analysis of mitochondrial diseases. Exogenous mitochondria from patients who are heteroplasmic for a specific mtDNA mutation can now be injected into ρ^o cells, and cybrids can be selected containing various proportions of mutant and wild-type mtDNAs, including lines with exclusively mutant and exclusively wild-type genomes. In this way, mtDNA genotypes can be correlated directly to biochemical, morphological and molecular biological phenotypes.

The ρ^o system has already been used successfully to analyse the mtDNA mutations in KSS [59], MERRF [93] and MELAS [104,106], and the analysis of other mtDNA mutations are sure to follow. Moreover, the ρ^o system can also be used to screen *in vitro*-mutagenized mtDNA mutations having functional consequences, to search for back-mutations that might complement a known mtDNA mutation, to study nuclear-mitochondrial communication and interactions, and to distinguish between mendelian- and maternally-inherited mutations. Finally, this system can be used to study various physical, genetic and biochemical interventions to treat and ultimately cure these devastating diseases.

ACKNOWLEDGEMENTS

I thank C.T. Moraes and S. DiMauro for critical comments in the preparation of this review. This work was supported by grants from the National Institutes of Health (PO1-NS11766, RO1-NS28828 and P50-AG08702), the Muscular Dystrophy Association, and Aaron Diamond Foundation, and a donation from Libero and Graziella Danesi, Milan, Italy.

REFERENCES

1. Anderson, S., Bankier, A.T., Barrel, B.G., de Bruijin, M.H.L., Coulson, A.R., Drouin, J., Eperon, I.C., Nierlich, D.P., Roe, B.A., Sanger, F., Schreier, P.H., Smith, A.J.H., Staden, R. and Young, I.G. (1981) Sequence and organization of the human mitochondrial genome. *Nature*, **290**, 457–465
2. Clayton, D.A. (1982) Replication of animal mitochondrial DNA. *Cell*, **28**, 693–705
3. Chang, D.D. and Clayton, D.A. (1984) Precise identification of individual promoters for transcription of each strand of human mitochondrial DNA. *Cell*, **36**, 635–643
4. Bogenhagen, D.F., Applegate, E.F. and Yoza, B.K. (1984) Identification of a promoter for transcription of the heavy strand of human mtDNA: in vitro transcription and deletion mutagenesis. *Cell*, **36**, 1105–1113
5. Attardi, G., Chomyn, A., King, M.P., Kruse, B., Polosa, P.L. and Murdter, N.N. (1989) Regulation of mitochondrial gene expression in mammalian cells. *Biochemical Society Transactions*, **18**, 509–513
6. Christianson, T.W. and Clayton, D.A. (1986) *In vitro* transcription of human mitochondrial DNA: accurate termination requires a region of DNA sequence that can function bidirectionally. *Proceedings of the National Academy of Sciences USA*, **83**, 6277–6281
7. Christianson, T.W. and Clayton, D.A. (1988) A tridecamer DNA sequence supports human mitochondrial RNA 3′-end formation in vitro. *Molecular and Cellular Biology*, **8**, 4502–4509
8. Kruse, B., Narasimhan, N. and Attardi, G. (1989) Termination of transcription in human mitochondria: identification and purification of a DNA binding protein factor that promotes termination. *Cell*, **58**, 391–397
9. Satoh, M. and Kuroiwa, T. (1991) Organization of multiple nucleoids and DNA molecules in mitochondria of a human cell. *Experimental Cell Research*, **196**, 137–140
10. Piko, L. and Matsumoto, L. (1976) Number of mitochondria and some properties of mitochondrial DNA in the mouse egg. *Developmental Biology*, **49**, 1–10
11. Piko, L. and Taylor, K.D. (1987) Amounts of mitochondrial DNA and abundance of some mitochondrial gene transcripts in early mouse embryos. *Developmental Biology*, **123**, 364–374

12. Michaels, G.S., Hauswirth, W.W. and Laipis, P.J. (1982) Mitochondrial DNA copy number in bovine oocytes and somatic cells. *Developmental Biology*, **94**, 246–251

13. Shuster, R.C., Rubenstein, A.J. and Wallace, D.C. (1988) Mitochondrial DNA in anucleate human blood cells. *Biochemical and Biophysical Research Communications*, **155**, 1360–1365

14. Giles, R.E., Blanc, H., Cann, H.M. and Wallace, D.C. (1980) Maternal inheritance of human mitochondrial DNA. *Proceedings of the National Academy of Sciences USA*, **77**, 6715–6719

15. Gyllensten, U., Wharton, D., Josefsson, A. and Wilson, A.C. (1991) Paternal inheritance of mitochondrial DNA in mice. *Nature*, **352**, 255–257

16. Wallace, D.C. (1987) Maternal genes: mitochondrial diseases. *Birth Defects*, **23**, 137–190

17. Clayton, D.A., Doda, J.N. and Friedberg, E.C. (1974) The absence of a pyrimidine dimer repair mechanism in mammalian mitochondria. *Proceedings of the National Academy of Sciences USA*, **71**, 2777–2781

18. Croizat, B. and Attardi, G. (1975) Selective in vivo damage by 'visible' light of BrdU-containing mitochondrial DNA in a thymidine kinase-deficient mouse cell with peristent mitochondrial enzyme activity. *Journal of Cell Science*, **19**, 69–84

19. Pettepher, C.C., LeDoux, S.P., Bohr, V.A. and Wilson, G.L. (1991) Repair of alkali-labile sites within the mitochondrial DNA of RINr 38 cells after exposure to the nitrosourea streptozotocin. *Journal of Biological Chemistry*, **266**, 3113–3117

20. Richter, C., Park, J.-W. and Ames, B.N. (1988) Normal oxidative damage to mitochondrial and nuclear DNA is extensive. *Proceedings of the National Academy of Sciences USA*, **85**, 6465–6467

21. Pavlakis, S.G., Rowland, L.P., DeVivo, D.C., Bonilla, E. and DiMauro, S. (1988) Mitochondrial myopathies and encephalomyopathies. In *Contemporary Neurology* (F. Plum ed.). Davis, Philadelphia, pp. 95–133

22. DiMauro, S., Zeviani, M., Rizzuto, R., Lombes, A., Nakase, H., Bonilla, E., Miranda, A. and Schon, E.A. (1988) Molecular defects in cytochrome *c* oxidase in mitochondrial disease. *Journal of Bioenergetics and Biomembranes*, **20**, 353–364

23. Holt, I.J., Harding, A.E. and Morgan-Hughes, J.A. (1988) Deletions of mitochondrial DNA in patients with mitochondrial myopathies. *Nature*, **331**, 717–719

24. Lestienne, P. and Ponsot, G. (1988) Kearns-Sayre syndrome with muscle mitochondrial DNA deletion. *Lancet*, **i**, 885(letter)

25. Zeviani, M., Moraes, C.T., DiMauro, S., Nakase, H., Bonilla, E., Schon, E.A. and Rowland, L.P. (1988) Deletions of mitochondrial DNA in Kearns-Sayre syndrome. *Neurology*, **38**, 1339–1346

26. Moraes, C.T., DiMauro, S., Zeviani, M., Lombes, A., Shanske, S., Miranda, A.F., Nakase, H., Bonilla, E., Wernec, L.C., Servidei, S., Nonaka, I., Koga, Y., Spiro, A., Brownell, K. W., Schmidt, B., Schotland, D.L., Zupanc, M.D., DeVivo, D.C., Schon, E.A. and Rowland, L.P. (1989) Mitochondrial DNA deletions in progressive external ophthalmoplegia and Kearns-Sayre syndrome. *New England Journal of Medicine*, **320**, 1293–1299

27. Holt, I.J., Harding, A.E. and Morgan-Hughes, J.A. (1989) Deletions of muscle mitochondrial DNA in mitochondrial myopathies: sequence analysis and possible mechanisms. *Nucleic Acids Research*, **17**, 4465–4469

28. Gerbitz, K.-D., Obermaier-Kusser, B., Zierz, S., Pongratz, D., Müller-Höcker, J. and Lestienne, P. (1990) Mitochondrial myopathies: divergence of genetic deletions, biochemical defects and the clinical syndromes. *Journal of Neurology*, **237**, 5–10

29. Goto, Y.-i., Koga, Y., Horai, S. and Nonaka, I. (1990) Chronic progressive external ophthalmo-plegia: a correlative study of mitochondrial DNA deletions and their phenotypic expression in muscle biopsies. *Journal of Neurological Sciences*, **100**, 63–69

30. Yamamoto, H., Tanaka, M., Katayama, M., Obayashi, T., Nimura, Y. and Ozawa, T. (1992) Significant existence of deleted mitochondrial DNA in cirrhotic liver surrounding hepatic tumor. *Biochemical and Biophysical Research Communications*, **182**, 913–920

31. Nelson, I., Degoul, F., Obermaier-Kusser, B., Romero, N., Borrone, C., Marsac, C., Vayssiere, J., Gerbitz, K., Fardeau, M., Ponsot, G. and Lestienne, P. (1989) Mapping of heteroplasmic mitochon-drial DNA deletions in Kearns-Sayre syndrome. *Nucleic Acids Research*, **17**, 8117–8124

32. Johns, D.R., Rutledge, S.L., Stine, O.C. and Hurko, O. (1989) Directly repeated sequences associ-ated with pathogenic mitochondrial DNA deletions. *Proceedings of the National Academy of Sciences USA*, **86**, 8059–8062

33. Schon, E.A., Rizzuto, R., Moraes, C.T., Nakase, H., Zeviani, M. and DiMauro, S. (1989) A direct repeat is a hotspot for large-scale deletions of human mitochondrial DNA. *Science*, **244**, 346–349

34. Mita, S., Rizzuto, R., Moraes, C.T., Shanske, S., Arnaudo, E., Fabrizi, G., Koga, Y., DiMauro, S. and Schon, E.A. (1990) Recombination via flanking direct repeats is a major cause of large-scale deletions of human mitochondrial DNA. *Nucleic Acids Research*, **18**, 561–567

35. Degoul, F., Nelson, I., Amselem, S., Romero, N., Obermaier-Kusser, B., Ponsot, G., Marsac, C. and Lestienne, P. (1991) Different mechanisms inferred from sequences of human mitochondrial DNA deletions in ocular myopathies. *Nucleic Acids Research*, **19**, 493–496
36. Rötig, A., Cormier, V., Koll, F., Mize, C.E., Saudubray, J.-M., Veerman, A., Pearson, H.A. and Munnich, A. (1991) Site-specific deletions of the mitochondrial genome in the Pearson marrow-pancreas syndrome. *Genomics*, **10**, 502–504
37. Shoffner, J.M., Lott, M.T., Voljavec, A.S., Soueidan, S.A., Costigan, D.A. and Wallace, D.C. (1989) Spontaneous Kearns-Sayre/chronic external ophthalmoplegia plus syndrome associated with a mitochondrial DNA deletion: a slip-replication model and metabolic therapy. *Proceedings of the National Academy of Sciences USA*, **86**, 7952–7956
38. Tanaka, M., Sato, W., Ohno, K., Yamamoto, T. and Ozawa, T. (1989) Direct sequencing of deleted mitochondrial DNA in myopathic patients. *Biochemical and Biophysical Research Communications*, **164**, 156–163
39. Moraes, C.T., Zeviani, M., Schon, E.A., Hickman, R.O. and Velck, B.W. (1991) Mitochondrial DNA deletion in a girl with manifestations of Kearns-Sayre and Lowe syndromes: an example of phenotypic mimicry? *American Journal of Medical Genetics*, **41**, 301–305
40. Zupanc, M., Moraes, C.T., Shanske, S., Langman, C.B., Ciafaloni, E. and DiMauro, S. (1991) Mitochondrial DNA deletion in patients with combined features of Kearns-Sayre and MELAS syndromes. *Annals of Neurology*, **29**, 680–683
41. Poulton, J., Deadman, M.E., Ramacharan, S. and Gardiner, R.M. (1991) Germ-line deletions of mtDNA in mitochondrial myopathy. *American Journal of Human Genetics*, **48**, 649–653
42. Larsson, N.-G., Eiken, H.G., Holme, E., Oldfors, A. and Tulinius, M.H. (1992) Lack of transmission of deleted mtDNA from a woman with Kearns-Sayre syndrome to her child. *American Journal of Human Genetics*, **50**, 360–363
43. Shanske, S., Moraes, C.T., Lombes, A., Miranda, A.F., Bonilla, E., Lewis, P., Whelan, M.A., Ellsworth, C.A. and DiMauro, S. (1990) Widespread tissue distribution of mitochondrial DNA deletions in Kearns-Sayre syndrome. *Neurology*, **40**, 24–28
44. Obermaier-Kusser, B., Müller-Höcker, J., Nelson, I., Lestienne, P., Enter, C., Riedele, T. and Gerbitz, K.-D. (1990) Different copy numbers of apparently identical deleted mitochondrial DNA in tissue from a patient with Kearns-Sayre syndrome. *Biochemical and Biophysical Research Communications*, **169**, 1001–1015
45. Larsson, N.-G., Holme, E., Kristiansson, B., Oldfors, A. and Tulinius, M. (1990) Progressive increase of the mutated mitochondrial DNA fraction in Kearns-Sayre syndrome. *Pediatric Research*, **28**, 131–136
46. McShane, M.A., Hammans, S.R., Sweeney, M., Holt, I.J., Beattie, T.J., Brett, E.M. and Harding, A.E. (1991) Pearson syndrome and mitochondrial encephalomyopathy in a patient with a deletion of mtDNA. *American Journal of Human Genetics*, **48**, 39–42
47. Schon, E.A., Bonilla, E., Miranda, A.F. and DiMauro, S. (1991) Molecular biology of mitochondrial diseases. *Molecular Genetic Approaches to Neuropsychiatric Disease*, 57–80
48. Poulton, J., Deadman, M.E. and Gardiner, R.M. (1989) Duplications of mitochondrial DNA in mitochondrial myopathy. *Lancet*, **i**, 236–240
49. Poulton, J., Deadman, M.E. and Gardiner, R.M. (1989) Tandem direct duplications of mitochondrial DNA in mitochondrial myopathy: analysis of nucleotide sequence and tissue distribution. *Nucleic Acids Research*, **17**, 10223–10229
50. Rötig, A., Bessis, J.-L., Romero, N., Cormier, V., Saudubray, J.-M., Narcy, P., Lenoir, G., Rustin, P. and Munnich, A. (1992) Maternally inherited duplication of the mitochondrial genome in a syndrome of proximal tubulopathy, diabetes mellitus, and cerebellar ataxia. *American Journal of Human Genetics*, **50**, 364–370
51. Moraes, C.T., Andreetta, F., Bonilla, E., Shanske, S., DiMauro, S. and Schon, E.A. (1991) Replication-competent human mitochondrial DNA lacking the heavy-strand promoter region. *Molecular and Cellular Biology*, **11**, 1631–1637
52. Johns, D.R. and Cornblath, D.R. (1991) Molecular insight into the asymmetric distribution of pathogenetic human mitochondrial DNA. *Biochemical and Biophysical Research Communications*, **174**, 244–250
53. Miyabayashi, S., Hanamizu, H., Endo, H., Tada, K. and Horai, S. (1991) A new type of mitochondrial DNA deletion in patients with encephalomyopathy. *Journal of Inherited Metabolic Diseases*, **14**, 805–812
54. Ballinger, S.W., Shoffner, J.M., Hedaya, E.V., Trounce, I., Polak, M.A., Koontz, D.A. and Wallace, D.C. (1992) Maternally transmitted diabetes and deafness associated with a 10.4 kb mitochondrial DNA deletion. *Nature and Genetics*, **1**, 11–15
55. Nakase, H., Moraes, C.T., Rizzuto, R., Lombes, A., DiMauro, S. and Schon, E.A. (1990) Trans-

cription and translation of deleted mitochondrial genomes in Kearns-Sayre syndrome: implications for pathogenesis. *American Journal of Human Genetics*, **46**, 418–427

56. Mita, S., Schmidt, B., Schon, E.A., DiMauro, S. and Bonilla, E. (1989) Detection of 'deleted' mitochondrial genomes in cytochrome *c* oxidase-deficient muscle fibers of a patient with Kearns-Sayre syndrome. *Proceedings of the National Academy of Sciences USA*, **86**, 9509–9513

57. Nakamura, S., Sato, T., Hirawake, H., Kobayashi, R., Fukuda, Y., Kawamura, J., Ujike, H. and Horai, S. (1990) In situ hybridization of muscle mitochondrial mRNA in mitochondrial myopathies. *Acta Neuropathologica*, **81**, 1–6

58. Shoubridge, E.A., Karpati, G. and Hastings, K.E.M. (1990) Deletion mutants are functionally dominant over wild-type mitochondrial genomes in skeletal muscle fiber segments in mitochondrial disease. *Cell*, **62**, 43–49

59. Hayashi, J.-I., Ohta, S., Kikuchi, A., Takemitsu, M., Goto, Y.-I. and Nonaka, I. (1991) Introduction of disease-related mitochondrial DNA deletions into HeLa cells lacking mitochondrial DNA results in mitochondrial dysfunction. *Proceedings of the National Academy of Sciences USA*, **88**, 10614–10618

60. Hammans, S.R., Sweeney, M.G., Holt, I.J., Cooper, J.M., Toscano, A., Clark, J.B., Morgan-Hughes, J.A. and Harding, A.E. (1992) Evidence for intramitochondrial complementation between deleted and normal mitochondrial DNA in some patients with mitochondrial myopathy. *Journal of Neurological Sciences*, **107**, 87–92

61. Moraes, C.T., Ricci, E., Petruzzella, V., Shanske, S., DiMauro, S., Schon, E.A. and Bonilla, E. (1992) Molecular analysis of the muscle pathology associated with mitochondrial DNA deletions. *Nature Genetics*, **1**, 359–367

62. Zeviani, M., Servidei, S., Gellera, C., Bertini, E., DiMauro, S. and DiDonato, S. (1989) An autosomal dominant disorder with multiple deletions of mitochondrial DNA starting at the D-loop region. *Nature*, **339**, 309–311

63. Zeviani, M., Bresolin, N., Gellera, C., Bordoni, A., Pannacci, M., Amati, P., Moggio, M., Servidei, S., Scarlato, G. and DiDonato, S. (1990) Nucleus-driven multiple large-scale deletions of the human mitochondrial genome: a new autosomal dominant disease. *American Journal of Human Genetics*, **47**, 904–914

64. Yuzaki, M., Ohkoshi, N., Kanazawa, Y. and Ohta, S. (1989) Multiple deletions in mitochondrial DNA at direct repeats of non-D-loop regions in cases of familial mitochondrial myopathy. *Biochemical and Biophysical Research Communications*, **164**, 1352–1357

65. Ciafaloni, E., Shanske, S., Apostolski, S., Griggs, R.L., Bird, T.D., Sumi, M. and DiMauro, S. (1991) Multiple deletions of mitochondrial DNA. *Neurology*, **41** (Suppl.1), 207

66. Ohno, K., Tanaka, M., Sahashi, K., Ibi, T., Sato, W., Yamamoto, T., Takahashi, A. and Ozawa, T. (1991) Mitochondrial DNA deletions in inherited recurrent myoglobinuria. *Annals of Neurology*, **29**, 364–369

67. Cormier, V., Rötig, A., Tardieu, M., Colonna, M., Saudubray, J.-M. and Munnich, A. (1991) Autosomal dominant deletions of the mitochondrial genome in a case of progressive encephalomyopathy. *American Journal of Human Genetics*, **48**, 643–648

68. Ikebe, S.-i., Tanaka, M., Ohno, K., Sato, W., Hattori, K., Kondo, T., Mizuno, Y. and Ozawa, T. (1990) Increase of deleted mitochondrial DNA in the striatum in Parkinson's disease and senescence. *Biochemical and Biophysical Research Communications*, **170**, 1044–1048

69. Ozawa, T., Tanaka, M., Ikebe, S.-i., Ohno, K., Kondo, T. and Mizuno, Y. (1990) Quantitative determination of deleted mitochondrial DNA relative to normal DNA in Parkinsonian striatum by a kinetic PCR analysis. *Biochemical and Biophysical Research Communications*, **172**, 483–489

70. Mann, V.M., Cooper, J.M. and Schapira, A.H.V. (1992) Quantitation of a mitochondrial DNA deletion in Parkinson's disease. *FEBS Letters*, **299**, 218–222

71. Yen, T.-C., Su, J.-H., King, K.-L. and Wei, Y.-H. (1991) Ageing-associated 5 kb deletion in human liver mitochondrial DNA. *Biochemical and Biophysical Research Communications*, **178**, 124–131

72. Cortopassi, G.A. and Arnheim, N. (1991) Detection of a specific mitochondrial DNA deletion in tissues of older humans. *Nucleic Acids Research*, **18**, 6927–6933

73. Corral-Debrinski, M., Stepien, G., Shoffner, J.M., Lott, M.T., Kanter, K. and Wallace, D.C. (1991) Hypoxemia is associated with mitochondrial DNA damage and gene induction. *Journal of the American Medical Association*, **266**, 1812–1816

74. Hattori, K., Tanaka, M., Sugiyama, S., Obayashi, T., Ito, T., Satake, T., Hanaki, Y., Asai, J., Nagano, M. and Ozawa, T. (1991) Age-dependent increase in deleted mitochondrial DNA in the human heart: possible contributory factor to presbycardia. *American Heart Journal*, **121**, 1735–1742

75. Ozawa, T., Tanaka, M., Sugiyama, S., Hattori, K., Ito, T., Ohno, K., Takahashi, A., Sato, W., Takada, G., Mayumi, B., Yamamoto, K., Adachi, K., Koga, Y. and Toshima, H. (1990) Multiple

mitochondrial DNA deletions exist in cardiomyocytes of patients with hypertrophic or dilated cardiomyopathy. *Biochemical and Biophysical Research Communications*, **170**, 830–836

76. Simonetti, S., Chen, X., DiMauro, S. and Schon, E.A. (1993) Accumulation of deletions in human mitochondrial DNA during normal aging: analysis by quantitative PCR.

77. Cortopassi, G.A., Shibata, D., Soong, N.W. and Arnheim, N. (1992) A pattern of accumulation of a somatic deletion of mitochondrial DNA in aging human tissues. *Proceedings of the National Academy of Sciences USA*, **89**, 7370–7374

78. Zhang, C., Baumer, A., Maxwell, R.J., Linnane, A.W. and Nagley, P. (1992) Multiple mitochondrial DNA deletions in an elderly human individual. *FEBS Letters*, **297**, 34–38

79. Wallace, D.C., Singh, G., Lott, M.T., Hodge, J.A., Schurr, T.G., Lezza, A.M.S., Elsas II, L.J. and Nikoskelainen, E.K. (1988) Mitochondrial DNA mutation associated with Leber's hereditary optic neuropathy. *Science*, **242**, 1427–1430

80. Singh, G., Lott, M.T. and Wallace, D.C. (1989) A mitochondrial DNA mutation as a cause of Leber's hereditary optic neuropathy. *New England Journal of Medicine*, **320**, 1300–1305

81. Holt, I.J., Miller, D.H. and Harding, A.E. (1989) Genetic heterogeneity and mitochondrial DNA heteroplasmy in Leber's hereditary optic neuropathy. *Journal of Medical Genetics*, **26**, 739–743

82. Huoponen, K., Vilkki, J., Aula, P., Nikoskelain, E.K. and Savontaus, M.-L. (1991) A new mtDNA mutation associated with Leber hereditary optic neuropathy. *American Journal of Human Genetics*, **48**, 1147–1153

83. Howell, N., Bindoff, L.A., McCullough, D.A., Kubacka, I., Poulton, J., Mackey, D., Taylor, L. and Turnbull, D.M. (1991) Leber hereditary optic neuropathy: identification of the same mitochondrial ND1 mutation in six pedigrees. *American Journal of Human Genetics*, **49**, 939–950

84. Howell, N., Kubacka, I., Xu, M. and McCullough, D.A. (1991) Leber hereditary optic neuropathy: involvement of the mitochondrial ND1 gene and evidence for an intragenic suppressor mutation. *American Journal of Human Genetics*, **48**, 935–942

85. Johns, D.R. and Neufeld, M.J. (1992) Cytochrome b mutations in Leber hereditary optic neuropathy. *Biochemical and Biophysical Research Communications*, **181**, 1358–1364

86. Brown, M.D., Voljavec, A.S., Lott, M.T., Torroni, A., Yang, C.-C. and Wallace, D.C. (1992) Mitochondrial DNA complex I and III mutations with Leber's hereditary optic neuropathy. *Genetics*, **130**, 163–173

87. Vilkki, J., Ott, J., Savontaus, M.-L., Aula, P. and Nikoskelainen, E.K. (1991) Optic atrophy in Leber hereditary optic neuroretinopathy is probably determined by an X-chromosomal gene closely linked to DXS7. *American Journal of Human Genetics*, **48**, 486–491

88. Bu, X. and Rotter, J.I. (1991) X chromosome-linked and mitochondrial gene control of Leber hereditary optic neuropathy: evidence from segregation analysis for dependence on X chromosome inactivation. *Proceedings of the National Academy of Sciences USA*, **88**, 8198–8202

89. Rosing, H.S., Hopkins, L.C., Wallace, D.C., Epstein, C.M. and Weidenheim, K. (1985) Maternally inherited mitochondrial myopathy and myoclonic epilepsy. *Annals of Neurology*, **17**, 228–237

90. Fukuhara, N., Tokigushi, S., Shirakawa, K. and Tsubaki, T. (1980) Myoclonus epilepsy associated with ragged-red fibers (mitochondrial abnormalities): disease entity or syndrome? Light and electron microscopic studies of two cases and review of the literature. *Journal of Neurological Science*, **47**, 117–133

91. Shoffner, J.M., Lott, M.T., Lezza, A.M.S., Seibel, P., Ballinger, S.W. and Wallace, D.C. (1990) Myoclonic epilepsy and ragged-red fiber disease (MERRF) is associated with a mitochondrial DNA tRNA^{Lys} mutation. *Cell*, **61**, 931–937

92. Yoneda, M., Tanno, Y., Horai, S., Ozawa, T., Miyatake, T. and Tsuji, S. (1990) A common mitochondrial DNA mutation in the t-RNA^{Lys} of patients with myoclonus epilepsy associated with ragged-red fibers. *Biochemistry International*, **21**, 789–796

93. Chomyn, A., Meola, G., Bresolin, N., Lai, S.T., Scarlato, G. and Attardi, G. (1991) *In vitro* genetic transfer of protein synthesis and respiration defects to mitochondrial DNA-less cells with myopathy-patient mitochondria. *Molecular and Cellular Biology*, **11**, 2236–2244

94. Zeviani, M., Amati, P., Bresolin, N., Antozzi, C., Piccolo, G., Toscano, A. and DiDonato, S. (1991) Rapid detection of the A$\rightarrow$G$^{(8344)}$ mutation of mtDNA in Italian families with myoclonus epilepsy and ragged-red fibers (MERRF). *American Journal of Human Genetics*, **48**, 203–211

95. Silvestri, G., Caifaloni, E., Shanske, S. and DiMauro, S. (1992) Clinical manifestations associated with the 'MERRF point mutation'. *Neurology*, **42** (Suppl 3), 417

96. Tanno, Y., Yoneda, M., Nonaka, I., Tanaka, K., Miyatake, T. and Tsuji, S. (1991) Quantitation of mitochondrial DNA carrying tRNA^{Lys} mutation in MERRF patients. *Biochemical and Biophysical Research Communications*, **179**, 880–885

97. Pavlakis, S.G., Phillips, P.C., DiMauro, S., DeVivo, D.C. and Rowland, L.P. (1984) Mitochondrial myopathy, encephalopathy, lactic acidosis, and stroke-like episodes: a distinctive clinical syndrome. *Annals of Neurology*, **16**, 481–487

98. Goto, Y.-i., Nonaka, I. and Horai, S. (1990) A mutation in the tRNA[Leu(UUR)] gene associated with the MELAS subgroup of mitochondrial encephalomyopathies. *Nature*, **348**, 651–653

99. Kobayashi, Y., Momoi, M.Y., Tominaga, K., Momoi, T., Nihei, K., Yanagisawa, M., Kagawa, Y. and Ohta, S. (1990) A point mutation in the mitochondrial tRNA[Leu(UUR)] gene in MELAS (mitochondrial myopathy, encephalopathy, lactic acidosis and stroke-like episodes). *Biochemical and Biophysical Research Communications*, **173**, 816–822

100. Goto, Y.-i., Nonaka, I. and Horai, S. (1991) A new mtDNA mutation associated with mitochondrial myopathy, encephalopathy, lactic acidosis and stroke-like episodes (MELAS). *Biochimica et Biophysica Acta*, **1097**, 238–240

101. Goto, Y.-i., Horai, S., Matsuoka, T., Koga, Y., Nihei, K., Kobayashi, M. and Nonaka, I. (1992) Mitochondrial myopathy, encephalopathy, lactic acidosis, and stroke-like episodes (MELAS): a correlative study of the clinical features and mitochondrial DNA mutation. *Neurology*, **42**, 545–550

102. Ciafaloni, E., Ricci, E., Shanske, S., Moraes, C.T., Silvestri, G., Hirano, M., Simonetti, S., Angelini, C., Donati, A., Garcia, C., Martinuzzi, A., Mosewich, R., Servidei, S., Zammarchi, E., Bonilla, E., DeVivo, D.C., Rowland, L.P., Schon, E.A. and DiMauro, S. (1992) MELAS: clinical features, biochemistry, and molecular genetics. *Annals of Neurology*, **31**, 391–398

103. Hess, J.F., Parisi, M.A., Bennett, J.L. and Clayton, D.A. (1991) Impairment of mitochondrial transcription termination by a point mutation associated with the MELAS subgroup of mitochondrial encephalomyopathies. *Nature*, **351**, 236–239

104. Chomyn, A., Martinuzzi, A., Yoneda, M., Daga, A., Hurko, O., Johns, D., Lai, S.T., Nonaka, I., Angelini, C. and Attardi, G. (1992) MELAS mutation in mtDNA binding site for transcription termination factor causes defects in protein synthesis and in respiration but no changes in levels of upstream and downstream mature transcripts. *Proceedings of the National Academy of Sciences USA*, **89**, 4221–4225

105. Kobayashi, Y., Momoi, M., Tominaga, K., Shimoizumi, H., Nihei, K., Yanagisawa, M., Kagawa, Y. and Ohta, S. (1991) Respiration-deficient cells are caused by a single point mutation in the mitochondrial tRNA-Leu(UUR) gene in mitochondrial myopathy, encephalopathy, lactic acidosis, and stroke-like episodes (MELAS). *American Journal of Human Genetics*, **49**, 590–599

106. King, M.P., Koga, Y., Davidson, M. and Schon, E.A. (1992) Defects in mitochondrial protein synthesis and respiratory chain activity segregate with the tRNA[Leu(UUR)] mutation associated with MELAS. *Molecular and Cellular Biology*, **12**, 480–490

107. Moraes, C.T., Ricci, E., Bonilla, E., DiMauro, S. and Schon, E.A. (1992) The mitochondrial tRNA[Leu(UUR)] mutation in MELAS: genetic, biochemical, and morphological correlations in skeletal muscle. *American Journal of Human Genetics*, **50**, 934–949

108. Schon, E.A., Koga, Y., Davidson, M., Moraes, C.T., and King, M.P. (1993) The mitochondrial tRNA[Leu(UUR)] mutation in MELAS: a model for pathogenesis. *Biochimica et Biophysica Acta*, **1101**, 206–209

109. Ciacci, F., Moraes, C.T., Silvestri, G., Shanske, S., Schon, E.A. and DiMauro, S. (1992) The 'MELAS-3,243' mutation in mtDNA is found in many patients with Progressive External Ophthalmoplegia (PEO). *Neurology*, **42** (Suppl. 3), 417

110. Zeviani, M., Gellera, C., Antozzi, C., Rimoldi, M., Villani, F., Tiranti, V. and DiDonato, S. (1991) Maternally inherited myopathy and cardiomyopathy: association with mutation in mitochondrial DNA tRNA[Leu(UUR)]. *Lancet*, **338**, 143–147

111. Holt, I.J., Harding, A.E., Petty, R.K.H. and Morgan-Hughes, J.A. (1990) A new mitochondrial disease associated with mitochondrial DNA heteroplasmy. *American Journal of Human Genetics*, **46**, 428–433

112. Tatuch, Y., Christodoulou, J., Feigenbaum, A., Clark, J.T.R., Wherret, J., Smith, C., Rudd, N., Petrova-Benedict, R. and Robinson, B.H. (1992) Heteroplasmic mtDNA mutation (T→G) at 8993 can cause Leigh's disease when the percentage of abnormal mtDNA is high. *American Journal of Human Genetics*, **50**, 852–858

113. Miranda, A.F., Ishii, S., DiMauro, S. and Shay, J.W. (1989) Cytochrome *c* oxidase deficiency in Leigh's syndrome: genetic evidence for a nuclear DNA-encoded mutation. *Neurology*, **39**, 697–702

114. Moraes, C.T., Shanske, S., Tritschler, H.-J., Aprille, J.R., Andreetta, F., Bonilla, E., Schon, E.A. and DiMauro, S. (1991) Mitochondrial DNA depletion with variable tissue expression: a novel genetic abnormality in mitochondrial diseases. *American Journal of Human Genetics*, **48**, 492–501

115. Tritschler, H.-J., Andreetta, F., Moraes, C.T., Bonilla, E., Arnaudo, E., Danon, M.J., Glass, S., Zelaya, B.M., Vamos, E., Telerman-Toppet, N., Kadenbach, B., DiMauro, S. and Schon, E.A. (1991) Mitochondrial myopathy of childhood associated with depletion of mitochondrial DNA. *Neurology*, **42**, 209–217

116. Boustany, R.N., Aprille, J.R., Halperin, J., Levy, H. and DeLong, G.R. (1983) Mitochondrial cytochrome deficiency presenting as a myopathy with hypotonia, external ophthalmoplegia, and lactic acidosis in an infant and as fatal hepatopathy in a second cousin. *Annals of Neurology*, **14**, 462–470
117. Dalakas, M., Illa, I., Pezeshkpour, G.H., Laukaitis, J.P., Cohen, B. and Griffin, J.L. (1990) Mitochondrial myopathy caused by long-term AZT (Zidovudine) therapy: management and differences from HIV-associated myopathy. *New England Journal of Medicine*, **322**, 1098–1105
118. Mhiri, C., Baudrimont, M., Bonne, G., Geny, C., Degoul, F., Marsac, C., Roullet, E. and Gherardi, R. (1991) Zidovudine myopathy: a distinctive disorder associated with mitochondrial dysfunction. *Annals of Neurology*, **29**, 606–614
119. Arnaudo, E., Dalakas, M., Shanske, S., Moraes, C.T., DiMauro, S. and Schon, E.A. (1991) Depletion of mitochondrial DNA in AIDS patients with zidovudine-induced myopathy. *Lancet*, **337**, 508–510
120. Chalmers, A.C., Greco, C.M. and Miller, R.G. (1991) Prognosis in AZT myopathy. *Neurology*, **41**, 1181–1184
121. Simpson, M.V., Chin, C.D., Keilbaugh, S.A., Lin, T.-S. and Prusoff, W.H. (1989) Studies on the inhibition of mitochondrial DNA replication by 3'-azido-3'-deoxythimidine and other dideoxy-nucleoside analogues which inhibit HIV-1 replication. *Biochemical Pharmacology*, **38**, 1033–1036
122. Chen, C.H. and Cheng, Y.C. (1992) The role of cytoplasmic deoxycytidine kinase in the mitochondrial effects of the antihuman immunodeficiency virus compound, 2',3'-dideoxycytidine. *Journal of Biological Chemistry*, **267**, 2856–2859
123. Whitcup, S.M., Butler, K.M., Pizzo, P.A. and Nussenblatt, R.B. (1992) Retinal lesions in children treated with dideoxyinosine. *New England Journal of Medicine*, **326**, 1226–1227
124. Hayakawa, M., Ogawa, T., Sugiyama, S., Tanaka, M. and Ozawa, T. (1991) Massive conversion of guanosine to 8-hydroxy-guanosine in mouse liver mitochondrial DNA by administration of azidothymidine. *Biochemical and Biophysical Research Communications*, **176**, 87–93
125. Lamperth, L., Dalakas, M.C., Dagani, F., Anderson, J. and Ferrari, R. (1991) Abnormal skeletal and cardiac muscle mitochondria induced by zidovudine (AZT) in human muscle *in vitro* and in an animal model. *Laboratory Investigations*, **65**, 742–751
126. Lewis, W., Papoian, T., Gonzalez, B., Louie, H., Kelly, D.P., Payne, R.M. and Grody, W. (1991) Mitochondrial ultrastructural and molecular changes induced by zidovudine in rat hearts. *Laboratory Investigations*, **65**, 222–236
127. Lewis, W., Gonzalez, B., Chomyn, A. and Papoian, T. (1992) Zidovudine induces molecular, biochemical, and ultrastructural changes in rat skeletal muscle mitochondria. *Journal of Clinical Investigations*, **80**, 1354–1360
128. Weissman, J.D., Constantinitis, I., Hudgins, P. and Wallace, D.C. (1992) ^{31}P magnetic resonance spectroscopy suggests impaired mitochondrial function in AZT-treated HIV-infected patients. *Neurology*, **42**, 619–623
129. King, M.P. and Attardi. G. (1988) Injection of mitochondria into human cells leads to a rapid replacement of the endogenous mitochondrial DNA. *Cell*, **52**, 811–819
130. King, M.P. and Attardi, G. (1989) Human cells lacking mtDNA: repopulation with exogenous mitochondria by complementation. *Science*, **246**, 500–503

3
Mitochondrial myopathies: clinical features, investigation, treatment and genetic counselling

S.R. Hammans and J.A. Morgan-Hughes

CLINICAL FEATURES

The mitochondrial myopathies are a clinically diverse group of diseases defined historically by the presence of structural mitochondrial abnormalities in biopsied skeletal muscle [1]. They have been increasingly recognized since the application of histochemistry and electron microscopy to muscle pathology in the 1960s. The histological hallmark of mitochondrial myopathy is the ragged red fibre (RRF), demonstrated with the modified Gomori trichrome stain [2,3], containing peripheral and intermyofibrillar accumulations of abnormal mitochondria. Although the diagnostic importance of muscle biopsy is undisputed, the presence of RRFs as a sole criterion for establishing the diagnosis has been challenged by recent histochemical, biochemical and genetic advances.

The first description of a mitochondrial myopathy was in a 35-year-old Swedish woman who presented with severe euthyroid hypermetabolism due to defective coupling of oxidation and phosphorylation in muscle mitochondria [4]. Only one other similar case has been described [5]. Heat intolerance, profuse sweating, polyphagia, polydipsia without polyuria and mild fatiguable muscle weakness were features of both cases. Peripheral accumulations of structurally abnormal mitochondria were observed in muscle fibres. Subsequently, it became apparent that RRFs were frequently associated with progressive external ophthalmoplegia (PEO) [3,6], often with proximal myopathy and weakness induced or increased by exertion, and also with other myopathic syndromes without ophthalmoplegia.

In addition to myopathic presentations, RRFs occur in patients with clinical syndromes predominantly or exclusively involving the central nervous system (CNS). These diseases, termed 'mitochondrial encephalomyopathies' have been increasingly identified and classified during the 1980s. Patients may present with psychomotor retardation, dementia, seizures, ataxia, movement disorders, stroke-like episodes, pigmentary retinopathy, deafness and peripheral neuropathy in various combination [1,7–9]. Subsets of clinical features within the spectrum of mitochondrial encephalomyopathies led to the identification of apparently specific clinicopathological entities, such as the Kearns–Sayre syndrome (KSS [10]), mitochondrial myopathy, encephalopathy, lactic acidosis and stroke-like episodes (MELAS [11]) and myoclonic epilepsy with ragged red fibres (MERFF [8]).

Involvement of tissues other than muscle, nerve and CNS has further broadened the spectrum of mitochondrial myopathies to include haematological, endocrine, renal, cardiac, gastrointestinal and metabolic manifestations. Multiorgan involvement emphasizes the importance of mitochondria in most tissues. Differential tissue involvement is determined by differing dependence on oxidative energy production and by the distribution and expression of underlying mitochondrial (mt) DNA defects. Developmental and tissue-specific expression of nuclearly encoded mitochondrial proteins, as well as immunological and environmental factors, may also play a role in the variability of organ involvement. This chapter describes some of the major clinical manifestations of mitochondrial disease in the different tissues.

Myopathies

SKELETAL

Limb weakness with or without exercise intolerance may be the sole clinical feature of mitochondrial disease. It may be associated with PEO at presentation or PEO may develop subsequently. Limb weakness is present in some but not all patients with predominant CNS involvement. In the study of Petty and colleagues [12], limb weakness was a presenting complaint in 18 of 66 patients, but on examination was present in 47 of 66. Weakness is mainly proximal, mild (Medical Research Council grade 4+) and most easily detectable in the arms. In 10 of 66 patients, weakness was grade 4– or greater, and muscle wasting was confined to this group. None was unable to walk because of muscle weakness alone. Fatigue on sustained or repeated contraction may be a diagnostic clue, being present in 50% of cases [12]. Weakness may occasionally have a facioscapulohumeral distribution [13–15]. Of 15 patients presenting with proximal weakness or exercise intolerance, 13 had a relatively benign clinical course with the clinical syndrome remaining restricted to a predominant proximal myopathy, with or without PEO [12]. Two patients developed CNS disease, evident within 5 years of presentation.

Harding and colleagues [16] found that six of 12 cases with proximal limb weakness alone had affected relatives. One patient with limb weakness alone harboured the bp 3243 tRNA$^{Leu(UUR)}$ mutation [17], initially described in association with the MELAS syndrome [18,19]. Other families with isolated myopathy have been described [20,21]. Multiple deletions of mtDNA have been described in two brothers with mitochondrial myopathy who presented with recurrent exertional myoglobinuria and exercise intolerance [22]. Apart from muscle weakness, no neurological abnormality was detected.

PEO and KSS   PEO is one of the most frequent manifestations of mitochondrial myopathy. It is insidious in onset with ptosis and limitation of eye movements in all directions of gaze, upgaze typically being the most affected. Often there is absent or transient diplopia, but gaze is frequently disconjugate. Onset may occur at any age but usually before 20 years, and the deficit may be static or progressive. Of 52 patients with PEO [12], three had ptosis alone, one had ophthalmoplegia without ptosis, and 48 had both. The ptosis was asymmetrical in 29 and unilateral in four. In 32 patients, gaze was limited to less than 10% of normal in any direction. Ocular movements were disconjugate in 18 cases. Nineteen experienced transient or persistent diplopia.

While PEO may be the sole clinical manifestation of mitochondrial myopathy, it is more commonly associated with limb weakness, fatigue and/or retinopathy. Encephalopathic features may also occur, including dementia, seizures, myoclonus or stroke-like episodes [15,17,23]. In a study of 66 patients [12], 36 had PEO and muscle weakness without major CNS disease. Of these, 22 had additional minor features such as salt and pepper retinopathy with preserved visual acuity (ten cases), mild limb or gait ataxia (nine), sensorineural deafness (five), extensor plantar responses (five) and infrequent seizures (two). The prognosis in this group was good; at a mean disease duration of 17 years, 16 patients had no significant disability and a further 15 were mildly disabled but still able to work. None was physically dependent on others. In contrast, patients with co-existing CNS disease were frequently disabled and prognosis was poor.

KSS is defined by the presence of PEO and pigmentary retinopathy with onset before 20, together with one or more of the following: ataxia, cerebrosinal fluid (CSF) protein concentration > 1 g/l and complete heart block [10]. Patients who fulfil these criteria may also develop other features associated with mitochondrial myopathy. Patients with PEO may have some but not all the elements of KSS, suggesting that PEO, PEO 'plus' and KSS are part of a continuous spectrum of the same disease. This hypothesis is supported by genetic data. Large single deletions of mitochondrial DNA (mtDNA) in muscle but not in blood have been observed in 80% of patients with KSS, and 70% of those with PEO [24–26]. Conversely, to date, all patients with single mtDNA deletions and neurological disease have had PEO, with two exceptions [27,28].

CARDIAC INVOLVEMENT

Petty *et al.* [12] found cardiac conduction defects in nine of 66 patients with mitochondrial myopathy, but identified no patients with symptomatic cardiomyopathy. The defects were: first-degree heart block in two cases; intraventricular conduction block in two cases; right bundle branch block in three cases; complete heart block and a pre-excitation syndrome were seen in single cases. Egger *et al.* [29] reported similar findings in 13 children. Of six patients with predominant myopathy studied by Tulinius and colleagues [30], three had hypertrophic non-obstructive cardiomyopathy; all three presented in infancy. Heart block is commonly associated with Kearns–Sayre syndrome [9] and is one of the criteria for its diagnosis [10], but Petty and colleagues [12] found that cardiac conduction defects were not associated with any specific CNS manifestation.

Occasionally familial mitochondrial myopathy shows prominent cardiomyopathy. Three unrelated families with seven of 22 children with the features of congenital cataract, cardiomyopathy and mitochondrial myopathy on histological examination have been described [31]. All seven children had cardiomegaly; hypertrophic obstructive cardiomyopathy was confirmed at catheterization in four. A family with adult onset myopathy and cardiomyopathy has been described in association with a tRNA$^{Leu(UUR)}$ mutation [32]. Affected members had echocardiographic features of non-obstructive concentric hypertrophy of the left ventricle. No clinical involvement of the nervous system was observed. An infant with rapidly fatal encephalopathy and cardiomyopathy has been described [33]. Another infant with severe cardiomyopathy died at 4 weeks. At necropsy, cardiac muscle fibres were enlarged with mitochondrial accumulations characteristic of histiocytoid cardiomyopathy [34]. Hypertrophic cardiomyopathy has been described in association with MELAS [35].

Cardiac manifestations usually present with or after other features, which give clinical clues to the presence of mitochondrial disease. Patients with cardiomyopathy as the major clinical problem may cause diagnostic difficulties [36], and may remain undiagnosed. Cardiac failure secondary to respiratory failure is an uncommon complication of mitochondrial myopathy [29,37], but respiratory support may lead to a dramatic improvement in symptoms (S.R. Hammans and J.A. Morgan-Hughes, unpublished data).

ZIDOVUDINE MYOPATHY

The cause of diffuse myopathy in HIV-infected patients is controversial, but evidence points to a causal role of zidovudine in at least some patients [38]. Histological changes within muscle of such patients include inflammation, abundant RRFs and proliferation of abnormal mitochondria. Arnaudo *et al.* [39] have identified depletion of muscle mtDNA in zidovudine-induced myopathy, thought to be related to inhibition of mitochondrial DNA polymerase γ by the nucleoside analogue zidovudine.

Of 48 HIV cases with various neuromuscular symptoms [40], 13 showed changes of a mitochondrial myopathy, all of whom had received long-term zidovudine (mean cumulative dose 498 g). The myopathy was painful, proximal and with pronounced wasting. Creatine kinase was normal or moderately elevated. The condition usually improved on withdrawal of the drug.

Mitochondrial encephalopathies

The term mitochondrial encephalomyopathy was introduced by Shapira *et al.* [7] to describe cases of complex multisystem disease associated with 'structurally and/or functionally abnormal mitochondria in the brain and/or muscle'. Rowland's group and others [8–11] have proposed that grouping of some features amongst the encephalopathies allows useful classification into a number of specific entities, such as KSS, MERRF and MELAS. Considerable overlap in clinical features lead to difficulty in classification of many patients, whereas others with encephalopathy do not satisfy the criteria of any of the three syndromes. This has led some authors [12,23] to question the usefulness of subclassification. The continuing elucidation of mitochondrial DNA mutations, which now allows a genetic diagnosis in more than half of our patients (unpublished data), promises a genetic-based classification of these disorders, but imperfect clinical–genetic correlation stimulates continuing debate (Tables 3.1 and 3.2).

MYOCLONIC EPILEPSY

The association of progressive myoclonic epilepsy (PME) with RRFs was first described by Tsairis *et al.* [41]. In 1980 Fukuhara *et al.* [8] proposed the syndrome of MERRF. It is clear that mitochondrial encephalopathy is an important cause of the syndrome of PME and PMA (progressive myoclonic ataxia) [42–44]. Other causes of PME/PMA include Unverricht–Lundborg disease, Lafora body disease, neuronal ceroid lipofuscinoses, sialidoses, spinocerebellar degenerations and rarer causes [44]. Unlike most other causes of PME/PMA, mitochondrial myopathy does not have a narrow age of onset or easily defined clinical course. Ages of onset between 5 and 75 have been described [45,46].

Table 3.1 Point mutations of mtDNA associated with mitochondrial disease

Phenotype	Mutation present	Gene	RRFs
MERRF [119], MERRF + PEO [17], MERRF + Leigh's syndrome [17], cervical lipomas [49]	A→G at bp 8344	tRNA^{Lys}	Usually
MELAS [18,19], other encephalopathies, PEO, myopathy [17]	A→G at bp 3243	tRNA$^{Leu(UUR)}$	Usually
MELAS [56]	T→C at bp 3271	tRNA$^{Leu(UUR)}$	Yes
Myopathy [133]	T→C at bp 3250	tRNA$^{Leu(UUR)}$	Yes
Fatal infantile cardiomyopathy and encephalopathy [33]	A→G at bp 4317	tRNA^{Ile}	Yes
Cardiomyopathy and myopathy [32]	A→G at bp 3260	tRNA$^{Leu(UUR)}$	Yes
Leber's optic neuropathy [134–137]	G→A at bp 11778	ND4	No
	T→C at bp 4160	ND1	
	G→A at bp 3460	ND1	
	G→A at bp 13708*	ND5	
Neurogenic muscle weakness, ataxia and retinitis pigmentosa [104], Leigh's syndrome [105]	T→G at bp 8993	ATPase 6	No

*May require other mtDNA mutations for disease expression.

Core features of the syndrome are myoclonus, ataxia and seizure activity. Associated features include deafness, limb weakness, dementia, headache, foot deformity, optic atrophy, peripheral neuropathy and cervical lipomas. Cases with the core features and some or all of the associated features together with the detection of RRFs on muscle biopsy are uncontroversially classified as within the syndrome of MERRF. Some 70–80% of these patients have the bp 8344 tRNA^{Lys} mutation [17,47,48]. Conversely most but not all patients with this mutation will have MERRF [49]. Some patients with the tRNA^{Lys} mutation have the clinical picture of PME but no RRFs [17,49]. Others have features of MERRF but with additional PEO [17], Leigh's syndrome [17,49] or stroke-like episodes [50,51]. Virtually all clinical features seen in other mitochondrial myopathies may also be seen in patients with MERRF, such as optic atrophy, involuntary movements, episodic hemicranial headache, cardiomyopathy, deafness, retinopathy, spasticity or psychomotor retardation [8,23,30,45]. Such associated features, in particular deafness, are useful clinical clues in the differential diagnosis of PME/PMA. Retinopathy has been described in a case of MERRF without the tRNA^{Lys} mutation [1,17,23], but so far in no cases with the mutation. The frequency of clinical features associated with the bp 8344 mutation in a series of the authors' patients is summarized in Table 3.3.

Myoclonus is often the presenting symptom in the MERRF syndrome, and may be induced by action, noise or photic stimulation. Apart from myoclonus, seizure type may vary. Drop attacks, focal seizures and photosensitive tonic–clonic seizures have been reported [45,52]. Clinical myopathy is often mild or absent. Dementia, when present, results in an insidious decline in cognitive function and is rarely an early feature.

Table 3.2 Phenotypes associated with mtDNA deletions and duplications

Phenotype	MtDNA defect		Inheritance
	Muscle	Blood	
PEO, PEO + other manifestations including stroke-like episodes, and KSS [25,26,46,125–127], migraine and stroke-like episodes [27], Leigh's syndrome [138]	Heteroplasmic large single deletions	Usually not detectable by Southern blot techniques [139]	Usually sporadic
Pearson's syndrome [74–78]	Heteroplasmic large single deletions	*	Sporadic
Diabetes mellitus and deafness [28]	Single large deletion involving O_L	Deletion present in lower quantity	Matrilineal
KSS [89]	Tandem heteroplasmic duplications	*	Sporadic
PEO + other manifestations [130]	Large multiple deletions, starting in D-loop region	Deletions not detectable by Southern blot	Autosomal dominant
KSS [140]	Large multiple deletions, most starting in non-D-loop regions; mtDNA depletion	Not examined	Autosomal dominant
PEO, encephalomyopathy, tubulopathy [141]	Tandem heteroplasmic duplication	Detected by PCR	Matrilineal
Recurrent ketoacidosis, encephalomyopathy [93]	Large multiple deletions, most starting in non-D-loop regions	*	Probably autosomal dominant
Ptosis, optic atrophy, peripheral neuropathy [142]	As above	Not examined	Two brothers ?autosomal recessive
Recurrent myoglobinuria [22]	As above	Deletions not detected by PCR	Two brothers ?autosomal recessive

Low abundance deletions demonstrated in Parkinson's disease [143], cardiomyopathy [144] and normal aging [145] are excluded from this summary.
*Similar findings to muscle.

Table 3.3 Frequency of clinical features associated with the tRNA[Lys] A→G[(8344)] and tRNA[Leu(UUR)] A→G[(3243)] mutations

	bp 3243 mutation (n = 24)		bp 8344 mutation (n = 18)	
	Number	%	Number	%
Ataxia*	13	54	16	89
Myoclonus*	3	12	15	83
Seizures	11	46	14	72
Optic atrophy*	1	4	6	33
Neuropathy	3	12	7	38
Limb weakness	17	70	9	56
Deafness	15	62	6	33
Dementia	8	33	4	11
Ophthalmoplegia or ptosis†	10	41	2	11
Stroke-like episodes†	9	37	1	6
Retinopathy†	8	33	0	0
Short stature†	13	54	0	0

*More frequent with the bp 8344 mutation.
†More frequent with the bp 3243 mutation.
$P<0.05$ Fisher's exact test. Based on a study of a series of patients including probands and examined symptomatic relatives.
Dystonia, chorea, diabetes mellitus, renal tubular acidosis, childhood anaemia, gastrointestinal pseudo-obstruction and cardiomyopathy were observed in single cases with the bp 3243 mutation. Leigh's syndrome, infertility, cataracts and cervical lipomas were observed in single cases with the bp 8344 mutation.

Electrophysiological findings in MERRF are variable. So *et al.* [53] reported atypical spike and wave discharges with an abnormal EEG background in nine of 13 cases, and focal epileptiform abnormalities in six patients (usually in the occipital regions). Giant cortical sensory evoked potentials (SEPs) were recorded in four of six cases studied. There were no features to distinguish cases of mitochondrial disease from other causes of PME. Similar results were reported by Rosing *et al.* [54].

Large MERRF kindreds with several affected members show exclusive maternal inheritance [45,48,54]. No example of paternal inheritance has been described. Apparently sporadic cases may be found to have mildly affected relatives with manifestations such as deafness or myoclonus. Prognosis is variable, but appears to be partly related to age of onset. Two patients described by Berkovic *et al.* [45] and two further English patients died aged 21 or younger; all had onset in the first decade of life. At the other extreme, some family members of MERRF kindreds with late onset disease have epilepsy well controlled on medication and are essentially asymptomatic. Most patients fall between these extremes and have slowly progressive disease leading to significant disability. There is some correlation between the proportion of mtDNA with the bp 8344 transition and clinical parameters. This was initially suggested by Shoffner *et al.* [55] after a study of a single family and the proportion of mutant mtDNA in muscle mtDNA. We have found that the proportion of mutant mtDNA in blood is related to both age of onset of disease and also to clinical severity [52]. This is of potential clinical application and adds to evidence of the causal nature of the mutation.

Table 3.4 Four cases histories showing the diversity of clinical course associated with the bp 3243 mutation

Case number	Sex, age	History	Examination	Investigations
Case 39	Male, age 49	1.75 m tall. No relevant family history. Ptosis noted at age 10. Muscle weakness and fatigue developed since age 16. Left hemiparesis aged 45, with full recovery within a few days. No further stroke-like episodes	Right-sided ptosis; bilateral external ophthalmoplegia. Moderate proximal weakness and mild transient left pyramidal signs	Carotid angiogram, echocardiogram and ECG normal. Muscle biopsy: RRFs; bp 3243 mutation present. Complex I–III defect with normal cytochrome aa_3
Case 120	Male, died age 29	Well until aged 29 when he developed focal seizures affecting his right limbs. 5 months later he developed focal seizures affecting his left side, with a left homonymous hemianopia and hemiparesis. Subsequent subacute deterioration and death	Before last deterioration he had a defect of comprehension, with occipital blindness. Pursuit eye movements were broken, with facial and neck weakness. Proximal limb weakness was noted but the only residue of previous deficits was left-sided sensory inattention	EEG: continuous slow-wave activity more marked over left hemisphere; MRI: bilateral abnormal occipital and parieto-temporal regions; muscle biopsy: RRFs; bp 3243 mutation present
Case 86	Female, died age 24	Small for age, but well until age 15 when she developed episodes of abdominal pain, vomiting and diarrhoea. She had a fit aged 18 and then an episode of status epilepticus with a residual right homonymous hemianopia. The same year she had a laparotomy for suspected intestinal obstruction but no cause found. At age 19 she became unsteady, suffered falls and became dysarthic. She developed dysphasia and suffered intellectual impairment	Dysphasic and intellectually impaired. Pigmentary retinopathy, jerky pursuit eye movements were noted. Her hands were held in a dystonic posture, but limbs showed distal weakness only. She was ataxic with absent ankle jerks and extensor plantar responses	CT: diffuse atrophy; basal ganglia calcification with diffuse low density especially posteriorly. EMG: predominant sensory axonal neuropathy; EEG: widespread slow activity; Muscle biopsy: normal, no RRFs; bp 3243 mutation present in muscle and blood
Case 74	Female, age 51	Difficulty walking from age 14, slowly worsening, prominent fatigue. SOB on exercise	Examination: mild ptosis, no gaze restriction. Proximal muscle weakness, hyporeflexia	Investigations: RRFs on biopsy; bp 3243 mutation present in muscle and blood; Complex I defect with low cytochrome aa_3

Abbreviations: ECG, electrocardiogram; EEG, electroencephalogram; MRI, magnetic resonance imaging; CT, computed tomography; EMG, electromyography.

STROKE-LIKE EPISODES

The acronym MELAS (mitochondrial myopathy, encephalopathy, lactic acidosis and stroke-like episodes) was proposed by Pavlakis and colleagues [11] to delineate another distinctive clinical syndrome associated with RRFs, characterized by normal early development, stunted growth, focal or generalized seizures and recurrent neurological deficit resembling strokes. Absence of ophthalmoplegia, pigmentary retinopathy and heart block, and lack of myoclonus, optic atrophy and sensory neuropathy distinguished MELAS from KSS and MERRF.

Stroke-like episodes may result in hemiparesis, hemianopia or cortical blindness. CT brain scan shows low-density areas which may affect both white and grey matter, but do not always conform to vascular territories [11], and are more common posteriorly. CT may show low-density lesions without clinical correlate. The course of MELAS is typically progressive but punctuated by events of hemicranial headache, recurrent vomiting and/or lactic acidosis, and stroke-like episodes. Strokes may be followed by complete recovery (see Table 3.4, case 39), but more typically lead to progressive encephalopathy which may be rapid (see Table 3.4, case 120). Strokes may be heralded by focal seizures, which may be continuous, recurrent or secondarily generalized. Cessation of ictal activity frequently reveals a hemiparesis, hemianopia or other focal cerebral deficit. Although normal early development was suggested by Pavlakis and colleagues [11] as a diagnostic criterion, delayed psychomotor development has been observed in patients with a clinical syndrome otherwise suggestive of MELAS in studies by ourselves and others [35].

In addition to stroke-like episodes, ataxia, cognitive impairment, focal or generalized seizures, deafness, short stature and limb weakness are common. Optic atrophy, movement disorders, peripheral neuropathy, myoclonus, pigmentary retinopathy, spasticity and PEO also occur. About 80% of patients with MELAS have the bp 3243 tRNA$^{Leu(UUR)}$ mutation [17–19]; some of the remainder may have another mutation in the tRNA$^{Leu(UUR)}$ gene (Table 3.4, [56]). Conversely, in one series of unselected patients with mitochondrial myopathies, only half (eight of 17) the cases with the bp 3243 mutation had the features of MELAS at the time of mtDNA analysis [17]. The incidence of clinical features in patients harbouring the bp 3243 mutation studied by the present authors is given in Table 3.3. The diversity of clinical features associated with the bp 3243 mutation is illustrated by case summaries of four patients (Table 3.4).

OTHER MITOCHONDRIAL ENCEPHALOPATHIES

Many patients with mitochondrial encephalopathies do not develop stroke-like episodes or myoclonus and accordingly are not classified as having the MERRF or MELAS syndromes. Of 28 encephalopathic patients with RRFs without deletions of mtDNA [17], seven had MERRF, eight had MELAS and two had KSS. The remaining 11 had other features commonly associated with mitochondrial encephalopathy such as ataxia, deafness, seizures, dementia, PEO, retinopathy and limb weakness. Four of these had the bp 3243 tRNA$^{Leu(UUR)}$ mutation. The absence of myoclonus or stroke-like episodes does not exclude mitochondrial disease in an undiagnosed encephalopathy.

OVERLAP SYNDROMES

Using standard criteria for inclusion into the three subcategories of mitochondrial encephalopathy KSS [10], MERRF [8] and MELAS [11], several reported patients fall into two groups. Examples are given in Table 3.5.

Table 3.5 Overlap syndromes in mitochondrial myopathies

Overlap syndrome	Reference
KSS ∩ MELAS	Holt *et al.* [25],
	Zupanc *et al.* [91]
PEO ∩ MERRF	Truong *et al.* [23] and Hammans *et al.* [17]
	Rowland *et al.* [15]
KSS ∩ MERRF	Truong *et al.* [23]
MERRF ∩ MELAS	Kuriyama *et al.* [50]
	Byrne *et al.* [51]

DEAFNESS

Sensorineural deafness is common in mitochondrial encephalomyopathies and often provides a valuable clinical clue in cases with multisystem involvement. Of a series of 66 cases of mitochondrial myopathy [12], 17 were deaf, including 12 of 18 patients with CNS disease. The anatomical basis of deafness is unclear. A patient with sensorineural hearing loss but with normal brainstem evoked responses indicated that the deafness was probably of cochlear origin [45].

Deafness may be the sole manifestation of mitochondrial disease, sometimes apparent in otherwise unaffected relatives of patients with mitochondrial encephalopathy [45,54]. Hearing loss is insidious and may be attributed to early onset presbyacusis. A pedigree reported by Ballinger and colleagues [28] had the clinical manifestations of deafness and diabetes mellitus. This was associated with a large mtDNA deletion which involved the origin of replication of the light-strand of mtDNA, and was maternally inherited. No RRF were observed on muscle biopsy, and no CNS features were evident.

A Chinese study described 36 pedigrees with aminoglycoside-induced deafness [57]. Ototoxicity was maternally inherited, suggesting that this predisposition is associated with particular mtDNA genotypes.

THE VISUAL SYSTEM

The retinal manifestations of mitochondrial myopathy were reviewed by Mullie *et al.* [58] and Petty *et al.* [12]. Twenty-four of 66 patients had one of three types of pigmentary retinopathy. In 20 cases there was a 'salt and pepper' retinopathy, with diffuse hypo- and hyper-pigmentation most marked in the equatorial region. This did not give rise to severe impairment of visual acuity unless there was associated optic atrophy or CNS disease; visual acuity was 6/6 or better in eight, and in one it was 6/12. Of the 11 patients with salt and pepper retinopathy and CNS disease, four had normal visual acuity and seven had acuities ranging from 6/9 to 6/36. Four of the latter had optic atrophy. Two patients had diffuse loss of retinal pigment epithelium (RPE) and two had retinitis pigmentosa of the bone corpuscle type. These four patients had severe visual loss (6/36, 3/36, perception of light in two), optic atrophy and clinical evidence of widespread CNS disease [58]. One had optic atrophy without clinically detectable retinopathy, and she also had severe CNS disease. Although prevalence of optic nerve disease without clinically detectable retinopathy was rare in this study, our own experience suggests it may be more common in patients with the syndrome of MERRF (Table 3.3).

In Petty and colleagues' study [12], electroretinography (ERG) was performed in 14 patients, 11 of whom had pigmentary retinopathy. ERG was abnormal in all of the latter, showing subnormal rod- and cone-mediated ERG functions in eight and depression of cone function alone in three. ERG was normal in two of the three cases without retinopathy, and abnormal in one of them.

The retina obtained at necropsy from a 14-year-old boy who died with Kearns–Sayre syndrome showed marked photoreceptor and pigment epithelial cell loss in the retinal periphery and around the optic nerve head [59]. There were structural changes in the RPE cells with enlarged mitochondria. The authors speculated that a breakdown in the energy-dependent relationship between the RPE and the photoreceptor layer was responsible for retinal degeneration. Another post-mortem study also concluded that the primary abnormality was in the RPE [60].

Rarely reported in association with mitochondrial myopathy are corneal opacities [10,61] and cataracts [31,61]. The latter have also been reported in association with autosomal dominant syndrome of mitochondrial myopathy with multiple deletions [62].

PERIPHERAL NEUROPATHY
Peripheral neuropathy is relatively common in mitochondrial myopathies, although it is often asymptomatic. It manifests usually with distal sensory loss, especially proprioception and vibration sense, hypo- or areflexia. Motor nerve involvement is usually subclinical, distal muscle wasting and weakness being rare [63]. In a study of 66 patients, 11 had clinical evidence of neuropathy such as sensory loss or areflexia. Twelve of 51 patients had absent or small sensory action potentials; motor nerve conduction velocities were normal or just below the control range in all but one in which they were markedly reduced to a median of 24 m/s. Electromyography (EMG) showed features of denervation in seven patients [12].

The predominant mechanism of peripheral nerve dysfunction is axonal degeneration. Pathological examination reveals myelinated fibre loss, active axonal degeneration, and small nerve clusters indicating axonal regeneration in biopsied peripheral nerves [64]. Necropsy studies have shown dorsal root ganglia or anterior horn neuronal loss with marked degeneration of the posterior columns [65–67]. However, evidence of additional Schwann cell pathology exists in occasional onion-bulb formation, segmental demyelination and abnormal mitochondria in Schwann cell cytoplasm [64,68–70]. Although rare, the presence of cases with demyelinating neuropathy [71] demonstrates that demyelination may occasionally be the predominant pathogenetic mechanism.

Mizusawa *et al.* [63] have reviewed three cases with mitochondrial myopathy with prominent peripheral nerve involvement. Two of these cases had multiple mtDNA deletions but the third had not been investigated genetically. All three cases had distal sensory loss in association with PEO and muscle weakness.

Non-neurological manifestations

BONE MARROW
Haematological manifestations are uncommon, at least in adults [12]. In a series of 17 children with mitochondrial myopathy, two presented with hypoplastic

anaemia in infancy [29]. Both cases had renal disease. A family with mitochondrial myopathy and sideroblastic anaemia has been described [72].

Pearson's syndrome [73] presents in infancy with refractory sideroblastic anaemia, thrombocytopenia, neutropenia and pancreatic exocrine dysfunction. Renal tubular acidosis and hepatic dysfunction may also occur. Despite repeated transfusions most patients die before the age of 3 years. The disease is associated with large deletions of mtDNA similar to those found in PEO and KSS [74–76]. Unlike most cases of PEO and KSS, deletions may be detected in the blood by Southern blot. Two patients who survived in initial complications of Pearson's syndrome subsequently developed features of Kearns–Sayre syndrome in childhood [77,78], suggesting that these disorders are age-dependent manifestations of the same disease. The two cases with transfusion-dependent hypoplastic anaemia in the series of Egger *et al.* [29] share some of the features of Pearson's syndrome. It appears that early onset mitochondrial myopathy may present with characteristic systemic manifestations such as sideroblastic anaemia, tubulopathy or pancreatic exocrine dysfunction representing the effects of disturbed mitochondrial function on different organs.

KIDNEY

Renal disease appears to be rare in adult patients. It was absent in all but one case of the 66 patients with mitochondrial myopathy described by Petty *et al.* [12]; the single case had renal failure of unknown aetiology. In contrast, six of 17 patients presenting in childhood had renal dysfunction [29], two with glomerulopathy, three with tubulopathy and one with nephronophthisis. A further child had four affected matrilineal relatives who died from renal failure. Tubular defects (including the Fanconi syndrome) appear to be the commonest manifestation of mitochondrial renal disease. Renal tubulopathy frequently occurs in Pearson's syndrome [76,77]. Apart from Pearson's syndrome, three children have been described with renal tubulopathy in association with large mtDNA deletions. One child presented with renal tubular acidosis and subsequently developed Kearns–Sayre syndrome [79]. Two further patients with Kearns–Sayre syndrome were described by Goto *et al.* [80]: one had tubular dysfunction mimicking Bartter's syndrome [81].

The Fanconi syndrome is common in the fatal infantile myopathy associated with cytochrome *c* oxidase (COX) deficiency [82,83]. Some of these cases were subsequently shown to have depletion of mtDNA in muscle or other organs. The Fanconi syndrome has also been reported in Leigh's syndrome due to COX deficiency [84], and has also been observed in three children of one family with mitochondrial myopathy [85], with unknown genetic defect.

ENDOCRINE

Of 66 patients, endocrinopathies occurred in four (two hypothyroid, two diabetic) [12]. In addition, two female patients had been investigated for infertility, and three further patients has short stature. Endocrine involvement appears more common in young patients; of 17 childhood onset cases all had reduced somatic growth [29], three had diabetes, one was hypothyroid. In these cases reduced growth with greatly retarded skeletal maturation was usually apparent within the first three years of life. Growth hormone stimulation tests were performed in five cases with absence of response in three. Treatment with growth hormone doubled the yearly growth rate in these patients, although height remained below the third centile. Short stature, with or without growth hormone deficiency, has been extensively reported [3,10,14,61,86,87] and appears to be a cardinal feature of early onset mitochondrial myopathy.

Insulin-dependent diabetes mellitus has been reported in association with multisystem disease [12,14,29,61,88–90]. Deafness and diabetes mellitus were the major features of the maternally inherited disease reported by Ballinger *et al.* [28]. Hypoparathyroidism is a less common but well-recognized complication [61,91–93]. Hypothyroidism [12,29], adrenocorticotrophin deficiency [66,94] and hyperaldosteronism [95] are rare associations. Infertility with ovarian failure [90,96] or hypothalamic dysfunction also occurs (authors' unpublished observations and [97]).

GASTROINTESTINAL TRACT AND LIVER
Gastrointestinal disturbances may be manifested by recurrent vomiting or pseudo-obstruction, and may occur particularly in patients with encephalopathy [11]. A syndrome consisting of mitochondrial myopathy, peripheral neuropathy, gastrointestinal disease and encephalopathy, termed MNGIE (myo-neuro-gastrointestinal encephalopathy) has been reported in three patients [98,99]. Gastrointestinal disease was manifested as intermittent diarrhoea. Similar features, with intestinal pseudo-obstruction, have been described in five other individuals [100], although the presence of mitochondrial abnormalities was not reported. A further patient with PEO had intestinal pseudo-obstruction [101], a manifestation also seen in one of our cases (case 86, Table 3.4).

Hepatocellular dysfunction occurs in Pearson's syndrome [76,77], and in the infantile mtDNA depletion syndromes [91,102]. In both of these disorders hepatic failure may lead to early death.

MULTIPLE SYMMETRICAL LIPOMATOSIS
Multiple symmetrical lipomatosis (MSL) is a striking but uncommon feature, characterized by large unencapsulated lipomas around the neck, shoulders and other axial regions. Berkovic and colleagues [103] described four patients, three with RRFs and one with COX-negative fibres. Neurological involvement ranged from a mild sensory neuropathy in one patient to the full MERRF syndrome. Two of the patients had the bp 8344 tRNA^{Lys} mutation. The index case of a family with the bp 8344 mutation [47,103] developed lipomas, as has one of our cases with this mutation (authors' unpublished data).

Newly recognized disorders associated with mitochondrial abnormalities

SPINAL MUSCULAR ATROPHY
Rowland and colleagues [15] described a boy with possible spinal muscular atrophy with delayed motor milestones and progressive muscular weakness. He was the child of Ashkenazi parents who were first cousins. Muscle biopsy showed changes of denervation in addition to RRFs.

MIGRAINE WITH RRFs AND A DELETION OF mtDNA
Bresolin *et al.* [27] recently identified a 5 kb deletion of muscle mtDNA in a 40-year-old woman with recurrent headaches since the age of 14 years. She suffered a single episode of unconsciousness at the age of 29 years which left her with a

persistent lower quadrantic field defect on the right. CT brain scan showed a small ischaemic lesion in the left occipital pole. Muscle strength was normal but a biopsy showed RRFs and COX-negative fibres.

RECURRENT KETOACIDOTIC COMA

Cormier and colleagues [93] describe a family with a probable autosomal dominant syndrome of multiple mtDNA deletions, in which the only clinically affected member presented with short stature, ataxia and recurrent episodes of drowsiness or ketoacidotic coma. A pyramidal syndrome, ataxia, PEO, deafness and hypoparathyroidism developed.

NEUROGENIC MUSCLE WEAKNESS, ATAXIA AND RETINITIS PIGMENTOSA (NARP)

A family with these cardinal features, and variable developmental delay, fits, dementia and sensory neuropathy, has been described [104]. RRFs were lacking. Blood and muscle were heteroplasmic for a mutation in a subunit of ATPase (Complex V). A further family had similar features, but also included the phenotype of Leigh's disease [105].

INVESTIGATION

Clinical assessment

Clinical features suggesting mitochondrial disease should be sought in both the patient and their available relatives. Asymptomatic relatives may have mild ptosis, limb weakness or fatigue, retinopathy, or hearing loss. Detailed ophthalmological or audiological assessment may be helpful. Mitochondrial myopathies may be sporadic, or show matrilineal, autosomal dominant or autosomal recessive modes of inheritance.

Laboratory investigation

Serum creatine kinase is usually normal or marginally increased [12], and is unhelpful in diagnosis. Mitochondrial dysfunction may result in increased concentrations of lactate and pyruvate or an abnormal lactate/pyruvate ratio in blood and cerebrospinal fluid. Resting blood lactate concentrations were less than twice normal in 35 of 50 cases [12]. Greater discrimination between patients and controls is achieved after aerobic exercise; peak exercise serum lactate was above 2 mM in 19 of 29 patients compared with a maximum of 1.8 mM in normal or disease controls [12].

Of 61 cases, EMG was normal in 16, showed myopathic change in 38 and features of denervation in seven [12], and therefore is not a specific or sensitive test for mitochondrial myopathy. MRI or CT scan of the brain may be normal, but in encephalopathies show a wide range of abnormalities including low-density lesions or atrophy of the cerebrum or cerebellum, and lucencies or calcification of the basal ganglia. Multifocal areas of hyperintense signal confined to the cortex of the cerebrum, cerebellum and immediately adjacent white matter on T_2-weighted MRI scan has been suggested as a specific finding in the MELAS syndrome [106]. In Leigh's syndrome, cranial MRI scan is more sensitive than CT, characteristically showing bilateral symmetrical lesions in the basal ganglia and the brainstem, and less commonly elsewhere in the brain [107–109].

Although some mtDNA mutations may be detected in blood, muscle biopsy is usually necessary to confirm the diagnosis. A closed biopsy is usually adequate to establish the diagnosis but open biopsy is necessary for detailed biochemical studies which require a minimum of 1–2 g of tissue. Diagnostic yield from muscle histology is increased by the use of histochemical stains, particularly those for succinate dehydrogenase and cytochrome oxidase [110]. A negative muscle biopsy does not exclude mitochondrial disease. If the clinical picture is suggestive, ultra-structural [45] and genetic studies should be performed, and a second biopsy for biochemical studies should be considered.

Although the diagnostic importance of muscle biopsy is well established, the definition of mitochondrial myopathy by morphological criteria alone is complicated by the fact that small numbers of RRFs and COX-negative fibres are occasionally observed in the elderly and in other muscle diseases such as muscular dystrophies, polymyositis or glycogen storage diseases [111,112]. Conversely, the diagnosis of mitochondrial myopathy has been made in the absence of RRFs but on the basis of a suggestive clinical course and COX-negative fibres on muscle biopsy [110], a family history [45,54], a biochemical defect [45] or mitochondrial DNA defects [17].

Screening leukocyte DNA for the bp 3243 and bp 8344 mutations has recently been proposed to allow diagnosis before muscle biopsy in patients with mitochondrial encephalopathy [17]. Absence of the mutations does not exclude the diagnosis. Three patients without RRFs, but with features suggestive of mitochondrial encephalopathy, had mutations at bp 8344 (one patient) or bp 3243 (two patients) in both blood and muscle. In two further patients, genetic diagnoses were established before biopsy. Point mutations occurring in NARP [104], myopathy/cardiomyopathy [32], optic neuropathy/myelopathy [113], infantile cardiomyopathy/encephalopathy [33] and Leber's hereditary optic neuropathy may also be detectable in blood (Table 3.1). mtDNA deletions in mitochondrial myopathies are rarely detectable in blood by Southern blot [26], but are present in blood in Pearson's syndrome [77]. For genetic counselling, leucocyte and (if available) muscle DNA should be analysed or stored.

Magnetic resonance spectroscopy (MRS) of muscle

Phosphorus MRS allows study of muscle energy metabolism *in vivo* [114–116]. Inorganic phosphate (P_i), phosphocreatine (PCr), adenosine mono-, di- or tri-phosphate (AMP, ADP, ATP), phosphomono- and di-esters and intracellular pH may be measured. The P_i/PCr ratio is the most useful derived parameter and may be monitored at rest, during exercise and recovery. An increased ratio is found in most patients with mitochondrial myopathies [116], with a prolonged recovery of the P_i/PCr ratio being the most sensitive indicator of mitochondrial dysfunction. In specialized centres, MRS is a useful tool in diagnosis of mitochondrial disease, and also to monitor therapeutic trials. However, the observed abnormalities are not specific to mitochondrial myopathies, as they may be reproduced by other pathologies where secondary mitochondrial dysfunction is associated, including inflammatory myopathies, hypothyroidism and muscular dystrophies. No correlation between MRS findings and the site of the respiratory chain defect can be made [114].

Table 3.6 Drug therapy in mitochondrial myopathies

Therapy	Dose
Ubidecarone (coenzyme Q_{10})	2 mg/kg per day [146], 80–160 mg/day [117], 300 mg/day [119]
Carnitine	50–200 mg/kg per day [117]
Riboflavin	100 mg/day [122]
Thiamine	400–800 mg/day [117]
Menadione	40–160 mg/day [117], 40–80 mg/day [121]
Prednisolone	Variable
Succinate	6 g/day [119]
Ascorbate	4 × 1 g/day [117,121]

TREATMENT

Lactic acidosis may be precipitated by exercise, alcohol or intercurrent infection. Patients should accordingly be advised to avoid excess alcohol or exercise. On theoretical grounds it is advisable to avoid sodium valproate (which sequesters carnitine), barbiturates (which inhibit the respiratory chain), tetracyclines and chloramphenicol (which inhibit mitochondrial protein synthesis)[117], although detrimental effects have not been proven.

Effective treatment remains limited, and is summarized in Table 3.6. Acute exacerbations of lactic acidosis may be improved by slow infusion of sodium bicarbonate. Carnitine supplements are indicated in patients with secondary carnitine deficiency. In an open trial of ubidecarone (CoQ_{10}), 16 of 44 patients showed a greater than 25% decrease in post-exercise lactate levels [118]. However, this subgroup of 'responders' was studied further in a blind placebo controlled trial for 3 months; no significant difference was observed between placebo and treated groups. Combined therapy with succinate and CoQ_{10} improved respiratory function in one patient with KSS [119]. Sustained improvement is reported in one case with Complex III deficiency treated with menadione (vitamin K_3) and ascorbic acid (vitamin C), monitored with ^{31}P-MRS [120, 121]. Treatment with riboflavin has been associated with improvement in a patient with Complex I-deficient myopathy [122]. Headaches and stroke-like episodes seem to improve rapidly on steroid treatment [123]. Gubbay and colleagues have described a case of steroid-responsive mitochondrial encephalopathy [124].

GENETIC COUNSELLING IN MITOCHONDRIAL MYOPATHIES

Advice to families of patients with mitochondrial myopathies should cater for their need to know the nature of the underlying defect, their prognosis and the recurrence risks. Harding *et al.* [16] performed a genetic study of 71 cases of mitochondrial myopathy and calculated overall recurrence risks. Subsequent description of associated mtDNA deletions and mutations has allowed genetic classification of mitochondrial myopathies. More than half of our own series of patients have one of the published mtDNA deletions or mutations. Knowledge concerning modes of

Table 3.7 Difficulties in calculating recurrence risks in mitochondrial myopathies

1	Disease may be sporadic or inherited in a matrilineal, autosomal dominant or recessive manner
2	Published pedigrees generally give incomplete information, hindering calculation of recurrence risks
3	Insufficient data are available on individual phenotype–genotype combinations
4	Only 50–60% of patients have known mutations or deletions of mtDNA
5	Most mtDNA mutations have age-depednent penetrance, as yet undefined
6	Affected patients have varying severity of disease
7	The degree of heteroplasmy in blood, muscle or placental tissue has not been shown to be closely related to clinical severity

inheritance of mtDNA defects often allows prediction of the heritability of the individual genotype/phenotype combinations to a greater accuracy than allowed by the data previously available. Consequently, molecular genetic analysis of family members is essential because of its implications for the likely mode of inheritance. Difficulties encountered in deriving genetic risks are summarized in Table 3.7. We summarize below, the present knowledge applicable to genetic counselling. The swift rate of progress in associating mitochondrial DNA mutations with disease threatens to render such advice rapidly obsolete.

Approach to counselling

The diagnosis of mitochondrial myopathy should be made secure by a full clinical evaluation, with appropriate investigation as previously outlined. A full evaluation of the family is essential to establish the mode of inheritance and may also lend support to the clinical diagnosis.

Genetic investigation

Analysis of muscle mtDNA is preferable to blood because large deletions are usually not detectable in blood by Southern blotting techniques, and mtDNA point mutations are generally present in higher proportions in muscle mtDNA than in blood. If a muscle biopsy has not been performed, analysis of blood mtDNA, especially in patients with encephalopathy, may allow genetic diagnosis without recourse to biopsy [17]. As the phenotypes associated with individual mtDNA mutations are diverse, it is our practice to screen new patients for most of the published mutations. The use of this approach is illustrated by an 18-year-old male who recently presented to us with PEO and deafness. A muscle biopsy was declined but blood mtDNA analysis showed the presence of the bp 3243 tRNA$^{Leu(UUR)}$ mutation.

In estimating recurrence risks in specific subgroups, errors are likely to be caused by the factors outlined in Table 3.7. Where given, quantitative estimates should be used cautiously and may require adjustment in the light of individual pedigree structure.

Deletions

SINGLE mtDNA DELETIONS

This genetic defect is associated with a wide variety of phenotypes (Table 3.2). Since the first description of mtDNA deletions [24], it has become clear that diseases associated with deletions are usually sporadic [25,26,46,125–127]. Exceptions to this rule are a mother and daughter reported by Ozawa *et al.* [80,128] with mtDNA deletions at different sites, and a family with diabetes and deafness [28]. Two of the 30 cases described by Holt *et al.* [25] had affected relatives. The daughter of the sister of one patient and the daughter of another patient reportedly had PEO. One was unavailable for study; the other had mild ptosis but no other abnormal physical signs, and muscle mtDNA analysis showed no evidence of a deletion. Well over a hundred cases with deletions have been reported and no other familial cases have been identified [25,26,46,125–127]. Poulton and colleagues [129] suggested that deleted mtDNAs were maternally transmitted in one family. This was based on finding small amounts of deleted mtDNA in the blood of the asymptomatic mother and sister of one patient using the polymerase chain reaction undetectable by Southern blotting. These observations have not been confirmed and their significance is difficult to assess given that maternal transmission of disease associated with a single large deletion has not been reported. Reproductive fitness is clearly affected by the disease and published observations do not include full pedigree data. However, recurrence risk of disease in the maternal line could be reasonably estimated as less than 5%, and less than 1% in the paternal line.

Duplications of mtDNA in two patients have been reported [89]. Both cases were sporadic.

MULTIPLE mtDNA DELETIONS

In the genetic analysis of mitochondrial myopathy, care should be taken to exclude the presence of multiple deletions which may be overlooked on Southern blot, appearing as a smear below the normal fragment. The phenotypes associated with multiple deletions are summarized in Table 3.2. The families described by Zeviani *et al.* [62,130] were characterized by PEO and limb weakness, with deafness and neuropathy in some. Affected members had RRFs on muscle biopsy and multiple deletions starting in the D-loop region. In contrast with single mtDNA deletions, these families showed autosomal dominant inheritance.

The case described by Cormier *et al.* [93] had a complicated clinical picture of ataxia and recurrent ketoacidotic comas. Genetic investigation revealed multiple deletions between the *COX II* and *Cyt b* genes not only in the proband but also in the asymptomatic mother and maternal aunt. Autosomal dominant inheritance was suggested, but maternal inheritance could not be excluded. Two brothers, described by Ohno *et al.* [22], presented with recurrent myoglobinuria and mitochondrial myopathy. Their mother was clinically normal but their father appears not to have been examined and inheritance in this family is unclear.

Point mutations of mtDNA (Table 3.1)

THE BP 8344 tRNA^{Lys} MUTATION

This mutation is commonly associated with the myoclonic epilepsy phenotype, with various additional features or exceptions described previously. The mutation

Table 3.8 Proportion of affected offspring of mothers with the A→G(8344) mutation

Description of pedigree	Symptomatic (inc. RRFs) children/total children	
	Symptomatic mother	Asymptomatic mother
Rosing *et al.* [47,54]	3/4,1/5,3/6,1/1,0(2)/4 total 8(10)/20	
Berkovic *et al.* [45]	3/3,2/2 total 5/5	
Zeviani *et al.* [48]	1/3,2/3,0(2)/2 total 3(5)/8	1(5)/6,0(1)/2,1(2)/3,0/2 total 2(8)/13
Hammans *et al.* [52]	1/2,3/5,3/4,1/2 total 8/13	1/2,0/11,0/2,2/3,2/3 total 5/21
Total	24(28)/46	7(8)/34

Each quotient refers to the offspring of an individual female within a pedigree.
Values in parentheses include asymptomatic cases shown to have RRFs in muscle.

is detectable in blood as well as muscle [17]. Large families established the MERRF syndrome as the prototype maternally inherited mitochondrial myopathy [45,47,54]. From these pedigrees and others [17,48] no examples of paternal inheritance are apparent. The following data are derived from published pedigrees (Table 3.8) and our own unpublished pedigrees, subject to errors outlined in Table 3.7. Our own studies include five families with the bp 8344 mutation [52]. Of 13 offspring of affected mothers, five are presently unaffected and are aged between 22 and 40, while affected offspring were all symptomatic by the age of 25. Only one child of an unaffected mother had no detectable mutation in blood. As patients may develop symptoms at any age up to at least 75 [46], such 'unaffected' patients are still at risk of developing the disease. It would seem likely that the offspring of an affected mother with the mutation have a risk of between 50% and 80% of developing disease by the age of 25. Should patients develop symptoms later, the clinical course may be less severe. Advice to asymptomatic females heteroplasmic for the mutation is more difficult. Of our patients, 21 children were born to unaffected women heteroplasmic for the mutation. Only five were symptomatic but some of the unaffected children were under the age of 10 [52]. We have found a correlation between the proportion of mutant mtDNA in blood and clinical parameters, but the relationship is imperfect and further data are required before potential clinical application.

THE BP 3243 tRNA^{Leu(UUR)} MUTATION

THE BP 3243 tRNA$^{\text{Leu(UUR)}}$ MUTATION

The study by Ciafaloni *et al.* [131] represents the sole published pedigree data concerning this mutation. In five pedigrees, eight of 13 offspring of clinically affected mothers had the disorder but some of those unaffected were under the age of 10. Seven of 23 offspring of clinically unaffected mothers also had the disease. Similar recurrence risks of developing the disease by age 25 are applicable to both the bp 3243 and bp 8344 mutations. A risk of 60–80% for offspring of affected females and approximately half this for offspring of unaffected females harbouring the mutation in blood or muscle is likely. No case of paternal transmission has been documented to date.

The clinical spectrum associated with the bp 3243 mutation is wide. We have studied two families with affected members in two and three generations, one with predominant limb weakness and the other with PEO and myopathy. The clinical syndrome, while varying in severity, may be qualitatively similar within families.

OTHER POINT MUTATIONS

Mitochondrial myopathies associated with point mutations described in single families may be maternally inherited [32] or sporadic [33].

Counselling in patients without known mtDNA mutations

Patients with mitochondrial myopathy without known mutations may have hitherto undescribed mtDNA mutations. It is possible that some may have nuclear DNA defects or are non-genetic phenocopies. Alternatively, such patients may have combinations of mtDNA and nuclear genetic 'defects' that are not individually pathogenic.

In the absence of a known genetic defect and without a mode of inheritance being apparent in an individual pedigree, some guidance may be derived from published data. The study of Harding *et al.* [16] surveyed 71 cases with a recurrence rate of 3% for sibs and 5.5% for offspring (over the age of 20) of index cases. The corresponding risk for offspring of females was 8.1%, whereas no examples of paternal transmission were observed. These risks are not directly applicable to patients without known mtDNA defects, as 28 of Harding's patients have since been shown to have mtDNA deletions in muscle and these patients are known to have a very low recurrence risk (see above). In addition, maternal transmission is much more common than paternal transmission (about 9:1). The true recurrence risk for offspring of affected females without known mutations is likely to be in the range 10–20% and 1–2% for offspring of males.

Prenatal and predictive testing

Prenatal testing has been performed twice in a family with a heteroplasmic mutation at bp 8993 [104,132] with the features of neurogenic muscle weakness, ataxia and retinitis pigmentosa. Prenatal testing was made possible by a close correlation within the family between degree of heteroplasmy and severity of disease. A high proportion of mutant mtDNA was seen in both chorionic villus samples. Both pregnancies were terminated and high proportions of mutant mtDNA were present in all fetal tissues examined.

In general, within mitochondrial myopathies, the relationship between the proportion of abnormal mtDNA in blood, placenta, muscle and other tissues (especially brain) is unclear. Heteroplasmy may quantitatively change *in utero* and subsequently during childhood. There is some relationship between degree of heteroplasmy in muscle in patients and their relatives and severity of disease [47,131], but this is imperfect. Further work is required to validate the relationship between the prevalence of normal and abnormal mtDNA and clinical expression of the disease both between tissue and during aging. Even with these data, a loose relationship between heteroplasmy and disease expression would prevent useful prenatal and predictive testing. Other nuclear and mitochondrial genes, as well as environmental influences, may modulate the expression of single-base mtDNA mutations.

In woman with known mutations associated with mitochondrial disease, recurrence risks for offspring are high. In addition, the patient is at risk of becoming increasingly physically and/or mentally disabled. This situation requires careful counselling and support. Ovum donation programmes offer one possible option to such patients, and mitochondrial DNA diseases represent a particularly compelling indication for this approach.

REFERENCES

1. Morgan-Hughes, J.A., Hayes, D.J., Clark, J.B. *et al.* (1982) Mitochondrial encephalomyopathies: biochemical studies in two cases revealing defects in the respiratory chain. *Brain*, **105**, 553–582
2. Engel, W.K. and Cunningham, G.G. (1963) Rapid examination of muscle tissue; an improved trichrome stain method for fresh-frozen biopsy sections. *Neurology, Minneapolis*, **13**, 919–923
3. Olson, W., Engel, W.K., Walsh, G.O. *et al.* (1972) Oculocraniosomatic neuromuscular disease with 'ragged-red' fibers; histochemical and ultrastructural changes in limb muscles of a group of patients with idiopathic progressive external ophthalmoplegia. *Archives of Neurology*, **26**, 193–211
4. Luft, R., Ikkos, D., Palmieri, G. *et al.* (1962) A case of severe hypermetabolism of nonthyroid origin with a defect in the maintenance of mitochondrial respiratory control; a correlated clinical, biochemical and morphological study. *Journal of Clinical Investigation*, **41**, 1776–1804
5. DiMauro, S., Bonilla, E., Lee, C.P. *et al.* (1976) Luft's disease: further biochemical and ultrastructural studies of skeletal muscle in the second case. *Journal of the Neurological Sciences*, **27**, 217–232
6. Morgan-Hughes, J.A. and Mair, W.G. (1973) Atypical muscle mitochondria in oculoskeletal myopathy. *Brain*, **96**, 215–224
7. Shapira, Y., Harel, S. and Russell, A. (1977) Mitochondrial encephalomyopathies: a group of neuromuscular disorders with defects in oxidative metabolism. *Israel Journal of Medical Sciences*, **13**, 161–164
8. Fukuhara, N., Tokiguchi, S., Shirakawa, K. *et al.* (1980) Myoclonus epilepsy associated with ragged-red fibres (mitochondrial abnormalities): disease entity or a syndrome? Light- and electron-microscopic studies of two cases and review of literature. *Journal of the Neurological Sciences*, **47**, 117–133
9. DiMauro, S., Bonilla, E., Zeviani, M. *et al.* (1985) Mitochondrial myopathies. *Annals of Neurology*, **17**, 521–538
10. Berenberg, R.A., Pellock, J.M., DiMauro, S. *et al.* (1977) Lumping or splitting? 'Ophthalmoplegia-plus' or Kearns–Sayre syndrome? *Annals of Neurology*, **1**, 37–54
11. Pavlakis, S.G., Phillips, P.C., DiMauro, S. *et al.* (1984) Mitochondrial myopathy, encephalopathy, lactic acidosis, and strokelike episodes: a distinctive clinical syndrome. *Annals of Neurology*, **16**, 481–488
12. Petty, R.K.H., Harding, A.E. and Morgan-Hughes, J.A. (1986) The clinical features of mitochondrial myopathy. *Brain*, **109**, 915–938
13. Hudgson, P., Bradley, W.G. and Jenkinson, M. (1972) Familial mitochondrial myopathy: disordered oxidative metabolism in muscle fibres. *Journal of the Neurological Sciences*, **16**, 343–370
14. Kamieniecka, Z. (1977) Myopathies with abnormal mitochondria. A clinical, histological, and electrophysiological study. *Acta Neurologica Scandinavica*, **55**, 57–75
15. Rowland, L.P., Blake, D.M., Hirano, M. *et al.* (1991) Clinical syndromes associated with ragged red fibres. *Revue Neurologique*, **147**, 467–473
16. Harding, A.E., Petty, R.K.H. and Morgan-Hughes, J.A. (1988) Mitochondrial myopathy: a genetic study of 71 cases. *Journal of Medical Genetics*, **25**, 528–535
17. Hammans, S.R., Sweeney, M.G., Brockington, M. *et al.* (1991) Mitochondrial encephalopathies: molecular genetic diagnosis from blood samples. *Lancet*, **337**, 1311–1313
18. Goto, Y., Nonaka, I. and Horai, S. (1990) A mutation in the tRNA$^{Leu(UUR)}$ gene associated with the MELAS subgroup of mitochondrial encephalomyopathies. *Nature*, **348**, 651–653
19. Kobayashi, Y., Momoi, M.Y., Tominaga, K. *et al.* (1990) A point mutation in the mitochondrial tRNA(Leu)(UUR) gene in MELAS (mitochondrial myopathy, encephalopathy, lactic acidosis and stroke-like episodes). *Biochemical Biophysical Research Communications*, **173**, 816–822
20. Koga, Y., Nonaka, I., Kobayashi, H. *et al.* (1988) Findings in muscle in complex I (NADH Coenzyme Q reductase) deficiency. *Annals of Neurology*, **24**, 749–756

21. Morgan-Hughes, J.A., Cooper, J.M., Schapira, A.H.V., Hayes, D.J. and Clark, J.B. (1987) The mitochondrial myopathies. Defects of the respiratory chain and oxidative phosphorylation system. In *The London Symposia* (eds. R.J. Ellingson, N.M.F. Murray and A.M. Halliday), Elsevier, Amsterdam, pp. 103–114
22. Ohno, K., Tanaka, M., Sahashi, K. *et al.* (1991) Mitochondrial DNA deletions in inherited recurrent myoglobinuria. *Annals of Neurology*, **29**, 364–369
23. Truong, D.D., Harding, A.E., Scaravilli, F. *et al.* (1990) Movement disorders in mitochondrial myopathies. A study of nine cases with two autopsy studies. *Movement Disorders*, **5**, 109–117
24. Holt, I.J., Harding, A.E. and Morgan-Hughes, J.A. (1988) Deletions of muscle mitochondrial DNA in patients with mitochondrial myopathies. *Nature*, **331**, 717–719
25. Holt, I.J., Harding, A.E., Cooper, J.M. *et al.* (1989) Mitochondrial myopathies: clinical and biochemical features of 30 patients with major deletions of muscle mitochondrial DNA. *Annals of Neurology*, **26**, 699–708
26. Moraes, C.T., DiMauro, S., Zeviani, M. *et al.* (1989) Mitochondrial DNA deletions in progressive external ophthalmoplegia and Kearns–Sayre syndrome. *New England Journal of Medicine*, **320**, 1293–1299
27. Bresolin, N., Martinelli, P., Barbiroli, B. *et al.* (1991) Muscle mitochondrial DNA deletion and ^{31}P-NMR spectroscopy alterations in a migraine patient. *Journal of the Neurological Sciences*, **104**, 182–189
28. Ballinger, S.W., Shoffner, J.M., Hadaya, E.V. *et al.* (1992) Maternally transmitted diabetes and deafness associated with a 10.4 kb mitochondrial DNA deletion. *Nature Genetics*, **1**, 11–15
29. Egger, J., Lake, B.D. and Wilson, J. (1981) Mitochondrial cytopathy. A multisystem disorder with ragged red fibres on muscle biopsy. *Archives of Diseases in Childhood*, **56**, 741–752
30. Tulinius, M.H., Holme, E., Kristiansson, B. *et al.* (1991) Mitochondrial encephalomyopathies in childhood. II. Clinical manifestations and syndromes. *Journal of Pediatrics*, **119**, 251–259
31. Sengers, R.C., Trijbels, J.M., Willems, J.L. *et al.* (1975) Congenital cataract and mitochondrial myopathy of skeletal and heart muscle associated with lactic acidosis after exercise. *Journal of Pediatrics*, **86**, 873–880
32. Zeviani, M., Gellera, C., Antozzi, C. *et al.* (1991) Maternally inherited myopathy and cardiomyopathy: association with mutation in mitochondrial DNA tRNA$^{Leu(UUR)}$. *Lancet*, **338**, 143–147
33. Tanaka, M., Ino, H., Ohno, K. *et al.* (1990) Mitochondrial mutation in fatal infantile cardiomyopathy. *Lancet*, **336**, 1452
34. Papadimitriou, A., Neustein, H.B., DiMauro, S. *et al.* (1984) Histiocytoid cardiomyopathy of infancy: deficiency of reducible cytochrome b in heart mitochondria. *Pediatric Research*, **18**, 1023–1028
35. Yoneda, M., Tanaka, M., Nishikimi, M. *et al.* (1989) Pleiotropic molecular defects in energy-transducing complexes in mitochondrial encephalomyopathy. *Journal of the Neurological Sciences*, **92**, 143–158
36. Wiles, C.M., Morgan-Hughes, J.A., Rose, P.E. *et al.* (1986) Breathlessness and weakness in a 22 year old woman. *Hospital Update*, **12**, 31–42
37. Byrne, E., Dennett, X., Trounce, I. *et al.* (1985) Mitochondrial myoneuropathy with respiratory failure and myoclonic epilepsy. *Journal of the Neurological Sciences*, **71**, 273–281
38. Dalakas, M., Illa, I., Pezeshkpour, G.H. *et al.* (1990) Mitochondrial myopathy caused by long-term zidovudine therapy. *New England Journal of Medicine*, **322**, 1098–1105
39. Arnaudo, E., Dalakas, M., Shanske, S. *et al.* (1991) Depletion of muscle mitochondrial DNA in AIDS patients with zidovudine-induced myopathy. *Lancet*, **337**, 508–510
40. Mhiri, C., Baudrimont, M., Bonne, G. *et al.* (1991) Zidovudine myopathy: a distinctive disorder associated with mitochondrial dysfunction. *Annals of Neurology*, **29**, 606–614
41. Tsairis, P., Engel, W.K. and Kark, P. (1973) Familial myoclonic epilepsy syndrome associated with skeletal muscle mitochondrial abnormalities. *Neurology, Minneapolis*, **23**, 408
42. Berkovic, S.F., Andermann, F., Carpenter, S. *et al.* (1986) Progressive myoclonic epilepsies: specific causes and diagnosis. *New England Journal of Medicine*, **315**, 296–305
43. Marseille Consensus Group (1990) Classification of progressive myoclonus epilepsies and related disorders. *Annals of Neurology*, **28**, 113–116
44. Marsden, C.D., Harding, A.E., Obeso, J.A. *et al.* (1990) Progressive myoclonic ataxia (the Ramsay Hunt syndrome). *Archives of Neurology*, **47**, 1121–1125
45. Berkovic, S.F., Carpenter, S., Evans, A. *et al.* (1989) Myoclonus epilepsy and ragged-red fibres (MERRF). 1. A clinical, pathological, biochemical, magnetic resonance spectrographic and positron emission tomographic study. *Brain*, **112**, 1231–1260
46. Geny, C., Cormier, V., Meyrignac, C. *et al.* (1991) Muscle mitochondrial DNA in encephalomyopathy and ragged red fibres: a Southern blot analysis and literature review. *Journal of Neurology*, **238**, 171–176

47. Shoffner, J.M., Lott, M.T., Lezza, A.M. *et al.* (1990) Myoclonic epilepsy and ragged-red fiber disease (MERRF) is associated with a mitochondrial DNA tRNA(Lys) mutation. *Cell,* **61,** 931–937

48. Zeviani, M., Servidei, S., Bresolin, N. *et al.* (1991) Rapid detection of the A→G[8344] mutation in Italian families with myoclonus, epilepsy and ragged red fibres (MERRF). *American Journal of Human Genetics,* **48,** 203–211

49. Berkovic, S.F., Shoubridge, E.A., Andermann, F. *et al.* (1991) Clinical spectrum of mitochondrial DNA mutation at base pair 8344. *Lancet,* **338,** 457

50. Kuriyama, M., Umezaki, H., Fukuda, Y. *et al.* (1984) Mitochondrial encephalomyopathy with lactate-pyruvate elevation and brain infarctions. *Neurology, Cleveland,* **34,** 72–77

51. Byrne, E., Trounce, I., Dennett, X. *et al.* (1988) Progression from MERRF to MELAS phenotype in a patient with combined respiratory complex I and IV deficiencies. *Journal of the Neurological Sciences,* **88,** 327–338

52. Hammans, S.R., Sweeney, M.G., Brockingham, M. *et al.* (1993) The mitochondrial DNA transfer RNA^{Lys} A→G[8344] mutation and the syndrome of myoclonic epilepsy with ragged red fibres (MERRF). *Brain,* **116,** 617–632

53. So, N., Berkovic, S.F., Andermann, F. *et al.* (1989) Myoclonus epilepsy and ragged-red fibres (MERRF). 2. Electrophysiological studies and comparison with other progressive myoclonus epilepsies. *Brain,* **112,** 1261–1276

54. Rosing, H.S., Hopkins, L.C., Wallace, D.C. *et al.* (1985) Maternally inherited mitochondrial myopathy and myoclonic epilepsy. *Annals of Neurology,* **17,** 228–237

55. Shoffner, J.M., Lott, M.T. and Wallace, D.C. (1991) MERRF: a model disease for understanding the principles of mitochondrial genetics. *Revue Neurologique,* **147,** 431–435

56. Goto, Y., Nonaka, I. and Horai, S. (1991) A new mtDNA mutation associated with mitochondrial myopathy, encephalopathy, lactic acidosis and stroke-like episodes (MELAS). *Biochimica et Biophysica Acta,* **1097,** 238–240

57. Hu, D.-N., Qiu, W.-Q., Wu, B.-T. *et al.* (1991) Genetic aspects of antibiotic induced deafness: mitochondrial inheritance. *Journal of Medical Genetics,* **28,** 79–83

58. Mullie, M.A., Harding, A.E., Petty, R.K.H. *et al.* (1985) The retinal manifestations of mitochondrial myopathy. A study of 22 cases. *Archives of Ophthalmology,* **103,** 1825–1830

59. McKechnie, N.M., King, M. and Lee, W.R. (1985) Retinal pathology in the Kearns–Sayre syndrome. *British Journal of Ophthalmology,* **69,** 63–75

60. Eagle, R.C., Hedges, T.R. and Yanoff, M. (1982) The atypical pigmentary retinopathy of Kearns–Sayre syndrome. *Ophthalmology,* **89,** 1433–1440

61. Toppet, M., Telerman Toppet, N., Szliwowski, H.B. *et al.* (1977) Oculocraniosomatic neuromuscular disease with hypoparathyroidism. *American Journal of Diseases in Childhood,* **131,** 437–441

62. Zeviani, M., Servidei, S., Gellera, C. *et al.* (1989) An autosomal dominant disorder with multiple deletions of mitochondrial DNA starting at the D-loop region. *Nature,* **339,** 309–311

63. Mizusawa, H., Ohkoshi, N., Watanabe, M. *et al.* (1991) Peripheral neuropathy of mitochondrial myopathies. *Revue Neurologique,* **147,** 501–507

64. Yiannikas, C., McLeod, J.G., Pollard, J.D. *et al.* (1988) Peripheral neuropathy associated with mitochondrial myopathy. *Australian Paediatric Journal,* **24,** Suppl., 62–63

65. Takeda, S., Wakabayashi, K., Ohama, E. *et al.* (1988) Neuropathology of myoclonus epilepsy associated with ragged red fibres (Fukuhara's disease). *Acta Neuropathologica (Berlin),* **75,** 433–440

66. Sasaki, H., Kuzuhara, S., Kanazawa, I. *et al.* (1983) Myoclonus, cerebellar disorder, neuropathy, mitochondrial myopathy, and ACTH deficiency. *Neurology, Cleveland,* **33,** 1288–1293

67. Croft, P.B., Cutting, J.C., Jewesbury, E.C.D. *et al.* (1977) Ocular myopathy (progessive external ophthalmoplegia) with neuropathic complications. *Acta Neurologica Scandinavica,* **55,** 169–197

68. Peyronnard, J.M., Charron, L., Bellavance, A. *et al.* (1980) Neuropathy and mitochondrial myopathy. *Annals of Neurology,* **7,** 262–268

69. Pezeshkpour, G., Krarup, C., Buchthal, F. *et al.* (1987) Peripheral neuropathy in mitochondrial disease. *Journal of the Neurological Sciences,* **77,** 285–304

70. Yiannikas, C., McLeod, J.G., Pollard, J.D. *et al.* (1986) Peripheral neuropathy associated with mitochondrial myopathy. *Annals of Neurology,* **20,** 249–257

71. Groothuis, D.R., Schulman, S., Wollman, R. *et al.* (1980) Demyelinating radiculopathy in the Kearns–Sayre syndrome: a clinicopathological study. *Annals of Neurology,* **8,** 373–380

72. Rawles, J.M. and Weller, R.O. (1974) Familial association of metabolic myopathy, lactic acidosis and sideroblastic anemia. *American Journal of Medicine,* **56,** 891–897

73. Pearson, H.A., Lobel, J.S., Kocoshis, S.A. *et al.* (1979) A new syndrome of refractory sideroblastic anaemia with vacuolisation of bone marrow precursors and exocrine pancreatic dysfunction. *Journal of Pediatrics,* **95,** 976–984

74. Rotig, A., Colonna, M., Blanche, S. *et al.* (1988) Deletion of blood mitochondrial DNA in pancytopenia. *Lancet*, **ii**, 567–568
75. Rotig, A., Colonna, M., Bonnefont, J.P. *et al.* (1989) Mitochondrial DNA deletion in Pearson's marrow/pancreas syndrome. *Lancet*, **i**, 902–903
76. Rotig, A., Cormier, V., Blanche, S. *et al.* (1990) Pearson's marrow-pancreas syndrome: a multisystem mitochondrial disorder in infancy. *Journal of Clinical Investigation*, **86**, 1601–1608
77. McShane, M.A., Hammans, S.R., Sweeney, M. *et al.* (1991) Pearson syndrome and mitochondrial encephalomyopathy in a patient with a deletion of mtDNA. *American Journal of Human Genetics*, **48**, 39–42
78. Larsson, N.-G., Holme, E., Kristiansson, B. *et al.* (1990) Progressive increase of the mutated mitochondrial DNA fraction in Kearns–Sayre syndrome. *Pediatric Research*, **28**, 131–136
79. Eviatar, L., Shanske, S., Gauthier, B. *et al.* (1990) Kearns–Sayre syndrome presenting as renal tubular acidosis. *Neurology*, **40**, 1761–1763
80. Goto, Y., Koga, Y., Horai, S. *et al.* (1990) Chronic progressive external ophthalmoplegia: a correlative study of mitochondrial DNA deletions and their phenotypic expression in muscle biopsies. *Journal of the Neurological Sciences*, **100**, 63–69
81. Goto, Y., Itami, N., Kajii, N. *et al.* (1990) Renal tubular involvement mimicking Bartter syndrome in a patient with Kearns–Sayre syndrome. *Journal of Pediatrics*, **116**, 904–910
82. DiMauro, S., Zeviani, M., Rizzuto, R. *et al.* (1988) Molecular defects in cytochrome oxidase in mitochondrial diseases. *Journal of Bioenegetics and Biomembranes*, **20**, 353–365
83. Zeviani, M., Nonaka, I., Bonilla, E. *et al.* (1985) Fatal infantile mitochondrial myopathy and renal dysfunction caused by cytochrome *c* oxidase deficiency: immunological studies in a new patient. *Annals of Neurology*, **17**, 414–417
84. DiMauro, S., Servidei, S., Zeviani, M. *et al.* (1987) Cytochrome *c* oxidase deficiency in Leigh syndrome. *Annals of Neurology*, **22**, 498–506
85. Van Biervliet, J.P., Bruinvis, L., Ketting, D. *et al.* (1977) Hereditary mitochondrial myopathy with lactic acidemia, a De Toni–Fanconi–Debre syndrome, and a defective respiratory chain in voluntary striated muscles. *Pediatric Research*, **11**, 1088–1093
86. D'Agostino, A.N., Ziter, F.A., Rallison, M.L. *et al.* (1968) Familial myopathy with abnormal muscle mitochondria. *Archives of Neurology*, **18**, 388–401
87. Shy, G.M., Gonatas, N.K. and Perez, M. (1966) Two childhood myopathies with abnormal mitochondria. I. Megaconial myopathy. II. Pleoconial myopathy. *Brain*, **89**, 133–158
88. McLeod, J.G., Baker, W., Shorey, C.D. *et al.* (1975) Mitochondrial myopathy with multisystem abnormalities and normal ocular movements. *Journal of the Neurological Sciences*, **24**, 39–52
89. Poulton, J., Deadman, M.E. and Gardiner, R.M. (1989) Duplications of mitochondrial DNA in mitochondrial myopathy. *Lancet*, **i**, 236–240
90. Lakin, M. and Locke, S. (1961) Progressive ocular myopathy, ovarian insufficiency and diabetes mellitus. Report of case. *Diabetes*, **10**, 228–231
91. Zupanc, M.L., Moraes, C.T., Shanske, S. *et al.* (1991) Deletion of mitochondrial DNA in patients with combined features of Kearns–Sayre and MELAS syndromes. *Annals of Neurology*, **29**, 680–683
92. Pellock, J.M., Behrens, M., Lewis, L. *et al.* (1978) Kearns–Sayre syndrome and hypoparathyroidism. *Annals of Neurology*, **3**, 455–458
93. Cormier, V., Rotig, A., Tardieu, M. *et al.* (1991) Autosomal dominant deletions of the mitochondrial genome in a case of progressive encephalopathy. *American Journal of Human Genetics*, **48**, 643–648
94. Drachman, D.A. (1968) Ophthalmoplegia plus: the neurodegenerative disorders associated with progressive external ophthalmoplegia. *Archives of Neurology*, **18**, 654–674
95. Simopoulos, A.P., Delea, C.S. and Bartter, F.C. (1971) Neurodegenerative disorders and hyperaldosteronism. *Journal of Pediatrics*, **79**, 633–641
96. Julien, J., Vital, C., Vallat, J.M. *et al.* (1973) Myopathie oculaire avec hypogonadisme primaire – anomalies mitochondriale en ultrastructure. *Revue Neurologique*, **128**, 365–377
97. Fitzsimons, R.B., Clifton-Bligh, P. and Wolfenden, W.H. (1981) Mitochondrial myopathy and lactic acidaemia with myoclonic epilepsy, ataxia and hypothalamic infertility: a variant of Ramsay-Hunt syndrome? *Journal of Neurology, Neurosurgery and Psychiatry*, **44**, 79–82
98. Bardosi, A., Creutzfeldt, W., DiMauro, S. *et al.* (1987) Myo-, neuro-, gastrointestinal encephalopathy (MNGIE syndrome) due to partial deficiency of cytochrome-c-oxidase. *Acta Neuropathologica (Berlin)*, **748**, 248–258
99. Blake, D., Lombes, A., Minetti, C. *et al.* (1990) MNGIE syndrome: a report of 2 new patients. *Neurology*, **40** (Suppl. I), 294 (Abstract)
100. Simon, L.T., Horoupian, D.S., Dorfman, L.J. *et al.* (1990) Polyneuropathy, ophthalmoplegia,

leukoencephalopathy, and intestinal pseudo-obstruction: POLIP syndrome. *Annals of Neurology*, **28**, 349–360

101. Cervera, R., Bruix, J., Bayes, A. *et al.* (1988) Chronic intestinal pseudo-obstruction and ophthalmoplegia in a patient with mitochondrial myopathy. *Gut*, **29**, 544–547

102. Boustany, R.M., Aprille, J.R., Halperin, J. *et al.* (1983) Mitochondrial cytochrome deficiency presenting as a myopathy with hypotonia, external ophthalmoplegia, and lactic acidosis as an infant and as fatal hepatopathy in a cousin. *Annals of Neurology*, **14**, 462–470

103. Berkovic, S.F., Andermann, F., Shoubridge, E.A. *et al.* (1991) Mitochondrial dysfunction in multiple symmetrical lipomatosis. *Annals of Neurology*, **29**, 566–569

104. Holt. I.J., Harding, A.E., Petty, R.K.H. *et al.* (1990) A new mitochondrial disease associated with mitochondrial DNA heteroplasmy. *American Journal of Human Genetics*, **46**, 428–433

105. Tatuch, Y., Christodoulou, J., Feigenbaum, A. *et al.* (1992) Heteroplasmic mtDNA mutation (T→G) at 8993 can cause Leigh disease when the percentage of abnormal mtDNA is high. *American Journal of Human Genetics*, **50**, 852–858

106. Matthews, P.M., Tampieri, D., Berkovic, S.F. *et al.* (1991) Magnetic resonance imaging shows specific abnormalities in the MELAS syndrome. *Neurology*, **41**, 1043–1046

107. Coster, R.V., Lombes, A., De Vivo, D.C. *et al.* (1991) Cytochrome *c* oxidase-deficient Leigh syndrome: phenotypic features and pathogenetic speculations. *Journal of the Neurological Sciences*, **104**, 97–111

108. Medina, L., Chi, T.L., De Vivo, D.C. *et al.* (1990) MR manifestations of biochemically characterized subacute necrotizing encephalomyelopathy (Leigh syndrome). *American Journal of Neuroradiology*, **11**, 379–384

109. Berkovic, S.F., Karpati, G., Carpenter, S. *et al.* (1987) Progressive dystonia with bilateral putaminal hypodensities. *Archives of Neurology*, **44**, 1184–1187

110. Turnbull, D.M., Johnson, M.A., Dick, D.J. *et al.* (1985) Partial cytochrome oxidase deficiency without subsarcolemmal accumulation of mitochondria in chronic external ophthalmoplegia. *Journal of the Neurological Sciences*, **70**, 93–100

111. Yamamoto, M., Koga, Y., Ohtaki, E. *et al.* (1989) Focal cytochrome c oxidase deficiency in various neuromuscular diseases. *Journal of the Neurological Sciences*, **91**, 207–213

112. Müller-Höcker, J. (1990) Cytochrome c oxidase deficient fibres in the limb muscle and diaphragm of man without muscular disease: an age related alteration. *Journal of the Neurological Sciences*, **100**, 14–21

113. Johns, D.R., Hurko, O., Attardi, G. *et al.* (1991) Molecular basis of a new mitochondrial disease: acute optic neuropathy and myelopathy. *Annals of Neurology*, **30**, 234 (Abstract)

114. Argov, Z., Bank, W.J., Maris, J. *et al.* (1985) Bioenergetic heterogenicity of human mitochondrial myopathies: phosphorus magnetic resonance spectroscopy study. *Neurology*, **37**, 257–262

115. Argov, Z. and Bank, W.J. (1991) Phosphorus magnetic resonance spectropscopy (^{31}P MRS) in neuromuscular disorders. *Annals of Neurology*, **30**, 90–97

116. Matthews, P.M., Allaire, C., Shoubridge, E.A. *et al.* (1991) In vivo muscle magnetic resonance spectroscopy in the clinical investigation of mitochondrial disease. *Neurology*, **41**, 114–120

117. Przyrembel, H. (1987) Therapy of mitochondrial disorders. *Journal of Inherited Metabolic Disease*, **10**, 129–146

118. Bresolin, N., Doriguzzi, C., Ponzetto, C. *et al.* (1990) Ubidecarenone in the treatment of mitochondrial myopathies: a multi-center double-blind trial. *Journal of the Neurological Sciences*, **100**, 70–78

119. Shoffner, J.M., Lott, M.T., Voljavec, A.S. *et al.* (1989) Spontaneous Kearns–Sayre/chronic external ophthalmoplegia plus syndrome associated with a mitochondrial DNA deletion: a slip-replication model and metabolic therapy. *Proceedings of the National Academy of Sciences of the USA*, **86**, 7952–7956

120. Eleff, S., Kennaway, N.G., Buist, N.R.M. *et al.* (1984) ^{31}P-NMR study of improvement in oxidative phosphorylation by vitamins K_3 and C in a patient with a defect in electron transport at complex III in skeletal muscle. *Proceedings of the National Academy of Sciences of the USA*, **81**, 3529–3533

121. Argov, Z., Bank, W.J., Maris, J. *et al.* (1986) Treatment of mitochondrial myopathy due to complex III deficiency with vitamins K_3 and C: a ^{31}P-NMR follow-up study. *Annals of Neurology*, **19**, 598–602

122. Arts, W.F.M., Scholte, H.R., Bogaard, J.M. *et al.* (1983) NADH-CoQ reductase deficient myopathy: successful treatment with riboflavin. *Lancet*, **2**, 581–582

123. Shapira, Y., Cederbaum, S.D., Cancilla, P.A. *et al.* (1975) Familial poliodystrophy, mitochondrial myopathy, and lactate acidemia. *Neurology*, **25**, 614–621

124. Gubbay, S.S., Hankey, G.J., Tan, N.T.S. *et al.* (1989) Mitochondrial encephalomyopathy with corticosteroid dependence. *Medical Journal of Australia*, **151**, 100–108

125. Nelson, I., Degoul, F., Obermaier Kusser, B. *et al.* (1989) Mapping of heteroplasmic mitochondrial DNA deletions in Kearns–Sayre syndrome. *Nucleic Acids Research*, **17**, 8117–8124
126. Gerbitz, K.D., Obermaier Kusser, B., Zierz, S. *et al.* (1990) Mitochondrial myopathies: divergences of genetic deletions, biochemical defects and the clinical syndromes. *Journal of Neurology*, **237**, 5–10
127. Degoul, F., Nelson, I., Lestienne, P. *et al.* (1991) Deletions of mitochondrial DNA in Kearns–Sayre syndrome and ocular myopathies: genetic, biochemical and morphological studies. *Journal of the Neurological Sciences*, **101**, 168–177
128. Ozawa, T., Yoneda, M., Tanaka, M. *et al.* (1988) Maternal inheritance of deleted mitochondrial DNA in a family with mitochondrial myopathy. *Biochemical Biophysical Research Communications*, **154**, 1240–1247
129. Poulton, J., Deadman, M.E., Ramacharan, S. *et al.* (1991) Germ-line deletions of mtDNA in mitochondrial myopathy. *American Journal of Human Genetics*, **48**, 649–653
130. Zeviani, M., Bresolin, N., Gellera, C. *et al.* (1990) Nucleus-driven multiple large-scale deletions of the human mitochondrial genome: a new autosomal dominant disease. *American Journal of Human Genetics*, **47**, 904–914
131. Ciafaloni, E., Ricci, E., Shanske, S. *et al.* (1992) MELAS: clinical features, biochemistry, and molecular genetics. *Annals of Neurology*, **31**, 391–398
132. Harding, A.E., Holt, I.J., Sweeney, M.G. *et al.* (1992) Prenatal diagnosis of mitochondrial DNA[8993] $^{T \to G}$ disease. *American Journal of Human Genetics*, **50**, 629–633
133. Goto, Y., Tojo, M., Tohyama, J. *et al.* (1992) A novel point mutation in the mitochondrial tRNA[Leu(UUR)] gene in a family with mitochondrial myopathy. *Annals of Neurology*, **31**, 672–675
134. Wallace, D.C., Singh, G., Lott, M.T. *et al.* (1988) Mitochondrial DNA mutation associated with Leber's hereditary optic neuropathy. *Science*, **242**, 1427–1430
135. Howell, N., Kubacka, I., Xu, M. *et al.* (1991) Leber hereditary optic neuropathy: involvement of the mitochondrial ND1 gene and evidence for an intragenic suppressor mutation. *American Journal of Human Genetics*, **48**, 935–942
136. Huoponen, K., Vilkki, J., Aula, P. *et al.* (1991) A new mtDNA mutation associated with Leber hereditary optic neuroretinopathy. *American Journal of Human Genetics*, **48**, 1147–1153
137. Johns, D.R. and Berman, J. (1991) Alternative, simultaneous complex I mitochondrial DNA mutations in Leber's hereditary optic neuropathy. *Biochemical Biophysical Research Communications*, **174**, 1324–1330
138. Yamamoto, M., Clemens, P.R. and Engel, A.G. (1991) Mitochondrial DNA deletions in mitochondrial cytopathies: observations in 19 patients. *Neurology*, **41**, 1822–1828
139. Poulton, J., Deadman, M.E., Turnbull, D.M. *et al.* (1991) Detection of mitochondrial DNA deletions in blood using the polymerase chain reaction: non-invasive diagnosis of mitochondrial myopathy. *Clinical Genetics*, **39**, 33–38
140. Otsuka, M., Niijima, K., Mizuno, Y. *et al.* (1990) Marked decrease of mitochondrial DNA with multiple deletions in a patient with familial mitochondrial myopathy. *Biochemical Biophysical Research Communications*, **167**, 680–685
141. Rotig, A., Bessis, J.-L., Romero, N. *et al.* (1992) Maternally inherited duplication of the mitochondrial genome in a syndrome of proximal tubulopathy, diabetes mellitus and cerebellar ataxia. *American Journal of Human Genetics*, **50**, 364–370
142. Yuzaki, M., Ohkoshi, N., Kanazawa, I. *et al.* (1989) Multiple deletions in mitochondrial DNA at direct repeats of non-D-loop regions in cases of familial mitochondrial myopathy. *Biochemical Biophysical Research Communications*, **164**, 1352–1357
143. Ikebe, S., Tanaka, M., Ohno, K. *et al.* (1990) Increase of deleted mitochondrial DNA in the striatum in Parkinson's disease and senescence. *Biochemical Biophysical Research Communications*, **170**, 1044–1048
144. Ozawa, T., Tanaka, M., Sugiyama, S. *et al.* (1990) Multiple mitochondrial DNA deletions exist in cardiomyocytes of patients with hypertrophic or dilated cardiomyopathy. *Biochemical Biophysical Research Communications*, **170**, 830–836
145. Cortopassi, G.A. and Arnheim, N. (1990) Detection of a specific mitochondrial DNA deletion in tissues of older humans. *Nucleic Acids Research*, **18**, 6927–6933
146. Scarlato, G., Bresolin, N., Moroni, I. *et al.* (1991) Multicenter trial with ubidecarone: treatment of 44 patients with mitochondrial myopathies. *Revue Neurologique*, **147**, 542–548

4
Biochemical and molecular features of deficiencies of Complexes I, II and III

A.H.V. Schapira and J.M. Cooper

INTRODUCTION

Functional defects of the respiratory chain may be found in any of its components and may be specific to one complex or involve multiple deficiencies. As an intact respiratory chain is required for optimal ATP production during aerobic respiration, it is not surprising that a severe deficiency of any of the complexes will result in cellular dysfunction. It is now clear that the site of the respiratory chain defect does not correlate with any specific clinical presentation. Nevertheless, certain phenotypes are more frequently associated with particular patterns of respiratory chain abnormalities. The investigation of these deficiencies at the functional and structural levels can potentially provide valuable insights into the nature of the molecular genetic defects found in association with them.

INVESTIGATIVE APPROACH

Various approaches have been used in the investigation of patients with a suspected respiratory chain defect, leaving the literature regarding this subject a little confusing. In order to help clarify this situation, some of the procedures applied in investigating the respiratory chain will be discussed, although it should be pointed out that this review is not exhaustive and many aspects of the analysis of mitochondrial function have been covered elsewhere [1].

The experimental approach taken for the identification of respiratory chain defects in humans is often governed by the nature and amount of tissue available.

BLOOD SAMPLES

One of the standard procedures used to confirm the presence of a respiratory chain abnormality is the aerobic exercise test. In its simplest form, this involves taking blood samples for pyruvate and lactate measurements before, during and after exercise. It is important that the period of exercise is monitored and standardized in order to keep within the aerobic threshold of control subjects. In

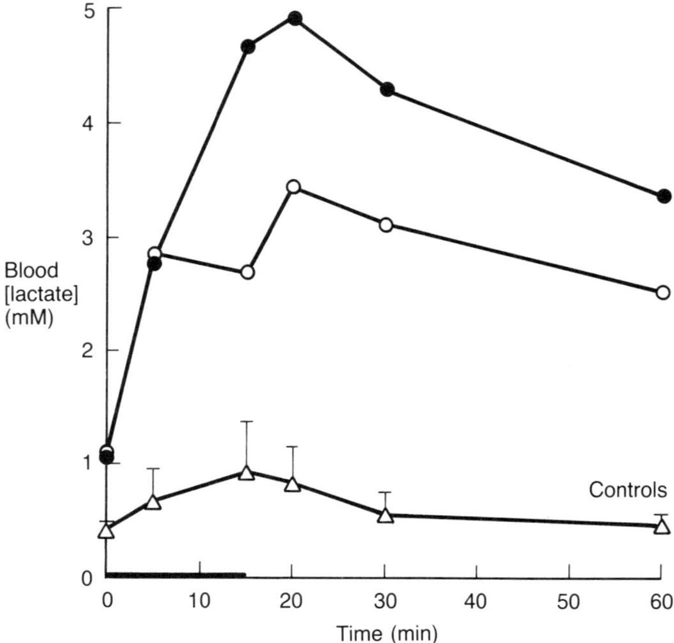

Figure 4.1 Blood lactate concentrations before, during and after a period
of aerobic exercise. Controls (mean ± S.D., *n* = 6) exercised for 15 min
(work = 4.3 ± 1.3 kilopond kilometer). A patient with a defect of Complex
I (O) exercised for 5 min (0.25 kpd km), and a patient with a defect of
Complexes II and III (●) exercised for 15 min (0.4 kpd km)

the majority of patients with respiratory chain abnormalities, the lactate levels
increase markedly above control values during exercise with a delayed recovery
to normal levels (Figure 4.1). The ratio of lactate/pyruvate also rises to abnormally
high levels. Both these parameters are good indicators for the presence of respiratory chain disorders. Although it is not possible to exercise infants, nor indeed
severely affected adults, such patients often have a resting lactic acidosis or raised
lactate/pyruvate ratios.

Both white cells and platelets contain mitochondria and are therefore potential
sources of material for mitochondrial analyses. This can involve screening for certain
mtDNA abnormalities [2], or the analysis of mitochondrial respiratory chain function
[3]. However, caution must be exercised when assessing the results from such samples
because the tissue distribution of the mitochondrial abnormality may not always
include these cell types. This is most clearly seen in the distribution of deletions of
mtDNA which are usually absent in blood samples but present in muscle [4].

MUSCLE BIOPSY

The detailed analysis of mitochondrial function can be determined using a skeletal
muscle biopsy. From a small biopsy (100–300 mg) several informative investigative

procedures can be performed including histochemistry, electron microscopy, DNA and limited biochemical and polypeptide analyses. From a larger biopsy (3–6 g) it is possible to perform a more extensive range of investigations involving mitochondrial isolation.

HISTOCHEMICAL AND ELECTRON MICROSCOPIC FEATURES

The presence of 'ragged red fibres' and/or cytochrome *c* oxidase-negative fibres is characteristic of abnormalities of mitochondrial respiratory chain dysfunction. However, not all patients with a respiratory chain abnormality show these features [5,6] and morphologically abnormal mitochondria have been observed in other disorders [7,8]. This section has been covered more extensively in Chapter 3.

BIOCHEMICAL ANALYSES

A wide variety of biochemical analyses have been performed on tissue samples removed at biopsy or autopsy. It is possible to analyse mitochondrial respiratory chain function in freshly obtained or frozen samples, although mitochondrial fractionation from freshly biopsied tissue is the method of choice.

MITOCHONDRIAL ISOLATION

Mitochondria can be isolated from almost any tissue. However, for diagnostic purposes skeletal muscle has been the tissue of choice mainly because of its accessibility and frequent involvement in the disease.

There are a number of techniques for mitochondrial isolation from various sized skeletal muscle samples and from both fresh and frozen muscle. Most of these procedures involve tissue homogenization followed by differential centrifugation resulting in a purification of mitochondria, although, without the use of density gradient centrifugation, all preparations will have some contamination from other organelles. The principles behind many of the techniques have been covered elsewhere [1].

Although high yields of pure intact mitochondria are not realistically attainable, higher yields should not be sought at the expense of significantly lower purities. In our laboratory, enough mitochondria of sufficient purity can be reproducibly obtained from 3–6 g of fresh muscle. Smaller sample sizes tend to result in higher and more variable degrees of contamination.

Mitochondrial-enriched fractions have been used, especially from smaller sample sizes, correcting for variations in purity by standardizing the results to a mitochondrial marker such as citrate synthase (CS). While this can be acceptable, it assumes that all the mitochondria are intact and have not lost proportions of their CS activity, which depends upon the severity and reproducibility of tissue homogenization. With the use of post-nuclear supernatants (PNS), there is potential for an even larger variability in the results. The number of mitochondria released from the cells can vary in both absolute and relative terms. Corrections made using CS activities assumes that there is uniform release of both CS (a soluble enzyme) and mitochondrial membranes. The use of whole tissue homogenates has the advantage that any analysis measures all the mitochondria present in the tissue, but has the disadvantage of high turbidity and decreased sensitivity. In many situations the best approach is a compromise. However, the important point to remember is the complete validation of the technique in terms of purity and reproducibility. Although these approaches are acceptable with the relevant safeguards, they inevitably introduce a greater degree of data scattering than with purified mitochondrial fractions.

POLAROGRAPHIC ANALYSES

Polarographic analysis of mitochondrial function is only feasible with either freshly isolated intact mitochondria or freshly harvested cells.

It is important to isolate the mitochondria immediately after the biopsy and to complete the analysis of the mitochondria within 3 or 4 h to avoid variability in the results. Various techniques of mitochondrial isolation have been used. However, a mixture of protease digestion and mechanical disruption of the skeletal muscle followed by differential centrifugation appears to be the method of choice. A micro-incubation chamber and a commercially available micro-electrode (Yellow Springs Instruments) can be used to measure oxygen utilization in a final volume of 200 µl, containing 100–200 µg of mitochondria. Measurement of the rate of oxygen utilization by mitochondria in the presence of a variety of substrates, both natural and artificial, enables a large proportion of mitochondrial functions to be analysed. Intact mitochondria are essential for this analysis. Being impermeable to most molecules, the normal rate of oxygen utilization by mitochondria using pyruvate, glutamate, palmitoylcarnitine and succinate not only requires a normal respiratory chain and oxidative phosphorylation system, but also many other mitochondrial systems including normal transporters (mono- and di-carboxylic acids), carnitine palmitoyl-transferase II, electron-transferring flavoprotein, ADP/ATP translocator, primary dehydrogenase (pyruvate dehydrogenase, glutamate dehydrogenase, glutamate–oxaloacetate transaminase) and tricarboxylic acid (TCA) cycle enzymes.

Although it is possible to purify mitochondria from cells in culture, it is difficult to obtain enough of sufficient quality to perform polarographic analysis. However, another approach has been developed which involves selective permeabilization of the plasma membrane with a detergent (usually digitonin) followed by washing of the cells and polarographic analysis. This involves fewer steps than mitochondrial fractionation and therefore sample recovery is less of a problem. This approach has been used successfully [9], although care must be taken to add sufficient detergent to solubilize the plasma membrane effectively but leave the mitochondrial membranes intact.

ENZYME ANALYSES

The activities of all five complexes of the respiratory chain and oxidative phosphorylation system can be analysed either individually or as functional units [10–13]. These analyses can be performed on all types of sample preparation, although Complex V (ATP synthase) activity can be difficult to differentiate from non-mitochondrial ATPase activities in preparations other than purified mitochondria. There are also methods for the determination of the activities of other mitochondrial enzymes (TCA cycle enzymes, primary dehydrogenases), although caution has to be exercised for some analyses because of the interference by other cytosolic or mitochondrial enzymes.

In contrast with the polarographic determination of mitochondrial respiratory chain function, the enzymic analysis of the mitochondrial respiratory chain complexes invariably involves artificial electron donors and/or acceptors, which are generally saturating to give maximal activities (V_{max}). Together the polarographic and enzymic determinations of mitochondrial function are powerful tools for identifying the site of any defect of mitochondrial function.

ANALYSIS OF COMPONENTS OF THE RESPIRATORY CHAIN

The mitochondrial respiratory chain and oxidative phosphorylation system is composed of five multisubunit protein complexes (Complexes I–V) and two mobile electron carriers, ubiquinone and cytochrome c. The number of protein subunits identified as part of this system has steadily risen over the years and now totals at least 80 different proteins (see Chapter 1). Associated with a number of these proteins are various redox centres which are either iron-sulphur centres or cytochrome s. These functional centres can be detected by either electron spin resonance (ESR) or light spectroscopy. Unfortunately, these analyses require a relatively large amount of material, preferably purified mitochondria.

CYTOCHROME ANALYSES

Cytochromes of the a, b and c type are components of the respiratory chain. Complex III contains cytochromes b-562, b-566 and c_1, Complex IV contains cytochromes a and a_3 and cytochrome c is a mobile electron carrier.

The cytochromes have a distinct visible spectrum when reduced; however, because the organelles produce a lot of light scattering and high levels of non-specific absorption, the difference (reduced minus oxidized) spectrum has to be determined. The peaks at 550 nm (cytochrome c), 556 nm (cytochrome c_1), 562 nm (cytochrome b) and 602 nm (cytochromes aa_3) can be used to calculate the cytochrome levels of the mitochondria [14]. Although the spectrum can be obtained at room temperature, the signal is enhanced at liquid nitrogen temperatures.

Samples can be reduced by either a non-specific reducing agent (dithionite) or a physiological substrate (succinate). Non-specific reducing agents must not be used when the sample contains haemoglobin or myoglobin because their different spectra interfere with the cytochrome spectra. When using a physiological reductant, any functional defect of the respiratory chain will affect the reduction of the cytochromes and may reflect the ability of the respiratory chain to reduce the cytochromes rather than the absolute level of the cytochrome.

IRON-SULPHUR CENTRES

A number of iron-sulphur centres present in Complexes I, II and III have characteristic ESR spectra [15,16], and the presence of these centres can be detected in mitochondria. The requirement for relatively large amounts of material, specialist equipment and the inability to detect all the iron-sulphur centres has limited the usefulness of this analysis. Using mitochondria isolated from liver obtained at *post mortem*, Moreadith *et al.* [17] and Ichiki *et al.* [18] analysed the ESR visible centres in two patients with a Complex I defect and revealed markedly diminished iron-sulphur centres in Complex I, but normal signals from other mitochondrial iron-sulphur centres consistent with a specific abnormality of Complex I function.

POLYPEPTIDE ANALYSES

Specific respiratory chain subunits can be detected using antisera raised to either the individual holo-complexes or to individual subunits which have been purified, for instance, from bovine heart mitochondria. These specific antisera can be used to: identify the specific components when used in conjunction with one- or two-dimensional SDS/PAGE followed by Western blotting; quantify the levels of the subunits by enzyme-linked immunosorbent assay (ELISA); localize the cellular distribution of the subunits using immunocytochemistry; and purify the respiratory chain complex/subunit using immunoprecipitation. However, with the use of antisera there are a number of drawbacks including: the specificity of the cross-reaction; the limited number of subunits that cross-react and can be detected; and the difficulty in comparing data between laboratories because of the variability in the specificity and subunit recognition between antisera preparations.

The five complexes of the respiratory chain and oxidative phosphorylation system can be separated from each other and most other cellular components using the technique known as blue native gel electrophoresis [19]. This separates the complexes while they are still in their native state, followed by a second-dimensional SDS/PAGE separation. After Coomassie staining, the relative levels of the complexes in the sample can be determined. This system has the potential of allowing analysis of the components of the five complexes in relatively small volumes of material, either purified mitochondria or whole tissue.

DNA ANALYSES

The fact that the mitochondrion contains its own DNA encoding 13 polypeptides, all of which are components of the respiratory chain and oxidative phosphorylation system, has focused much attention on this genome. Major alterations in mtDNA (large deletions and duplications) and point mutations affecting restriction endonuclease sites (either loss or gain of a site) have been detected by restriction fragment length polymorphism (RFLP) analysis. However, in the majority of cases, this can now be performed more easily using the polymerase chain reaction (PCR).

A number of methods are available for the detection of novel mutations of mtDNA. These include a combination of PCR and sequencing of selected regions of the mtDNA, single-stranded conformational polymorphism (SSCP) [20] and chemical cleavage mismatch (CCM) [21].

MITOCHONDRIAL BIOGENESIS

The translation products of the mitochondrial DNA transcripts can be determined using either mitochondria freshly isolated from patient biopsy samples or in cells cultured from the patient. This can be used to identify any abnormality that affects mtDNA translation, which could be due to abnormal rates of translation (i.e. due to defects of the translating machinery, tRNA, rRNA genes or RNA polymerase etc.) or abnormal products (i.e. fusion proteins). The results of such studies have shown decreased synthesis of the larger mtDNA-encoded proteins in patients with a point mutation of a tRNA gene [22,23]. Hayashi *et al.* [24] have also been able to demonstrate the translation of a predicted fusion protein resulting from a deletion of mtDNA in a ρ^o: patient cell cybrid cell line.

TISSUE CULTURE

Cells cultured from patients with respiratory chain abnormalities have been increasingly used to study mitochondrial abnormalities. Although some mtDNA mutations, most notably deletions, were often found to disappear in a mixed culture, if individual cell were clonally selected, the mutation could be preserved in selected lines. Clonal cell culture has therefore proved to be a useful model of mtDNA abnormalities.

The development of cells devoid of mtDNA (ρ^o cells) [25] and their fusion with enucleated cells from patients with mitochondrial DNA abnormalities has enabled the investigation of the relationship between the mtDNA mutation, the functional abnormality and the role played by the nucleus. These cell fusion experiments have been successfully performed using cells expressing either tRNA^{Leu} [22] or tRNA^{Lys} [26] and a mtDNA deletion [24]. In these experiments the mitochondrial respiratory chain abnormality clearly persists in the presence of a normal nuclear environment, demonstrating that the defect is caused by the abnormal mtDNA. Conversely, cell fusion experiments with cells from a patient with mtDNA depletion clearly show that the abnormality in this disease is due to a defect in a nuclear factor [27].

DISEASES OF THE MITOCHONDRIAL RESPIRATORY CHAIN

The classification of patients with defects of the mitochondrial respiratory chain can be by several different criteria: clinical presentation (myopathy, CPEO, encephalopathy, MELAS, MERRF); functional deficiency (localization of the enzyme defect) or genetic defect (mtDNA or nuclear abnormality). The clinical and genetic classifications have been covered in previous chapters. In this chapter we will concentrate on those patients with abnormalities of Complexes I–III.

Complex I deficiency

MITOCHONDRIAL MYOPATHY
Deficiency of Complex I activity, alone or in combination with reduced Complex IV function, is commonly found in patients with mitochondrial myopathy. A defect in Complex I can be identified from polarographic data, enzyme data or both. Polarographically a specific Complex I defect is characterized by a decreased rate of oxygen utilization with NAD^+-linked substrates (pyruvate or glutamate), but normal rates with succinate. The use of two distinct NAD^+-linked substrates (pyruvate and glutamate) allows for the exclusion of a defect of either

the dicarboxylic acid carrier or the primary dehydrogenase. Confirmation of the Complex I defect should be made by the spectrophotometric analysis of the rotenone-sensitive NADH–CoQ$_1$ reductase activity.

Clinical Presentation

There is no single clinical syndrome that characterizes Complex I deficiency. Patients may present at any age with clinical symptoms confined to a single organ or with multisystem failure.

The fatal infantile form of Complex I deficiency presents with a combination of cardiorespiratory and/or hepatorenal failure in association with hypotonia, acidosis and encephalopathy [28,29]. Affected infants invariably die within the first few weeks of life. Not surprisingly the biochemical defect appears widespread in those patients who come to post-mortem examination. Family history in this rare disorder may be absent, although two siblings with fatal infantile Complex I deficiency have been described [28].

Several patients with pure myopathy and Complex I deficiency have been described [30,31]. Symptoms usually date from early childhood or adolescence and are characterized by muscle fatigue, weakness and exercise intolerance. There may be a history of acute episodes of headache, nausea, vomiting and collapse provoked by alcohol or exercise. These are characteristic features of lactic acidosis during which arterial pH has been recorded to fall as low as 6.67 [32]. Examination may show generalized muscle thinning without focal wasting and a moderate proximal myopathy with fatigue. Ophthalmoplegia is uncommon. Muscle biopsy shows the typical changes of mitochondrial myopathy with the proportion of ragged red fibres roughly correlated to clinical weakness and biochemical deficiency, which in several cases has also invoked a partial deficiency of Complex IV [31,33,34]. Patients falling into the category of pure myopathy with Complex I deficiency who have been seen at the National Hospital, Queen Square, have remained only moderately affected and have not as yet developed any other features characteristic of some of the other mitochondrial phenotypes. The inheritance pattern is variable and may be absent or involve affected mothers [33,35,36] or siblings [35,37] including one pair of identical twins [35].

Complex I deficiency is frequently accompanied by central nervous system (CNS) dysfunction and is probably the most common respiratory chain defect associated with the encephalopathies of the MELAS and MERRF type. Biochemical analysis in patients with either the MELAS or MERRF phenotypes has often shown a defect of Complex I alone, or in combination with a defect of Complexes III and/or IV [38–43]. Thus the encephalopathy of Complex I deficiency will present with the features associated with these phenotypes or may even occur in isolation. The decrease in mtDNA-encoded polypeptide synthesis resulting from the MELAS and MERRF mutations may affect Complex I activity the most, as this complex has the highest proportion of such subunits. Specific Complex I deficiency has also been found in patients with CPEO and mtDNA deletions [44]. In these cases, the deleted region involved only Complex I genes and the intervening tRNAs. They are the only examples where the deleted mtDNA genes correspond exactly to the biochemical defect.

Family history is often positive in those bearing one of the mtDNA mutations, e.g. tRNA$^{\text{Leu(UUR)}}$ or tRNA^{Lys}, but sporadic cases have also been described. There are no particular histological features of muscle or other tissues which distinguish Complex I-deficient patients from those with other respiratory chain defects.

Movement disorders including dystonia and choreoathetosis have been documented in Complex I deficiency [45], and the recent discovery of a Complex I defect in Parkinson's disease (see Chapter 12) highlights this association.

Polypeptide Analysis
Analysis of Complex I, III and IV subunits by Western blotting has been performed in several patients with a defect in Complex I function. Deficiencies of the 75 kDa and 13 kDa iron-sulphur proteins were detected in post-mortem samples of liver and heart from a patient with fatal infantile Complex I deficiency [17]. A specific deficiency of the nuclear-encoded 24 kDa protein has been identified in two patients with pure myopathy [32,46] and in three patients with encephalopathy. One patient with MERRF and two patients with MELAS were subsequently found to have the tRNA^{Lys} mutation and the tRNA$^{Leu(UUR)}$ mutation respectively (see Figure 3.2) [47,48]. One of the MELAS patients and the patient with MERRF also had deficiency of the 13 kDa iron-sulphur protein. These specific deficiencies were found in association with a generalized reduction in other cross-reactive Complex I subunits and, in four cases, a specific decrease in subunit II of cytochrome oxidase. There are several possible explanations for these results, including reduced synthesis or increased turnover of the subunit specifically affected. A defect of the corresponding nuclear gene has not as yet been identified in these patients. The decreases in subunit II of Complex IV in some of these patients is in keeping with the presence of the tRNA$^{Leu(UUR)}$ or tRNA^{Lys} mutations. The relevance of the nuclear-encoded 24 kDa polypeptide deficiency is more difficult to interpret in this context. The decrease in this subunit may simply reflect increased turnover of a polypeptide resulting from impaired holoenzyme assembly. An alternative explanation would require bigenomic defects operating simultaneously. Such a mechanism might involve one or other of the mutations regulating phenotypic expression – in this case the tRNA$^{Leu(UUR)}$ would be the determinant in view of its strong association with MELAS.

LEBER'S HEREDITARY OPTIC NEUROPATHY (LHON)
The commonest mutations in LHON involve Complex I genes. LHON is characterized by a maternal pattern of inheritance and is now associated with an array of mtDNA mutations (see Chapter 10). Many of those mutations considered primary, e.g. 11778, 4160, 3460, are located within Complex I genes. Respiratory chain enzyme analysis in platelet mitochondria has demonstrated severe and selective deficiency of Complex I activity in those that carry the ND1 mutations at positions 3460 and 4160 [49,50]. The 11778 ND4 mutation is also associated with a mild but selective deficiency of platelet and skeletal muscle [51] mitochondrial Complex I activity when factors such as cigarette smoking (which also depresses Complex I activity) are taken into account [52].

PARKINSON'S DISEASE (PD)
The observations that there is a *c.* 40% decrease in Complex I activity in the substantia nigra of patients dying with PD [53,54] together with the fact that the Complex I inhibitor 1-methyl-4-phenyl-1,2,3,6 tetrahydropyridine (MPTP) can cause parkinsonism strongly suggests that Complex I dysfunction may be a key factor in the cause of PD (see Chapter 12).

AUTOIMMUNITY IN COMPLEX I DEFICIENCY

To date only one case of mitochondrial myopathy with mitochondrial autoantibodies has been described [47]. The patient was a 15-year-old boy with typical features of the MELAS phenotype: short stature, bilateral deafness, myopathy, ataxia, seizures, stroke-like episodes and dementia in association with a lactic acidosis. The clinical evolution was rapid and the patient died within 4 years of the onset of his symptoms. Muscle biopsy showed a high proportion of ragged red fibres and a few scattered cytochrome oxidase-negative fibres. Biochemistry showed a severe deficiency confined to Complex I but in association with a marked decrease in cytochrome aa_3 levels. Immunoblotting with respiratory chain antibodies showed selective deficiencies of the nuclear-encoded 24 kDa iron-sulphur protein of Complex I and the mitochondrial-encoded subunit II of Complex IV (Figure 4.2). The patient was found to carry the 3243 tRNA$^{Leu(UUR)}$ mtDNA mutation associated with the MELAS phenotype.

The patient's serum cross-reacted strongly with a specific mitochondrial 41 kDa polypeptide (Figure 4.3). This polypeptide was located within the mitochondrial matrix and present in heart, skeletal muscle, liver, kidney and brain from human, bovine, rabbit and rat tissues. This indicated that the polypeptide was highly conserved, and presumably functionally important, but not a component of the inner membrane respiratory chain. In addition, the polypeptide was not a recognized component of the M2 antigens associated with primary biliary cirrhosis.

The relevance of the patient's autoantibody to his disease pathogenesis is unclear. The limitation of the immune response to a single mitochondrial polypeptide suggests specificity and not a general response to proteins leaking from damaged mitochondria. The ability of the immune system to gain access to matrix proteins may involve mis-targeting of a proportion of the 41 kDa protein to the surface membrane or cross-reactivity with a protein bearing a similar epitope. The response of some patients with mitochondrial encephalomyopathy to steroids [55] suggests that an immune response may be relevant to disease pathogenesis in a proportion of cases.

Complex II deficiency

Selective deficiency of Complex II is uncommon and only approximately ten patients with such a defect have been identified to date. These patients occupy the clinical spectrum common to other respiratory chain defects with onset in infancy or adulthood with myopathy and, in half the patients, encephalopathy including one with Kearns–Sayre syndrome (KSS) [56]. Muscle biopsy shows ragged red fibres together with lipid storage in several cases. Histochemical staining with succinate dehydrogenase may be severely decreased or absent [5,57].

A defect of Complex II is characterized polarographically by a decreased rate of oxygen utilization with succinate, and enzymically by a decreased succinate–CoQ_2 reductase activity. The patient with KSS described by Rivner *et al.* [56] had decreased activities of succinate–cytochrome *c* reductase, succinate dehydrogenase and succinate–ubiquinone reductase, although Western blotting with antibodies to Complex II did not demonstrate any abnormality. Several of the other cases reported to date have relied on succinate–cytochrome *c* reductase activity for the diagnosis of Complex II deficiency. This enzyme does not reflect

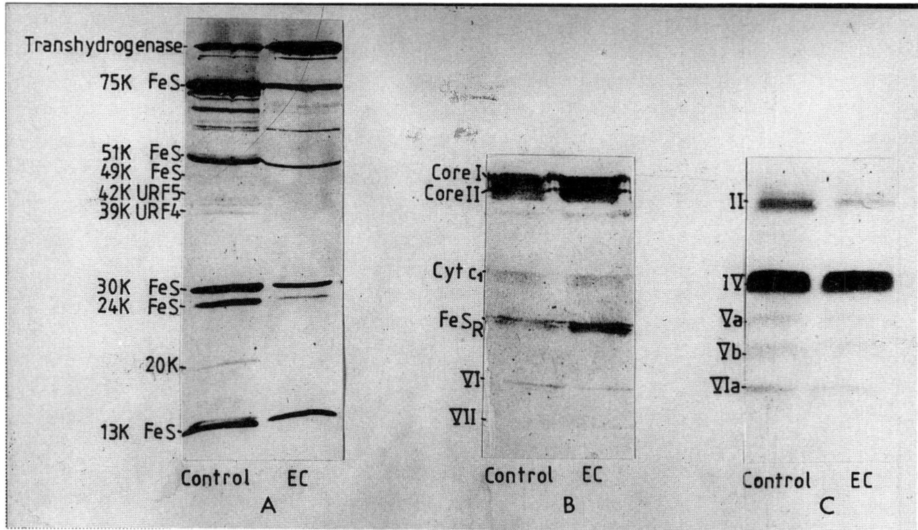

Figure 4.2 Immunoblots to detect mitochondrial respiratory chain components in muscle mitochondria isolated from control (C) and patient (E.C.) exhibiting mitochondrial autoantibodies: 12–16% linear gradient SDS/polyacryamide gels for Complex I (A) and III (B) and 12–19% gel for Complex IV (C) polypeptides. Holoenzyme antibodies to Complexes I, III and IV were used on the respective blots together with the following subunit-specific antisera: 30 kDa, 24 kDa, 13 kDa, iron-sulphur proteins (Complex I); Rieske iron-sulphur protein (Complex III) and subunit II (Complex IV) (Reproduced from ref. [47] with permission)

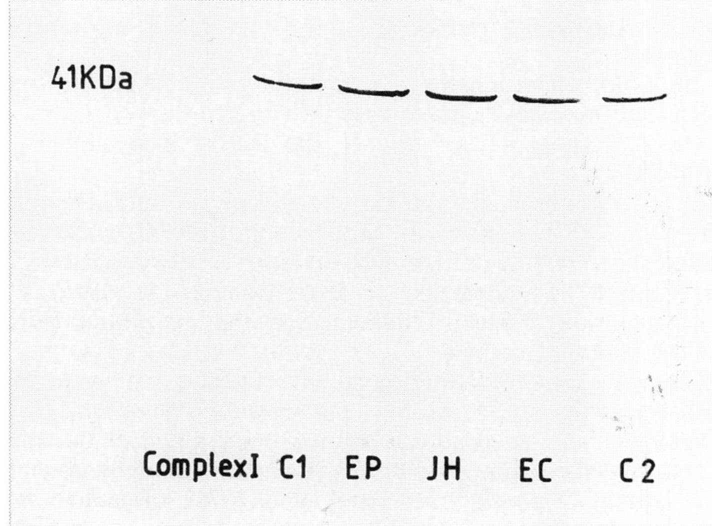

Figure 4.3 Immunoblot with serum from patient (E.C.) against skeletal muscle mitochondria from two controls (C1, C2), a patient with Complex III deficiency (E.P.), a patient with Complex I deficiency (J.H.) and the patient (E.C.) (Reproduced from ref. [47] with permission)

the activity of Complex II alone, and data to support a deficiency of this protein should be sought from the activities of succinate dehydrogenase, succinate–ubiquinone reductase and immunoblotting.

Complex III deficiency

Deficiency of Complex III activity is seen most often as part of a more comprehensive respiratory chain dysfunction, usually in conjunction with defects of Complex I and/or IV. Isolated Complex III deficiency has, however, been described in a few patients (see Kennaway [58] for review). Limb weakness and fatigue have been described in five patients, ophthalmoplegia in six and CNS involvement in ten. The latter varied from ataxia and myoclonus (MERRF) to mental retardation and seizures.

Morphological changes in patients with Complex III deficiency are similar to those found in other respiratory chain defects, i.e. ragged red fibres and a variable proportion of cytochrome oxidase-negative fibres. Polarography shows decreased oxygen utilization with pyruvate, glutamate and succinate, but normal oxygen uptake with ascorbate plus $NN'N'N'$-tetramethyl-phenylenediamine (TMPD). Enzyme analysis is required to define the defect further, with a deficiency of ubiquinol– cytochrome c reductase localizing the defect to Complex III. Analysis of cytochrome levels in such patients can be very informative. Severe decrease in the concentration of cytochrome b has been identified in several cases [59–62] including one 4-year-old child with histiocytoid cardiomyopathy [63]. The decrease in cytochrome b was also associated with deficiency of other cytochromes in some of these patients.

POLYPEPTIDE ANALYSIS
Immunoblotting of skeletal muscle mitochondria with antisera to Complex III in a patient with limb weakness and exercise intolerance showed severe deficiencies of cytochrome b, core proteins I and II, the Rieske iron-sulphur protein and subunit IV [60].

Mitochondria were obtained from the skeletal muscle and liver of a newborn infant who died within 3 days of birth with a severe lactic acidosis [64]. Biochemical analysis of this patient showed a severe and specific deficiency of Complex III and a 75% decrease in cytochrome b levels in both tissues. Immunoblotting showed a selective deficiency of the Rieske iron-sulphur protein in muscle mitochondrial fractions.

A further case of Complex III deficiency involved the first report of a defect of mitochondrial protein transport causing a mitochondrial myopathy [5]. The patient had limb weakness and exercise intolerance but no encephalopathic features. There were no ragged red fibres on biopsy, but succinate dehydrogenase staining was virtually absent. Electron microscopy and X-ray dispersive microanalysis showed intramitochondrial iron-rich inclusions in skeletal muscle. Polarography and enzyme analysis showed mild deficiencies of Complex I and IV, but severe defects of Complex II and III. The levels of cytochromes b, c_1 and c were normal, but cytochrome aa_3 was slightly decreased. Immunoblotting of muscle mitochondria showed the virtual absence of the Rieske iron-sulphur protein, but normal

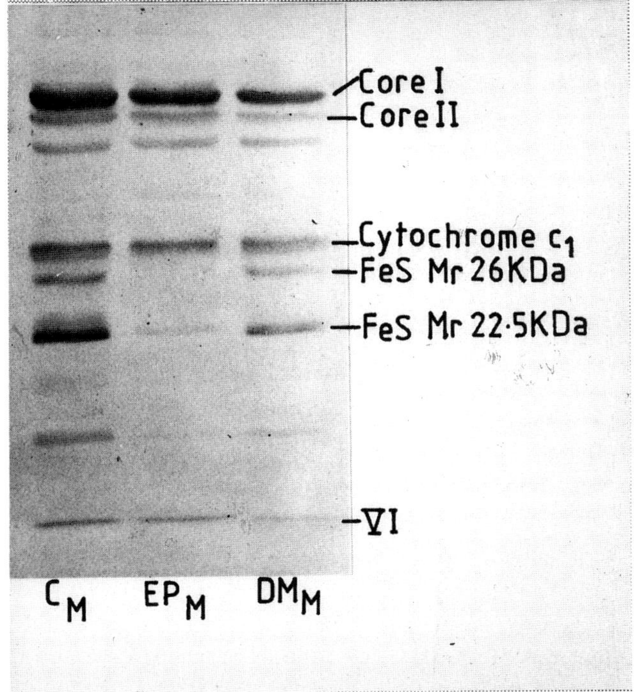

Figure 4.4 Immunoblots using antibody to the holoenzyme of Complex III and subunit-specific antiserum to the Rieske iron-sulphur (FeS) protein against 100 µg of muscle mitochondria from a control, and two patients with mitochondrial myopathy (Reproduced from ref. [5] with permission)

levels of core proteins I and II, cytochrome c_1 and subunit VI (Figure 4.4, patient EP). Immunoblots of the patient's muscle homogenate, however, showed normal amounts of Rieske protein. These results demonstrated an import defect of the Rieske protein. Complex II immunoblots of the same patient showed a severe deficiency of the 27.2 kDa iron-sulphur protein in both mitochondrial and homogenate fractions. This probably indicates either reduced synthesis or increased turnover of this protein. The precise molecular basis of the respiratory chain defects in this patient remain undefined. A mutation in the mitochondrial targeting sequence of the patient's Rieske protein or a mitochondrial receptor-based abnormality might explain the results.

REFERENCES

1. Sherratt, H.S.A., Watmough, N.J., Johnson, M. and Turnbull, D.M. (1988) Methods used for the study of normal and abnormal skeletal muscle mitochondria. *Methods of Biochemical Analysis*, **33**, 243–335

2. Hammans, S.R., Sweeney, M.G., Brockington, M., Morgan-Hughes, J.A. and Harding, A.E. (1991) Mitochondrial encephalopathies: molecular genetic diagnosis from blood samples. *Lancet*, **337**, 1311–1313

3. Krige, D., Carroll, M.T., Cooper, J.M., Marsden, C.D. and Schapira, A.H.V. (1992) Platelet mitochondrial function in Parkinson's disease. *Annals of Neurology*, **32**, 782–788

4. Holt, I.J., Harding, A.E. and Morgan-Hughes, J.A. (1988) Deletions of mitochondrial DNA in patients with mitochondrial myopathies. *Nature*, **331**, 717–719

5. Schapira, A.H.V., Cooper, J.M., Morgan-Hughes, J.A., Landon, D.N. and Clark, J.B. (1990) Mitochondrial myopathy with a defect of mitochondrial protein transport. *New England Journal of Medicine*, **323**, 37–42

6. Holt, I.J., Harding, A.E., Petty, R.K. and Morgan-Hughes, J.A. (1990) A new mitochondrial disease associated with mitochondrial DNA heteroplasmy. *American Journal of Human Genetics*, **46**, 428–433

7. Cullen, M.J. and Fulthorpe, J.J. (1975) Stages in fiber breakdown in Duchenne muscular dystrophy. An electron microscopic study. *Journal of the Neurological Sciences*, **24**, 179–200

8. Fardeau, M. (1970) Ultrastructural lesions in progressive muscular dystrophies: a critical study of their specificity. In *Muscle Diseases* (eds N. Canal, G. Scarlato and J.W. Walton), Excerpta Medica, Amsterdam, pp. 98–108

9. Granger, D.L. and Lehninger, A.L. (1982) Sites of inhibition of mitochondrial electron transport in macrophage-injured neoplastic cells. *The Journal of Cell Biology*, **95**, 527–535

10. King, T.E. (1967) Preparation of succinate cytochrome c reductase, and the cytochrome b-c_1 particle, and reconstitution of succinate cytochrome c reductase. *Methods in Enzymology*, **10**, 216–225

11. Wharton, D. and Tzagoloff, A. (1967) Cytochrome oxidase from beef heart mitochondria. *Methods of Enzymology*, **10**, 245–250

12. Ragan, C.I., Wilson, M.T., Darley-Usmar, V.M. and Lowe, P.N. (1987) Subfractionation of mitochondria, and isolation of the proteins of oxidative phosphorylation. In *Mitochondria, a Practical Approach* (eds V.M. Darley-Usmar, D. Rickwood and M.T. Wilson), IRL Press, London, pp. 79–112

13. Hatefi, Y. and Stiggall, D.A. (1978) Preparation and properties of NADH:cytochrome c oxidoreductase (complex I–III). *Methods in Enzymology*, **530**, 5–10

14. Wilson, D.F. and Epel, D. (1968) The cytochrome system of sea urchin sperm. *Archives of Biochemistry and Biophysics*, **126**, 23–90

15. Ohnishi, T., Ragan, C.I. and Hatefi, Y. (1985) EPR studies of iron-sulphur clusters in isolated subunits and subfractions of NADH ubiquinone oxidoreductase. *The Journal of Biological Chemistry*, **260**, 2782–2788

16. Ohnishi, T. (1987) Structure of the succinate-ubiquinone oxidoreductase (complex II). *Current Topics in Bioenergetics*, **15**, 37–65

17. Moreadith, R.W., Cleeter, M.J.W., Ragan, C.I. *et al.* (1987) Congenital deficiency of two polypeptide subunits of the iron-protein fragment of mitochondrial complex I. *Journal of Clinical Investigation*, **79**, 463–467

18. Ichiki, T., Tanaka, M., Kobayashi, M. *et al.* (1989) Disproportionate deficiency of iron-sulphur clusters and subunits of complex I in mitochondrial encephalopathy. *Pediatric Research*, **25**, 194–201

19. Schägger, H. and von Jagow, G. (1991) Blue native electrophoresis for isolation of membrane protein complexes in enzymatically active form. *Analytical Biochemistry*, **199**, 223–231

20. Soumalainen, A., Ciafolini, E., Koga, Y., Peltonen, L., DiMauro, S. and Schon, E.A. (1992) Use of a single strand conformation polymorphism analysis to detect point mutations in human mitochondrial DNA. *Journal of Neurological Sciences*, **111**, 222–226

21. Cotton, R.G.H., Rodrigues, N.R. and Cambell, R.D. (1988) Reactivity of cytosine and thymine in single-base-pair mismatches with hydroxylamine and osmium tetroxide and its application to the study of mutations. *Proceedings of the National Academy of Sciences USA*, **85**, 4397–4401

22. Chomyn, A., Martinuzzi, A., Yoneda, M. *et al.* (1992) MELAS mutation in mtDNA binding site for transcription termination factor causes defects in protein synthesis and in respiration but no change in levels of upstream and downstream mature transcripts. *Proceedings of the National Academy of Sciences USA*, **89**, 4221–4225

23. Shoffner, J.M., Lott, M.T., Lezza, A.M.A. *et al.* (1990) Myoclonic epilepsy and ragged red fibre disease (MERRF) is associated with a mitochondrial DNA tRNA[Lys] mutation. *Cell*, **61**, 931–937

24. Hayashi, J-I., Ohta, S., Kikuchi, A., Takemitsu, M., Goto, Y-I. and Nonaka, I. (1991) Introduction of disease-related mitochondrial DNA deletions into HeLa cells lacking mitochondrial DNA results in mitochondrial dysfunction. *Proceedings of the National Academy of Sciences USA*, **88**, 10614–10618

25. King, M.P. and Attardi, G. (1989) Human cells lacking mtDNA: repopulation with exogenous mitochondria by complementation. *Science*, **246**, 500–503
26. Chomyn, A., Meola, G., Bresolin, N., Lai, S.T., Scarlato, G. and Attardi, G. (1991) In vitro genetic transfer of protein synthesis and respiration defects to mitochondrial DNA-less cells with myopathy-patient mitochondria. *Molecular and Cellular Biology*, **11**, 2236–2244
27. Bodnar, A.G., Cooper, J.M., Holt, I.J., Leonard, J.V. and Schapira, A.H. V. (1993) Nuclear complementation restores mtDNA levels in cultured cells from a patient with mtDNA depletion. *American Journal of Human Genetics*, **53**, 663–669
28. Zheng, X., Shoffner, J.M., Lott, M.T. *et al.* (1989) Evidence in a lethal infantile mitochondrial disease for a nuclear mutation affecting respiratory complexes I and IV. *Neurology*, **39**, 1203–1209
29. Robinson, B.H., Ward, J., Goodyer, P. and Baudet, A. (1986) Respiratory chain defects in the mitochondria of cultured skin fibroblasts from three patients with lactic acidaemia. *Journal of Clinical Investigation*, **77**, 1422–1427
30. Morgan-Hughes, J.A., Hayes, D.J., Cooper, J.M. and Clark, J.B. (1985) Mitochondrial myopathies; deficiencies localized to complex I and complex III of the mitochondrial respiratory chain. *Biochemical Society Transactions*, **13**, 648–650
31. Roodhooft, A.M., Van Acker, K.J., Martin, J.J., Ceuterick, C., Scholte, H.R. and Luyt-Houwen, E.M. (1986) Benign mitochondrial myopathy with deficiency of NADH-CoQ reductase and cytochrome c oxidase. *Neuropediatrics*, **17**, 221–226
32. Schapira, A.H.V., Cooper, J.M., Morgan-Hughes, J.A. *et al.* (1988) Molecular basis of mitochondrial myopathies: polypeptide analysis in complex I deficiency. *Lancet*, **i**, 500–503
33. Morgan-Hughes, J.A., Cooper, J.M., Holt, I.J., Harding, A.E., Schapira, A.H.V. and Clark, J.B. (1990) Mitochondrial myopathies; clinical defects. *Biochemical Society Transactions*, **18**, 523–525
34. Koga, Y., Nonaka, I., Kobayashi, M., Tojyo, M. and Nihei, K. (1988) Findings in muscle in complex I (NADH) coenzyme Q reductase deficiency. *Annals of Neurology*, **24**, 749–756
35. Morgan-Hughes, J.A., Cooper, J.M., Schapira, A.H.V., Hayes, D.J. and Clark, J.B. (1987) The mitochondrial myopathies. Defects of the mitochondrial respiratory chain and oxidative phosphorylation system. In *The London Symposia* (eds R.J. Ellingson, N.M.F. Murray and A.M. Halliday), North Holland, Amsterdam, pp. 103–114
36. Scholte, H.R., Busch, H.F.M., Luyt-Houwen, I.E.M. *et al.* (1987) Defects in oxidative phosphorylation. Biochemical investigations in skeletal muscle and expression of the lesion in other cells. *Journal of Inherited Metabolic Disorders*, **10**, Suppl. 1, 81–97
37. Watmouth, M.J., Bindoff, L.A. and Birch-Machin, M.A. (1990) Impaired mitochondrial β-oxidation in a patient with an abnormality of the respiratory chain. *Journal of Clinical Investigation*, **85**, 177–184
38. Martinuzzi, A., Bartomomei, L., Carrozzo, R. *et al.* (1992) Correlation between clinical and molecular features in two MELAS families. *Journal of the Neurological Sciences*, **113**, 222–229
39. Ciafaloni, E., Ricci, E., Servidei, S. *et al.* (1991) Widespread tissue distribution of a transfer RNA-LEU(UUR) mutation in the mitochondrial DNA of a patient with MELAS syndrome. *Neurology*, **41**, 1663–1665
40. Moraes, C.T., Ricci, E., Bonilla, W., DiMauro, S. and Schon, E.A. (1992) The mitochondrial tRNA(UUR) mutation mitochondrial encephalomyopathy, lactic acidosis, and strokelike episodes (MELAS): genetic, biochemical, and morphological correlations in skeletal muscle. *American Journal of Human Genetics*, **50**, 934–949
41. Bindoff, L.A., Desnuelle, C., Birch-Machin, A. *et al.* (1991) Multiple defects of the mitochondrial respiratory chain in a mitochondrial encephalopathy (MERRF): a clinical, biochemical and molecular study. *Journal of the Neurological Sciences*, **102**, 17–24
42. Silvestri, G., Moraes, C.T., Shanske, S., Oh, S.J. and DiMauro, S. (1992) A new mtDNA mutation in the tRNA(Lys) gene associated with myoclonic epilepsy and ragged-red fibers (MERRF). *American Journal of Human Genetics*, **51**, 1213–1217
43. Seibel, P., Degoul, F., Bonne, G. *et al.* (1991) Genetic, biochemical and pathophysiological characterization of a familial mitochondrial encephalomyopathy (MERRF). *Journal of the Neurological Sciences*, **105**, 217–224
44. Holt, I.J., Harding, A.E., Cooper, J.M. *et al.* (1989) Mitochondrial myopathies: clinical and biochemical features of 30 patients with major deletions of muscle mitochondrial DNA. *Annals of Neurology*, **26**, 699–708
45. Truong, D.D., Harding, A.E., Scaravilli, F. *et al.* (1990) Movement disorders in mitochondrial myopathies: a report of nine cases with two autopsy studies. *Movement Disorders*, **5**, 109–117
46. Morgan-Hughes, J.A., Schapira, A.H.V., Cooper, J.M. and Clark, J.B. (1988) Molecular defects of NADH–ubiquinone oxidoreductase (complex I) in mitochondrial diseases. *Journal of Bioenergetics and Biomembranes*, **20**, 365–383
47. Schapira, A.H.V., Cooper, J.M., Manneschi, L. *et al.* (1990) A mitochondrial encephalomyopathy

with specific deficiencies of two respiratory chain polypeptides and a circulating autoantibody to a mitochondrial matrix protein. *Brain*, **113**, 419–432

48. Cooper, J.M., Mann, V.M., Krige, D. and Schapira, A.H.V. (1992) Human mitochondrial complex I dysfunction. *Biochimica et Biophysica Acta*, **1101**, 192–203

49. Howell, N., Bindoff, L.A., McCullough, D.A. *et al.* (1991) Leber's hereditary optic neuropathy: identification of the same mitochondrial ND1 mutation in six pedigrees. *American Journal of Human Genetics*, **49**, 939–950

50. Parker, W.D., Oley, C.A. and Parks, J.A. (1989) A defect in mitochondrial electron transport activity (NADH – coenzyme Q oxidoreductase) in Leber's hereditary optic neuropathy. *New England Journal of Medicine*, **320**, 1331–1333

51. Larsson, N.-G., Andersen, O., Holme, E., Oldfors, A. and Wahlström, J. (1991) Leber's hereditary optic neuropathy and complex I deficiency in muscle. *Annals of Neurology*, **30**, 701–708

52. Smith, P.R., Cooper, J.M., Govan, G.G., Harding, A.E. and Schapira, A.H.V. (1993) Smoking and mitochondrial function: a model for environmental toxins. *Quarterly Journal of Medicine*, **86**, 657–660

53. Schapira, A.H.V., Cooper, J.M., Dexter, D., Jenner, P., Clark, J.B. and Marsden, C.D. (1989) Mitochondrial complex I deficiency in Parkinson's disease. *Lancet*, **i**, 1269

54. Mann, V.M., Cooper, J.M., Krige, D., Daniel, S.E., Schapira, A.H.V. and Marsden, C.D. (1992) Brain, skeletal muscle and platelet homogenate mitochondrial function in Parkinson's disease. *Brain*, **115**, 333–342

55. Skoglund, R.R. (1979) Reversible ataxia, mitochondrial myopathy and lactic acidaemia. *Neurology*, **29**, 717–720

56. Rivner, M.H., Shamsnia, M., Swift, T.R. *et al.* (1989) Kearns-Sayre syndrome and complex II deficiency. *Neurology*, **39**, 693–696

57. Haller, R.G., Henriksson, K.G., Jorfeldt, L. *et al.* (1990) Muscle succinate dehydrogenase deficiency; exercise pathophysiology of a novel mitochondrial myopathy. *Neurology*, **40**, Suppl. 1, 413

58. Kennaway, N.G. (1988) Defects in the cytochrome bc_1 complex in mitochondrial diseases. *Journal of Bioenergetics and Biomembranes*, **20**, 325–352

59. Morgan-Hughes, J.A., Darveniza, P., Kahn, S.N. *et al.* (1977) A mitochondrial myopathy characterised by a deficiency in reducible cytochrome b. *Brain*, **100**, 617–640

60. Darley-Usmar, V.M., Kennaway, N.G., Buist, N.R.M. and Capaldi, R.A. (1983) Deficiency in ubiquinone cytochrome *c* reductase in a patient with mitochondrial myopathy and lactic acidosis. *Proceedings of the National Academy of Sciences USA*, **80**, 5103–5106

61. Kennaway, N.G., Buist, M.R.M., Darley-Usmar, V.M. *et al.* (1984) Lactic acidosis and mitochondrial myopathy associated with deficiency of several components of complex III of the respiratory chain. *Pediatric Research*, **18**, 991–999

62. Hayes, D.J., Lecky, B.R.F., Landon, D.N. *et al.* (1984) A new mitochondrial myopathy; biochemical studies revealing a deficiency in the cytochrome $b\text{-}c_1$ complex (complex III) of the respiratory chain. *Brain*, **107**, 1165–1177

63. Papadimitrou, A., Neustein, H.B., DiMauro, S. *et al.* (1984) Histiocytoid cardiomyopathy of infancy; deficiency of reducible cytochrome b in heart mitochondria. *Pediatric Research*, **18**, 1023–1028

64. Birch-Machin, M.A., Shepherd, I.M., Watmough, N.J. *et al.* (1989) Fatal lactic acidosis in infancy with a defect of complex III of the respiratory chain. *Pediatric Research*, **25**, 553–559

5
Cytochrome oxidase deficiency: progress and problems

Salvatore DiMauro, Michio Hirano, Eduardo Bonilla, Carlos T. Moraes and Eric A. Schon

INTRODUCTION

This is the eighth in a series of reviews by our group on cytochrome oxidase (COX) deficiency which started in 1983, when the different clinical presentations began to be delineated [1–7]. The number and frequency of these updates reflect the interest of both clinicians and basic scientists in COX deficiency and the rapid pace at which information on this topic has been accumulating. Much recent knowledge has come not from the study of patients with presumably primary COX deficiencies but, rather, from the widespread use of COX histochemistry in muscle biopsies from patients with diverse mitochondrial encephalomyopathies and from normal aging individuals. The finding, in most of these cases, of a 'chequerboard' pattern (that is, the coexistence of COX-positive and COX-negative fibres in cross-sections of muscle) together with normal or only partially reduced enzyme activity in muscle homogenates has considerably complicated the classification of COX deficiencies.

In this review, we will consider the COX deficiencies within the framework of a classification of the mitochondrial encephalomyopathies based on genetic criteria [8] (Table 5.1). Although the molecular basis of many forms of COX deficiency is still unknown, this classification allows rational pathogenetic mechanisms to be proposed for some COX deficiencies and genetic mechanisms to be postulated for others. As molecular probes are available for each of the 13 subunits of COX, experimental verification of the hypothetical mechanisms should be forthcoming. Thus *defects in nuclear DNA (nDNA)* could explain both tissue-specific and generalized forms of COX deficiency that are transmitted by mendelian inheritance and do not affect other enzymes of the respiratory chain. Unfortunately, as detailed below, this hypothesis has not yet been borne out by experimental evidence. On the other hand, *defects in mitochondrial DNA (mtDNA)* and *defects in intergenomic communication* can explain the usually incomplete and non-specific COX deficiencies in syndromes associated with mtDNA deletions or point mutations and mtDNA depletion. However, controversies abound, as the following discussion will illustrate.

Table 5.1 Mitochondrial diseases

Site of defect	Heredity	Clinical features	Biochemistry
Nuclear DNA (nDNA)			
Tissue-specific gene	Mendelian	Tissue-specific syndrome	Tissue-specific monoenzymopathy
Non-tissue-specific gene	Mendelian	Multisystemic disorder	Generalized monoenzymopathy
Protein import controlling gene	Mendelian	?	?
Mitochondrial DNA (mtDNA)			Generalized monoenzymopathy (structural genes)
Point mutations	Maternal	Multisystemic, heterogeneous	Generalized multienzymopathy (tRNA genes)
Deletions or duplications	Sporadic	PEO; KSS; Pearson	Generalized (±) multienzymopathy
nDNA/mtDNA communication			
Multiple mtDNA deletions	Mendelian (AD)	PEO ± other features	Generalized multienzymopathy
mtDNA depletion	Mendelian (AR)	Myopathy ± nephropathy; hepatopathy; encephalopathy	Tissue-specific multienzymopathy

THE ENZYME

COX (cytochrome c oxidase, EC 1.9.3.1), the last component of the respiratory chain (Complex IV), catalyses the transfer of reducing equivalents from cytochrome c to molecular oxygen. The energy released in this exergonic reaction sustains a transmembrane proton gradient, which is then utilized by mitochondrial ATP synthase (Complex V) to generate ATP. The holoenzyme contains two haem a moieties (a and a_3) and two copper atoms (Cu_A and Cu_B) bound to a multi-subunit protein frame embedded in the mitochondrial inner membrane [9,10]. In mammals, the apoprotein is composed of 13 different subunits. The three larger polypeptides (COX I to III) are associated with the prosthetic groups and form the catalytic core of the enzyme; haems and coppers are localized to subunits I and II and subunit III is responsible for the proton pumping activity [9]. They are encoded by mtDNA and are synthesized in mitochondria; both their coding sequences and primary structures have been established in humans [11]. The ten smaller subunits (IV, Va, Vb, VIa, VIb, VIc, VIIa, VIIb, VIIc and VIII, according to the nomenclature of Kadenbach *et al.* [12]), are encoded by nDNA and are synthesized in the cytoplasm. All except subunits VIa, VIb, VIc and VIII are synthesized as precursors carrying N-terminal basic presequences which allow them to be targeted to mitochondria and transported into the organelles [13]. Seven subunits (IV, VIa, VIc, VIIa, VIIb, VIIc and VIII) have similar tripartite structures with a hydrophilic N-terminus, a hydrophobic central domain and a hydrophilic C-terminus, consistent with transmembrane α-helices. The other three nuclear subunits have no obvious hydrophobic domains and are probably extrinsic to the inner membrane bilayer [9].

The functions of the nDNA-encoded subunits (which are missing in prokaryotes) have not been fully elucidated, but they may modulate the catalytic activity or the biosynthesis of COX, which is the central enzyme in the control of oxidative phosphorylation. Thus some or all nDNA-encoded COX subunits may function as regulators of COX activity, optimizing it to the metabolic requirements of different tissues. In agreement with this concept, tissue-specific isozymes have been documented for two human subunits, VIa and VIIa; the existence of isoforms had been suggested by differences in electrophoretic mobility, amino acid sequences and antibody specificity [12], and it has been confirmed by direct sequence of distinct cDNAs specifying 'liver' isoforms (expressed ubiquitously) and 'muscle' isoforms (expressed only in skeletal and cardiac muscles) [14–17]. Studies of human muscle cultures have shown that subunit VIa is also developmentally regulated; transcription of the liver isoforms decreases gradually during myogenesis while transcription of the muscle isoform increases [18]. In contrast, both subunits of COX VIIa are co-expressed throughout myogenesis [18]. The genes encoding the other eight human nDNA-encoded COX subunits have been isolated and sequenced and no evidence for tissue-specific isoforms has been provided by Northern blot analyses [19–25]. The expression of COX VIII as a single form in primates contrasts with the existence of at least two tissue-specific isoforms in other species [24]. This result is especially surprising, because sequence homology between COX genes for the same isoform in different mammals is greater than between COX genes for different isoforms in the same species. It is uncertain how nuclear-encoded COX subunits modulate COX activity. Mechanisms being considered include: (1) tissue-specific subunits could confer different electron-transfer activities on the holoenzymes; (2) the regulatory effect of ATP and ADP concentrations on COX activity could occur through binding of the nucleotides to specific subunits; (3) some subunits may regulate the amount of enzyme synthesized either by being present in limiting amounts or by controlling directly the rate of transcription or translation of mtDNA-encoded subunits; (4) some subunits may be important for the correct folding of mtDNA-encoded subunits [9]. None of these mechanisms, however, has been documented convincingly.

THE DISEASES

COX deficiencies attributed to defects in nuclear DNA

At present there are no molecular data to support this attribution, which is based on circumstantial evidence. Defects of nuclear genes encoding COX subunits which are expressed in a tissue-specific manner would be expected to cause disorders that are confined to one or a few tissues, are transmitted by mendelian inheritance, and affect exclusively COX activity (Table 4.1). Some diseases of infancy and fewer disorders of adult life approximate these criteria.

FATAL INFANTILE MYOPATHY

First described by Van Biervliet *et al.* [26] in 1977, this disorder has been reported in approximately 15 patients [27–35]. Hypotonia, weakness and lactic acidosis usually become evident in the first weeks of life, and death from respiratory failure occurs before 1 year of age (Table 5.2). Although heart, liver and brain are clinically spared, about half of the patients have renal disease with glycosuria, phosphaturia and generalized aminoaciduria (DeToni–Fanconi syndrome). Pedigree analysis in informative families suggests autosomal recessive transmission (equal distribution in the two sexes, siblings often affected, parents sometimes consanguineous and asymptomatic). Muscle biopsies show ragged red fibres (RRF), a hallmark of mitochondrial proliferation, and, histochemically, COX activity is undetectable in muscle fibres but normal in intrafusal fibres of muscle spindles and in the walls of intramuscular arteries [32,36]. There is an increased number of undifferentiated type IIC fibres [29] and satellite cells [37], and cultured myoblasts grow poorly and lack COX activity whereas fibroblasts appear normal [36].

In agreement with the clinical findings, biochemical analysis in one patient showed that the enzyme defect was confined to skeletal muscle, sparing heart, liver and brain [31]. Partial COX deficiency was also documented in the kidney of patients with myopathy and nephropathy [27,29]. It should be noted, however, that the lack of clinical cardiopathy and the normal COX activity found in the heart of these patients contrast with the knowledge that both muscle-specific COX subunits, VIa and VIIa, are also expressed in the heart. Another biochemical conundrum concerns the specificity of the COX deficiency. While only COX activity was lacking in muscle of most patients [27–29,31–35], other respiratory chain components, especially cytochrome *b*, were also decreased in some patients [26,30]. This biochemical heterogeneity could reflect genetic heterogeneity. In particular, multiple respiratory chain defects are seen in muscle of patients with fatal infantile myopathy and mtDNA depletion, a disorder virtually indistinguishable on clinical grounds from fatal infantile myopathy and COX deficiency (see below). In patients with multiple respiratory chain defects, it is therefore important to investigate mtDNA content in muscle.

Table 5.2 Clinical and laboratory features in 15 patients with fatal infantile myopathy and COX deficiency

Features	Frequency
Sex (M:F)	8:7
Age at onset	Birth – 3 months
Age at death	7 days – 15 months
Generalized weakness	15/15
Respiratory insufficiency	13/13
Renal dysfunction	11/15
Positive family history	8/15 (three sets of sibs)
Lactic acidosis	14/14

The patients are described in refs [26–33] (patient 2), [34] (patient 1–4), [35] (patients 1 and 2).

Immunological studies using antibodies raised against human heart COX holoenzyme showed markedly decreased amounts of cross-reacting material (CRM) in muscle of patients, by both ELISA and immunocytochemistry [31,32,36]. Electrophoretic analysis of the residual enzyme after immunoprecipitation failed to show any alteration in the subunit pattern [31]. In a further attempt to define the subunit composition of the presumably mutant enzyme, different investigators have employed a battery of antibodies against individual COX subunits obtained in Dr Kadenbach's laboratory and made available to the scientific community. The results have been controversial. In normal muscle sections, Romero *et al.* [38] showed that a monoclonal antibody to COX IV reacted only with type I fibres and a polyclonal antibody to COX Vab reacted preferentially with type II fibres. In muscle biopsies from four patients with fatal infantile myopathy and COX deficiency, Tritschler *et al.* [34] found an almost complete lack of immunoreactivity only with monoclonal antibodies to COX VIIa,b. However, using the same set of antibodies, Muller-Hocker *et al.* [39] found a lack of immunoreactivity for COX II/III and VIIb,c in muscle specimens from two siblings with a combined defect in COX and cytochrome *b*. This discrepancy could be due to genetic heterogeneity or to lack of specificity of the immunocytochemical results. Lack of specificity is suggested by molecular genetic analysis; direct sequencing of the cDNAs for the muscle-specific isoforms of COX VIa and VIIa in our patients lacking immunoreactive COX VIIa,b has failed to show any mutation (M. Hirano, unpublished results).

MYOPATHY AND CARDIOPATHY
In a few children there was a combination of generalized weakness and intractable cardiopathy leading to cardiac failure and death before 1 year of age [40–43]. COX deficiency, however, may be a more common cause of infantile cardiomyopathy because, in a survey of 28 children with mitochondrial diseases and diverse clinical presentations, Munnich *et al.* [44] found two cases of severe cardiopathy associated with COX deficiency in muscle. In the patient described by Zeviani *et al.* [42], COX deficiency was documented in both skeletal and cardiac muscles, and immunologically reactive enzyme was present in apparently normal concentrations in both tissues. In another patient, COX and Complex I were decreased in heart and muscle, and immunocytochemistry with Kadenbach's battery of antibodies

revealed decreased reactivity for COX II/III, Vab and VIIbc [43]. As mutations of mtDNA have been reported in patients with myopathy and cardiopathy [45–47], it cannot be excluded that some patients, especially those with combined defects of the respiratory chain, might have defects of the mitochondrial genome. However, mutations in the nuclear genes encoding either one of the two muscle/heart-specific COX isozymes, COX VIa and VIIa, could underlie this syndrome in some cases, and molecular studies will have to be performed to verify this hypothesis.

BENIGN INFANTILE MYOPATHY

Children with this disorder also present with severe weakness and often need assisted ventilation and gastric tube feeding early in life, but improve spontaneously and are usually normal by 2 or 3 years of age, although the tempo of recovery varies from patient to patient. This condition, first described by DiMauro *et al.* in 1983 [48], has been reported in eight additional patients [34,36,49–51]. A tenth child had identical clinical course and muscle pathology, although the biochemical defect was not documented [52]. Lactic acidosis, which may be even more severe initially than in the fatal form, also remits spontaneously. Although potentially benign, this myopathy is life-threatening in the first months of life; therefore it would be important to identify pathological or biochemical features capable of differentiating fatal infantile from benign myopathies.

The spontaneous clinical recovery in these children correlates with a gradual return of COX activity in muscle, which can be demonstrated both histochemically and biochemically. In our original patient [48], COX activity increased from 6 to 33 to 174% of normal in biopsies taken at 2, 7 and 36 months. Histochemistry showed that only scattered fibres stained for COX activity in the first biopsy, but the number of COX-positive fibres increased with time [48]. Immunocytochemistry with antibodies against the holoenzyme showed the presence of CRM even in fibres lacking COX activity, and the presence of inactive enzyme was confirmed by ELISA [49]. As in the fatal form, COX activity is normal in intrafusal fibres and in the vascular walls [34,36]. Immunocytochemistry of early biopsies with antibodies against individual COX subunits gave surprising results: not only was immunoreactivity lacking for COX VIIab (as in the fatal form) but also for COX II, a mtDNA-encoded subunit [34]. Immunoreactivity to these subunits returned to normal in parallel with the return of histochemical COX activity [34]. Although the different immunoreactivity (confirmed in four patients with fatal myopathy and four with the benign form) provides a useful tool in the differential diagnosis of the two disorders, the pathogenetic significance of these findings is uncertain [34].

The reversibility of COX deficiency suggests at least two hypothetical explanations. If the genetic error affects nDNA, the mutation may involve a subunit that is not only tissue-specific but also developmentally regulated. Mutations of a fetal or neonatal muscle isozyme would be corrected when the mature isozyme starts to be expressed. As mentioned above, developmental regulation of expression has been documented in muscle culture for COX VIIa [18]. However, as the 'immature' isozyme of COX VIIa appears to be the ubiquitous liver form, a mutation would be expected to affect multiple tissues. In addition, evidence from other enzyme systems suggests that the conversion from fetal to mature isoforms occurs early in fetal life rather than postnatally [53]. If, on the other hand, the genetic error affects mtDNA, there could be a gradual selection of fibres containing a majority of wild-type mitochondrial genomes over those that contain mostly mutant mtDNA. In favour of this hypothesis is the histochemical observation that fibres seem to be affected in an all-or-none fashion: it is the number of normal fibres that increases with time, not the intensity of the enzyme reaction in each fibre [48]. Against the hypothesis is the strict tissue-specificity of the defect and the lack of evidence of maternal inheritance, although the muscle biopsy from the mother of one of our patients showed a partial defect in COX [48] and scattered COX-deficient fibres (E. Bonilla, unpublished work), while the father's biopsy was normal. Sequencing of the COX II gene in patients has shown no mutations (M. Hirano, unpublished results).

MYOPATHIES OF ADULT LIFE

The most striking case was that of a 27-year-old woman who had life-long exercise intolerance, with exertional dyspnoea, tachycardia and premature fatigue of exercising muscles [54]. Although symptoms worsened with age, she could work as a registered nurse. Family history was non-informative: her parents and five siblings were well. She was short and had mild neck proximal limb weakness, but there were no signs of visceral or brain involvement. Muscle biopsy showed slight increase in red staining with the modified Gomori trichrome and increased number of lipid droplets but no frank RRF. Biochemical studies showed marked reduction of COX activity and less severe decrease in succinate–cytochrome c reductase activity in both crude muscle extracts and isolated mitochondria, and decreased cytochrome aa_3 content in mitochondrial fractions. The amount of immunologically detectable COX was markedly decreased by immunotitration. Exercise studies illustrated the functional consequences of blocks in the respiratory chain, with low maximal oxygen uptake and impaired muscle oxygen extraction. The increases in cardiac output and ventilation during exercise, normally closely coupled to muscle metabolic rate, were greatly exaggerated relative to oxygen uptake and carbon dioxide production [54].

Prominent involvement of respiratory muscles with respiratory insufficiency dominated the clinical picture in two patients with partial COX deficiency and RRF in their muscle biopsies [55]. Although myopathy was the main complaint, brain involvement with sensorineural hearing loss and complex partial seizures complicated the clinical picture in a 52-year-old man. There was partial COX deficiency in muscle with a commensurate decrease in CRM measured by ELISA [56]. Muscle and brain also appeared to be selectively affected in an 18-year-old man with exercise intolerance, moderate weakness and mental retardation [57]. Immunoblots suggested defects of COX II, VI and VII. Administration of coenzyme Q_{10} improved exercise tolerance and decreased the exaggerated lactate response to exercise [57].

COMBINED DEFECTS OF THE RESPIRATORY CHAIN

Because mtDNA encodes subunits of four of the five respiratory chain complexes, combined defects can be explained by alterations of mtDNA affecting its overall function, such as deletions, point mutations in tRNA genes and depletion (see below). However, mutations of nDNA can also cause multiple defects. Potential mechanisms include: (1) mutations of regulatory genes controlling more than one complex; (2) mutations in subunits shared by two or more complexes; (3) mutations affecting the phospholipid composition of the inner mitochondrial membrane; (4) defects in mitochondrial protein import affecting subunits of multiple complexes.

A combined defect in Complex I and IV affecting muscle, heart and liver but sparing brain and kidney was reported in an infant with a disorder characterized by growth retardation, diffuse weakness, progressive cardiomyopathy, hepatic insufficiency, severe lactic acidosis and RRF [58]. A mutation in the mature isoform of a nDNA-encoded tissue-specific developmentally-regulated subunit shared by COX and Complex I was postulated. Conversely, it was suggested that a mutation affecting the fetal isoform of the same subunit might explain the spontaneously reversible myopathy described by Roodhooft *et al.* [59], which was also characterized by a combined defect in Complex I and IV.

A combined defect in COX and Complex III was reported in an infant who died at 5 months with generalized weakness, lactic acidosis and cardiorespiratory insufficiency [60]. It was postulated that the primary defect may involve a nDNA-encoded protein controlling mtDNA translation or processing of the polycistronic mRNA.

Defects in genes encoding COX subunits that are expressed in all tissues would be expected to cause multisystemic disorders characterized, biochemically, by severe and specific loss of COX activity in all tissues (Table 5.1). The disorder that best approximates this situation is Leigh syndrome (subacute necrotizing encephalomyelopathy). COX deficiency is not the only biochemical defect associated with Leigh syndrome, which has been described in patients with pyruvate dehydrogenase complex (PDHC) deficiency [61], Complex I deficiency [62], a point mutation in the ATPase 6 gene of mtDNA [63], and in a few patients with the MERRF mutation at nt 8344 of mtDNA [64].

LEIGH SYNDROME

Typically, Leigh syndrome is a devastating encephalopathy of infancy or childhood characertized by psychomotor regression which usually becomes manifest around 1 year of age and is accompanied by a spectrum of neurological signs including ataxia, optic atrophy, ophthalmoplegia, ptosis, nystagmus, dystonia, tremor, pyramidal signs and respiratory abnormalities. The characteristic neuropathological lesions are symmetrical areas of necrosis involving preferentially mid-brain, pons, basal ganglia, thalamus and optic nerves. Microscopically, there is cystic cavitation, vascular proliferation, neuronal loss and demyelination. These lesions can be demonstrated as characteristically distributed signal abnormalities by cranial computed tomography (CT) or magnetic resonance imaging (MRI) scans, thus allowing premortem diagnosis. Muscle histochemistry is normal, although electron microscopy may show an increased number of mitochondria. Histochemical studies have shown diffusely decreased but not absent COX activity in all fibres and also in muscle spindles and vascular walls [33,36,65,66]. As shown by immunocytochemistry, there was uniformly decreased reactivity for all COX subunits, nDNA- as well as mtDNA-encoded, whereas in contrast, immunoreactivity to the mtDNA-encoded subunit 1 of Complex I was preserved [65].

COX deficiency in a patient with Leigh syndrome was first described by Willems *et al.* in 1977 [67] and has been documented in about 50 patients [68–73], 34 of whom have been recently reviewed by Van Coster *et al.* [71]. The enzyme defect is generalized in most cases, but normal activities were found in liver in one case [67], liver and cultured fibroblasts in another [74] and in fibroblasts from two patients [71]. The defect is partial (which may explain the lack of RRF) and residual activity varies in different tissues; in a study of five patients, we found that COX activity was about 35% of normal in brain, 15% in muscle and less than 10%

in liver [74]. Abnormal susceptibility of the enzyme to inhibition by oxidized cytochrome *c* was reported in fibroblasts from one patient [75], but we found normal affinity of COX for reduced cytochrome *c* in muscle and brain from three other patients [65]. The amount of CRM was normal or variably decreased in different patients by ELISA [71,74]. Immunoblot experiments using antibodies against the holoenzyme or individual COX subunits showed bands of decreased intensity but the pattern and relative concentration of the subunits appeared to be preserved [71,74,76–78]. In fibroblasts, the decrease in immunoreactive protein was roughly proportional to the decrease in COX activity and allowed separation of the patients into two groups, one with severe and the other with partial deficiency of COX activity, cytochrome aa_3 and immunoreactive protein [77].

A review of the family histories of 31 patients, 18 boys and 13 girls, described in eight publications [67–71,74,79,80] showed four pairs of siblings (including a pair of identical twins), a history of affected siblings in four cases, and parental consanguinity in five families, clearly suggesting autosomal recessive inheritance. Evidence for a nuclear defect was also provided by experiments in which COX-deficient fibroblasts from a patient with Leigh syndrome were fused with a variant strain of HeLa cells [81]. Prolonged cultivation of the hybrids in appropriate media led to preferential loss of HeLa cell mtDNA. COX activity was normal in clones that had lost almost all the HeLa cell mtDNA, suggesting that the enzyme defect was corrected by a nDNA-encoded factor from the HeLa parent cell [81]. The expression of COX deficiency in cultured skin fibroblasts in most (but not all) patients suggests that prenatal diagnosis may be feasible in families with one affected child. In fact, prenatal diagnosis has been successfully performed in four families using chorionic villi or cultured amniocytes; in all four cases, a first child had died of Leigh syndrome with COX deficiency, which was expressed in fibroblasts [82,83].

To verify the hypothesis that the genetic error in Leigh syndrome may affect one of the eight non-tissue-specific nuclear COX subunits, we have performed Northern blot analysis of RNA isolated from brain, muscle and cultured fibroblasts of patients: mRNAs for COX subunits I, II, IV, Va, Vb, VIc and VIII were normal in size and amount [65]. These findings, together with the normal subunit patterns obtained by immunoblot, suggest that COX deficiency in Leigh syndrome may not be due to a defect in the enzyme *per se*, but rather to the mutation of a nuclear regulatory gene controlling the assembly or stability of the enzyme [65,77]. In yeast, as many as 14 different nuclear genes are needed for proper COX assembly [84]; analogous genes probably exist in humans and they are attractive candidates for the genetic lesion or lesions causing Leigh syndrome.

OTHER ENCEPHALOPATHIES

COX deficiency was shown in muscle biopsies from two unrelated patients with Alpers disease (progressive sclerosing poliodystrophy), a neurodegenerative disorder of infancy dominated by intractable epilepsy and associated with liver disease [85]. The description of decreased activity of COX in platelets from five of six patients with Alzheimer's disease [86] is intriguing, but this study will have to be confirmed and extended to the brain.

Partial COX deficiency in muscle was documented in five of eight patients with mitochondrial neurogastrointestinal encephalomyopathy (MNGIE; in the first report by Bardosi *et al.* [87] the acronym stood for myoneurogastrointestinal encephalomyopathy), an autosomal recessive multisystem disorder characterized by peripheral neuropathy, ophthalmoplegia, gastrointestinal dysmotility with intestinal pseudo-obstruction, mitochondrial myopathy and, often, MRI evidence of leucodystrophy [87]. The syndrome is genetically heterogeneous, with Southern blot evidence of multiple mtDNA deletions in four of the eight patients. The pathogenesis of COX deficiency, which is also likely to be heterogeneous, remains to be clarified in patients without mtDNA deletions [88].

COX deficiencies associated with defects of mtDNA

In the four years since the description of the first pathogenetic mutation in human mtDNA [89], progress in this new area of human pathology has been astounding. Several comprehensive reviews of mitochondrial genetic diseases have appeared [90–92] and some are included in this volume. We will focus on COX deficiency, which has served as a very useful 'marker' of pathology in muscle biopsies from patients with mtDNA defects [93]. Conversely, studies of the correlations between mtDNA mutations and biochemical defects have shed light on the pathogenesis of partial COX deficiencies.

DELETIONS OF mtDNA

Even before the discovery of mtDNA deletions, a large literature had accumulated describing focal defects of COX by histochemistry in muscle biopsies from patients with Kearns–Sayre syndrome (KSS) or isolated progressive external ophthalmoplegia (PEO) [94–103]. Most of these patients probably had large-scale deletions, as several retrospective studies have later documented. The common association of focal COX deficiency with ophthalmoplegia and the rare occurrence of COX-negative fibres in a wide variety of non-mitochondrial neuromuscular diseases [103] suggested that these were not non-specific secondary changes. In agreement with this concept, the enzyme defect was expressed almost invariably in RRFs, but non-RRFs could also be COX-negative, suggesting that COX deficiency preceded mitochondrial proliferation and was closely related to the aetiology of these disorders. Histochemistry of longitudinal sections showed that the COX-negativity was segmental and did not extend the entire length of an affected fibre. Biochemical studies in whole muscle extracts or mitochondrial fractions showed either normal COX activity or extremely variable reduction in different patients; the activities of NADH–cytochrome c reductase (Complex I and III) and succinate–cytochrome c reductase (Complex II and III) were also decreased, although generally less markedly and in fewer patients [94,96,97,99–101,104–107]. A comparison of two muscle biopsies taken 7 years apart in a patient with KSS showed a marked increase in the number of COX-negative fibres and worsening of the enzyme defect (from 70% of normal in the first biopsy to 29% in the second), a clear indication that COX deficiency (or more general respiratory chain impairment) is progressive in these disorders [99]. Immunocytochemical studies using antibodies against individual COX subunits showed a consistent decrease in reactivity for COX II/III in COX-negative fibres, while staining with antibodies against nDNA-encoded COX subunits was variably decreased and showed an inconsistent pattern in different fibres [108].

Since the first description by Holt *et al.* in 1988 [89], large-scale deletions of mtDNA have been associated with three main clinical syndromes: KSS [109], sporadic PEO with RRFs [110], and Pearson's marrow-pancreas syndrome [111,112]. Both retrospective and prospective studies in patients with these syndromes confirmed that the histochemical hallmark of mtDNA deletions in muscle is the chequerboard distribution of COX-negative fibres [66]. Biochemical studies also confirmed the variably severe defect of COX activity, which was usually associated with defects in other respiratory chain complexes, but no consistent correlation was found between site or size of the deletion and biochemical data [110,113–119]. Moraes *et al.* [110] noted that only enzymes containing mtDNA-encoded subunits had decreased activities, whereas enzymes entirely encoded by nDNA were normal, and suggested that any deletion in the mtDNA could impair mitochondrial protein synthesis, probably because even the smallest deletion comprises one or more tRNA genes. This concept was bolstered by the demonstration that translation rather than transcription of deleted genomes was impaired in cultured muscle and fibroblasts from patients with KSS [120]. However, a few investigators suggested that there was a correlation between location of the deletions and biochemical defects, at least in some patients [113,114,119,121]. A direct relationship between proportion of deleted genomes and severity of COX deficiency was suggested by three studies [122–124], one of which correlated the percentage of deleted mtDNAs in single teased fibres with the sum of the lengths of COX-negative segments [123].

The combination of histochemistry, *in situ* hybridization and immunocytochemistry has allowed the study of the relationship between deletions and focal COX deficiency at the single-fibre level. Mita *et al.* [125] first documented that COX-deficient fibres had increased numbers of deleted mtDNAs and mRNAs and decreased numbers of wild-type genomes. They also found that, in COX-negative segments, immunoreactivity to COX II was lost whereas immunostaining for COX IV was preserved. As the deletion did not encompass the COX II gene, this seemed to reinforce the hypothesis that deletions cause a generalized impairment of mitochondrial protein synthesis [125]. A certain degree of controversy was generated by subsequent studies employing similar techniques [126–130]. There is general agreement that deleted mtDNAs accumulate in COX-negative segments of muscle fibres, but the proportion of wild-type genomes in the same areas varied in different studies. Shoubridge *et al.* [126] found normal amounts of wild-type mitochondrial mtRNAs and suggested that the deleted genomes somehow interfere with the expression of the normal ones. In another study, the proportion of deleted mtDNAs in COX-negative fibres was normal, decreased or even increased [127], while it was decreased in all other reports [128–130]. These discrepancies may be due to methodological differences or they may reflect genetic heterogeneity among patients. At variance with the immunocytochemical data of Mita *et al.* [125], Collins *et al.* [131] and Oldfors *et al.* [129] found that not only mtDNA-encoded subunits but also nDNA-encoded subunits of COX were lacking in COX-negative areas. An interesting observation was that deletions encompassing COX genes caused more severe COX deficiency within RRFs [128] and, conversely, the ratio between COX-negative fibres and RRFs was lowest in patients with deletions not affecting COX genes [121,129,130].

After combining data from the literature with the results of quantitative analysis in single fibres from seven patients with mtDNA deletions, Moraes *et al.* [130] have arrived at the following general conclusions: (1) RRFs are typically COX-negative, contain high levels of deleted mtDNAs and low levels of wild-type genomes, and lack mtDNA-encoded polypeptides; (2) wild-type and deleted mtDNAs are codominant in determining COX activity in a muscle segment, and the expression of COX deficiency depends on the ratio of deleted and wild-type genomes; (3) the threshold ratio causing segmental COX deficiency is related to the location of the deletion, as reflected by the lower COX-negative/RRF ratios in patients whose deletions do not encompass COX genes; (4) the presence of COX-positive RRFs is probably due to a decrease in the ratio of deleted/wild-type mtDNAs, allowing normal mtDNAs to provide missing tRNAs for the mutant genomes (intramitochondrial complementation); (5) intramitochondrial complementation for COX activity may well be impeded when COX genes are deleted, thus explaining the lower COX-negative/RRF ratios in patients with these types of deletions; (6) transcription of mutant mtDNA is not affected in RRFs lacking mtDNA-encoded polypeptides, suggesting that mitochondrial translation is the cause of oxidative phosphorylation defects associated with mtDNA deletions.

Some of these concepts have been verified in tissue culture systems. When various amounts of deleted mtDNAs from a patient were introduced into mtDNA-less HeLa cells, it was found that cybrid clones containing more than 60% deleted genomes showed decreased mitochondrial translation, decreased COX activity and reduced cell growth [132]. Similar findings were obtained in 'triplasmic' hybrids derived from the fusion of HeLacot cells and a heteroplasmic KSS fibroblast clone, although the threshold of deleted mtDNAs needed to cause COX deficiency and inhibition of mitochondrial translation seemed to be higher in this system, above 79% [133].

POINT MUTATIONS OF mtDNA

Typically, neither RRFs nor COX-negative fibres are seen in muscle biopsies from patients with point mutations in structural genes, such as those associated with Leber's hereditary optic neuropathy (LHON) [134] or the mutation in the ATPase 6 gene associated with the neurogenic atrophy, ataxia, retinitis pigmentosa (NARP) syndrome [135]. Likewise, no defect in COX activity was found in biochemical studies of muscle from patients with LHON [136]. Combined defects in Complexes I and III, but not Complex IV, were described in muscle biopsies from patients with the ATPase 6 mutation and Leigh syndrome by Shoffner *et al.* [137], and 'mildly decreased COX activity' was reported in another patient [138]; we found normal activities of Complexes I to IV in seven patients [118].

Point mutations in tRNA genes cause syndromes that are usually associated with RRFs and, less consistently, with COX-negative fibres in muscle biopsies. These similarities between tRNA point mutations and deletions are attributed to a common pathogenetic mechanism involving impairment of mtDNA translation.

In the myoclonic epilepsy with RRF (MERRF) syndrome, usually due to a point mutation at nt 8344 of the tRNA^{Lys} gene [139], COX deficiency has been reported in muscle biopsies from many patients, either alone [140,141] or, more often, combined with defects of other respiratory chain complexes [142–146]. The enzyme defect worsened with time in repeat biopsies [145]. In one post-mortem study [141], COX deficiency was generalized, partial and variably expressed in different tissues. Kinetic analysis showed decreased K_m for reduced cytochrome *c* in brain and muscle. Immunotitration suggested a moderate decrease in CRM in all tissues tested, and immunoblot with antibodies to subunits II and IV indicated a selective defect in COX II [141]. Histochemistry and immunocytochemistry showed that RRFs were COX-deficient; in these fibres, immunoreactivity to COX II was much less intense than immunoreactivity to COX IV, whereas in COX-positive fibres immunoreactivity to the two subunits was similar [141]. Northern blot analysis showed that the mRNAs for COX subunits II, III, IV and VIII were of normal size and normally abundant.

Tissue culture studies documented a pathogenetic relationship between the mutation and COX deficiency. Cybrid clones derived from the fusion of myoblasts from a MERRF patient and mtDNA-less cell lines showed marked impairment of mitochondrial protein synthesis, respiration and COX activity [147]. Analysis of myoblasts cloned from muscle biopsies of four MERRF patients showed severe depression of overall mitochondrial translation in homoplasmic clones [148]. The threshold for biochemical expression in heteroplasmic clones with different proportions of mutant mtDNA appeared to be approximately 85%; lower percentages of mutant genomes were compatible with normal protein synthesis and COX activity [148].

In patients with mitochondrial encephalomyopathy, lactic acidosis and stroke-like episodes (MELAS), usually associated with a point mutation at nucleotide 3243 of the mitochondrial tRNA$^{Leu(UUR)}$ gene [149], muscle biochemistry had shown predominantly Complex I deficiency [150], but recent studies have described multiple respiratory chain defects, including COX deficiency [150–152]. By histochemistry, however, RRFs were rarely devoid of COX activity and some even showed increased staining [128]; most RRFs and some non-RRFs showed partial COX deficiency [128,153]. Likewise, immunocytochemistry showed that COX-deficient fibres had decreased reactivity to mtDNA-encoded polypeptides but not to nDNA-encoded ones [153]. However, the histochemistry was different in a subset of patients with the same mutation but with a distinct clinical presentation, dominated by PEO [154]. In these patients, muscle biopsies showed numerous COX-negative fibres, similar to what is found in patients with PEO and deletions of mtDNA. Single-fibre polymerase chain reaction (PCR) followed by restriction fragment length polymorphism (RFLP) analysis showed that COX-deficient RRFs have higher proportions of mutant genomes than non-RRFs in patients with typical MELAS [153], but the proportions of mutant mtDNAs are highest in the COX-negative RRFs of patients with PEO (V. Petruzzella and E.A. Schon, unpublished work). The cumulative evidence derived from morphological studies [128,153] and analysis of cybrid cell lines [155,156] suggests that in MELAS, as in MERRF and mtDNA deletions, there is severe impairment of mtDNA translation above a certain threshold proportion of mutant genomes.

The same pathogenetic mechanism probably operates in patients with other tRNA mutations, although detailed studies have not been conducted. For example, both RRFs and COX-negative fibres were seen in muscle biopsies from patients with a maternally inherited form of cardiopathy and myopathy associated with a mutation at nucleotide 3260 of the tRNA$^{Leu(UUR)}$ gene [46].

COX deficiencies due to defects in intergenomic communication

In these disorders, it is assumed that primary defects of nDNA, transmitted by mendelian inheritance, affect mtDNA directly, causing qualitative or quantitative mtDNA alterations. COX deficiency is a common manifestation and its severity depends on the nature and degree of the mtDNA defect.

Multiple deletions of mtDNA characterize a group of patients with autosomal dominant PEO, weakness of limb and respiratory muscles, cataracts, hearing loss and early death [157–159]. Depression and mental retardation accompanied PEO in a Finnish family [160]. The distribution of the deletions in different tissues probably varies from family to family; only muscle seemed to be affected in the Italian family originally described by Zeviani *et al.* [157] and Servidei *et al.* [159], while multiple deletions were documented in all tissues in the Finnish patient [160]. Muscle biopsies show variable amounts of RRFs and ultrastructurally abnormal mitochondria [159,160]. Histochemically, most RRFs and some non-RRFs were COX-negative [159], the same pattern seen in patients with PEO and single mtDNA deletions. The only difference was that in patients with multiple deletions, an unusual number of fibres or fibre segments (which could be COX-positive or COX-negative) were non-reactive for the succinate dehydrogenase (SDH) reaction [159]. Because SDH is entirely encoded by nDNA, focal SDH deficiency cannot be explained by mtDNA deletions and has been attributed to some, as yet undefined, nuclear influence [159]. Biochemical studies of muscle extracts have documented a partial defect of COX and less marked deficiencies of other respiratory chain complexes [159,160]. COX deficiency was confirmed in isolated muscle mitochondria where, by immunotitration, there was also an apparent decrease in CRM [159]. The degree of the enzyme defect increased with age and clinical severity and was roughly proportional to the amount of deleted genome [159]. Despite the clinical and molecular evidence of multisystem involvement in the Finnish patient, the activities of COX and other complexes were not decreased in non-muscle tissues [160].

Depletion of mtDNA was identified by Moraes *et al.* [161] as the molecular basis for the severe COX deficiency described in muscle of two siblings with fatal infantile myopathy and in liver of a second cousin with fatal hepatopathy [162,163]. A second child with liver disease has been described [88], but myopathy appears to be a more common manifestation of mtDNA depletion [161,164–166]. Two forms of myopathy can be distinguished, differing in severity and associated with severe or partial mtDNA depletion.

Severe mtDNA depletion causes diffuse myopathy manifesting in the first weeks of life and leading to respiratory insufficiency and death before 1 year of age [142,161,165,166]. There is severe lactic acidosis, and muscle biopsy shows numerous RRFs. A few patients also have renal dysfunction with DeToni–Fanconi–Debre syndrome, making the differential diagnosis from the fatal infantile myopathy with COX deficiency (see above) problematic on purely clinical grounds. The few informative family histories suggest autosomal recessive inheritance.

In the patients with partial mtDNA depletion, onset of myopathy is later, between 5 and 12 months, and the course is slower, but two children died at 11 months and 3 years, and one was tetraplegic and ventilator-dependent at 3½ years [164]. Blood lactates were normal. Muscle biopsies showed RRFs, but one biopsy obtained early in the course of the disease showed only non-specific changes. Three of four children in one family were affected and the parents were asymptomatic and had normal biopsies, suggesting autosomal recessive inheritance.

Biochemical studies showed combined defects of respiratory chain complexes containing mtDNA-encoded subunits more marked in patients with severe depletion than in those with partial mtDNA depletion. In patients with severe depletion, there was an excellent correlation between degree of depletion and COX deficiency [161]. The decrease in mtDNA was documented by densitometry of Southern blots (using 18S ribosomal DNA as a normalizing control) and confirmed by immunocytochemistry with anti-DNA antibodies and by *in situ* hybridization

[161,164,165]. The depletion of mtDNA was virtually complete (83–98%) in the severely affected infants and varied between 66% and 92% in the less severely affected children. As predicated by the scarcity of mtDNA, immunocytochemistry showed lack of reactivity with antibodies against mtDNA-encoded subunits, such as COX II and ND1, and normal immunostain with antibodies against nDNA-encoded subunits, such as COX IV [161,164]. The difference between biopsies from patients with severe and partial mtDNA depletion was that the immuno-cytochemical reaction was absent from all fibres in the severe form but only in a subpopulation of fibres in the partial form [164].

A secondary form of mtDNA depletion is probably the cause of the mitochondrial myopathy seen in patients with acquired immunodeficiency syndrome (AIDS) after prolonged therapy with zidovudine (AZT) [167,168]. This conclusion was suggested by Southern blots showing up to 78% depletion of muscle mtDNA in treated but not untreated AIDS patients and by the finding that the depletion was reversible upon discontinuation of AZT [169]. Biochemical studies showed decreased activities of mitochondrial enzymes containing mtDNA-encoded subunits, especially COX [168]. Histochemical analysis showed that all RRFs and many non-RRFs were COX-deficient, and that COX-deficient fibres also contained areas of myofibrillar loss and numerous cytoplasmic bodies [170]. The defect in COX accompanying partial mtDNA depletion has been confirmed in rats treated with AZT [171], while mtDNA concentration did not change significantly and COX activity decreased only slightly in human muscle cultures exposed to the same drug [172].

Other COX deficiencies

The first report of COX deficiency was in *trichopoliodystrophy* (Menkes disease), an X-linked recessive disorder of copper metabolism characterized by infantile seizures, developmental regression, hair abnormalities, tortuous arteries, fragile bones, hypopigmentation and temperature instability [173]. These diverse manifestations have been attributed to secondary deficiencies of copper-dependent enzymes, including COX. Just as copper concentration varies from tissue to tissue [174], so probably does COX activity, and, even in the same tissue, such as muscle, COX deficiency may be expressed by some patients [175] but not others: we found normal COX activity in muscle specimens from two patients with Menkes disease.

COX activity was selectively decreased in the cerebellum of an animal model of Menkes disease, the macular mottled mouse, and this was confirmed by both immunocytochemistry and immuno-electron microscopy [176]. Selective COX deficiency and selective decrease in cytochrome aa_3 was also induced in muscle and heart of rats fed a copper-deficient diet [177].

A partial, but apparently significant and selective, defect in brain COX was induced in rats by infusion of sodium azide; because of the behavioural deficits that accompanied the treatment, this was considered an animal model of Alzheimer's disease [178].

Inhibition of COX by the antimitotic drug adriamycin was documented in isolated bovine heart mitochondria and attributed to drug-induced clustering of cardiolipin molecules in the inner membrane [179]. Although this mechanism could explain the cardiomyopathy that develops in children during treatment with adriamycin, we failed to document COX deficiency in the hearts removed at

Table 5.3 COX deficiencies (enzyme histochemistry)

Phenotype	Muscle fibres	Arterial walls	Muscle spindles
Fatal myopathy	Absent	Normal	Normal
Benign myopathy (initial stage)	Absent	Normal	Normal
Leigh syndrome	Decreased	Decreased	Decreased
KSS	Mosaic	Normal	Mosaic
MERRF	Mosaic	Normal/decreased	ND
MELAS	Decreased	Increased	ND
mtDNA depletion (severe)	Absent	Normal	Normal
mtDNA depletion (partial)	Mosaic	Normal	Normal

Mosaic refers to a mixture of COX-negative and COX-positive fibres. ND, not determined.

Table 5.4 COX deficiencies (immunohistochemistry)

Phenotype	Complex IV		Complex I NDI	mtDNA anti-DNA antibodies
	II	VIIa,b		
Fatal myopathy	+	−	+	+
Benign myopathy (early stage)	−	−	+	+
Leigh syndrome	±	±	+	+
KSS	−*	±*	−*	+
MERRF	−*	±*	−*	+
MELAS	+	+	+	+
mtDNA depletion (severe)	−	±	−	−
mtDNA depletion	−*	±*	−*	−*

The asterisk signifies that the change is seen only in COX-deficient fibres.

cardiac transplantation from three children with cardiomyopathy (S. DiMauro, unpublished work).

Both COX activity and cytochrome aa_3 concentration were markedly decreased in the liver of baboons after chronic ethanol consumption and the defects were associated with an altered phospholipid composition of the mitochondrial inner membrane [180]. However, studies of oxidative metabolism in muscle mitochondria from patients with chronic alcoholism did not reveal any abnormalities, and both COX activity and cytochrome aa_3 concentration were normal [181].

CONCLUSIONS

Despite the remarkable progress in our understanding of the structural and functional characteristics of human COX, and the availability of cDNAs for every subunit of the enzyme, we know relatively little about the aetiology and pathogenesis of human COX deficiencies. Recent advances in mtDNA genetics have largely clarified the pathogenesis of the partial and non-specific COX deficiency observed in patients with mtDNA deletions, point mutations in tRNA genes and mtDNA depletion. However, we still do not understand the molecular basis for the apparently specific COX deficiencies associated with tissue-specific or generalized syndromes such as the fatal infantile or benign myopathies and Leigh syndrome. We have failed to identify genetic defects in tissue-specific COX subunits in the myopathies or defects of the ubiquitous COX subunits in Leigh syndrome. These negative studies suggest that most COX deficiencies may be due to genetic errors affecting the control of COX biosynthesis and assembly rather than individual COX subunits. Thus efforts to clone the human equivalents of the 'control' genes identified in yeast will have not only scientific but also clinical value. In the meantime, however, the wealth of morphological data on muscle biopsies provides useful diagnostic clues for the clinician faced with the diagnosis of 'COX deficiency'. These data have been discussed in the individual sections and are summarized in Table 5.3 and 5.4

ACKNOWLEDGEMENTS

Part of the work described here was supported by Center Grant NS 11766 from the National Institute for Neurological Diseases and Stroke, by a grant from the Muscular Dystrophy Association, and by a generous donation from Libero and Graziella Danesi (Milan, Italy). M.H. was supported by a postdoctoral research fellowship from the Muscular Dystrophy Association and C.T.M. by the Brazilian Research Council (CNPq).

REFERENCES

1. DiMauro, S., Hays, A.P. and Eastwood, A.B. (1983) Different clinical expressions of cytochrome c oxidase deficiency. In *Mitochondrial Pathology in Muscle Diseases* (eds G. Scarlato and C. Cerri), Piccin, Padova pp. 111–129
2. DiMauro, S., Zeviani, M., Bonilla, E. *et al.* (1985) Cytochrome c oxidase deficiency. *Biochemical Society Transactions*, **13**, 651–653
3. DiMauro, S., Zeviani, M., Servidei, S. *et al.* (1986) Cytochrome oxidase deficiency: clinical and biochemical heterogeneity. *Annals of the New York Academy of Sciences*, **488**, 19–32
4. DiMauro, S., Zeviani, M., Servidei, S. *et al.* (1988) Biochemical and molecular aspects of cytochrome c oxidase deficiency. In *Advances in Neurology: Molecular Genetics of Neurological and Neuromuscular Disease*, (eds S. DiDonato, S. DiMauro, A. Mamoli and L.P. Rowland), Raven Press, New York, pp. 93–105
5. DiMauro, S., Zeviani, M., Rizzuto, R. *et al.* (1988) Molecular defects in cytochrome oxidase in mitochondrial diseases. *Journal of Bioenergetics and Biomembranes*, **20**, 353–364
6. Schon, E.A., Bonilla, E., Lombes, A. *et al.* (1988) Clinical and biochemical studies on cytochrome oxidase deficiencies. *Annals of the New York Academy of Sciences*, **550**, 348–359
7. DiMauro, S., Lombes, A., Nakase, H. *et al.* (1990) Cytochrome c oxidase deficiency. *Pediatric Research*, **28**, 536–541
8. DiMauro, S., Moraes, C.T. and Schon, E.A. (1991) Mitochondrial encephalomyopathies: problems of classification. In *Progress in Neuropathology* (eds T. Sato and S. DiMauro), Raven Press, New York, pp. 113–127
9. Capaldi, R.A. (1990) Structure and function of cytochrome c oxidase. *Annual Reviews of Biochemistry*, **59**, 569–596
10. Cooper, C.E., Nicholls, P. and Freedman, J.A. (1991) Cytochrome c oxidase: structure, function, and membrane topology of the polypeptide subunits. *Biochemistry and Cell Biology*, **69**, 586–607

11. Anderson, S., Bankier, A.T., Barrel, B.G. *et al.* (1981) Sequence and organization of the human mitochondrial genome. *Nature*, **290**, 457–465
12. Kadenbach, B., Kuhn-Nentwig, L. and Buge, U. (1987) Evolution of a regulatory enzyme: cytochrome c oxidase (complex IV). *Current Topics in Bioenergetics*, **15**, 113–161
13. Schatz, G. (1991) The mitochondrial protein import machinery. In *Mitochondrial Encephalomyopathies* (eds T. Sato and S. DiMauro), Raven Press, New York
14. Fabrizi, G.M., Rizzuto, R., Nakase, H., Mita, S., Kadenbach, B. and Schon, E.A. (1989) Sequence of a cDNA specifying subunit VIa of human cytochrome c oxidase. *Nucleic Acids Research*, **17**, 6409
15. Fabrizi, G.M., Sadlock, J., Hirano, M. *et al.* (1992) Differential expression of genes specifying two isoforms of subunit VIa of human cytochrome c oxidase. *Gene*, **119**, 307–312
16. Fabrizi, G.M., Rizzuto, R., Nakase, H. *et al.* (1989) Sequence of a cDNA specifying subunit VIIa of human cytochrome c oxidase. *Nucleic Acids Research*, **17**, 7107
17. Arnaudo, E., Hirano, M., Seelan, R.S. *et al.* (1992) Tissue-specific expression and chromosome assignment of genes specifying two isoforms of subunit VIIa of human cytochrome c oxidase. *Gene*, **119**, 299–305
18. Taanman, J.W., Herzberg, N.H., De Vries, H., Bolhuis, P.A. and Van den Bogert, C. (1992) Steady-state transcript levels of cytochrome c oxidase genes during human myogenesis indicate subunit switching of subunit VIa and co-expression of subunit VIIa isoforms. *Biochimica et Biophysica Acta*, **1139**, 155–162
19. Zeviani, M., Nakagawa, M., Herbert, J. *et al.* (1987) Isolation of a cDNA clone encoding subunit IV of human cytochrome c oxidase. *Gene*, **55**, 205–217
20. Rizzuto, R., Nakase, H., Zeviani, M., DiMauro, S. and Schon, E.A. (1988) Subunit Va of human and bovine cytochrome c oxidase is highly conserved. *Gene*, **69**, 245–256
21. Zeviani, M., Sakoda, S., Sherbany, A.A. *et al.* (1988) Sequence of cDNAs encoding subunit Vb of human and bovine cytochrome c oxidase. *Gene*, **65**, 1–11
22. Taanman, J.W., Schrage, C., Ponne, N., Bolhuis, P., deVries, H. and Agsteribbe, E. (1989) Nucleotide sequence of cDNA encoding subunit VIb of human cytochrome c oxidase. *Nucleic Acids Research*, **17**, 1766
23. Sadlock, J.E., Lightowlers, R.N., Capaldi, R.A. and Schon, E.A. (1993) Isolation of a cDNA specifying subunit VIIb of human cytochrome c oxidase. *Biochimica et Biophysica Acta*, **1172**, 223–225
24. Rizzuto, R., Nakase, H., Darras, B. *et al.* (1989) A gene specifying subunit VIII of human cytochrome c oxidase is localized to chromosome 11 and is expressed in both muscle and non-muscle tissues. *Journal of Biological Chemistry*, **264**, 10595–10600
25. Carrero-Valenzuela, R.D., Quan, F., Lightowlers, R., Kennaway, N.G., Litt, M. and Forte, M. (1991) Human cytochrome c oxidase subunit VIb: characterization and mapping of a multigene family. *Gene*, **102**, 229–236
26. Van Biervliet, J.P.A.M., Bruinvis, L., Ketting, D. *et al.* (1977) Hereditary mitochondrial myopathy with lactic acidemia, a DeToni–Fanconi–Debre syndrome, and a defective respiratory chain in voluntary striated muscles. *Pediatric Research*, **11**, 1088–1093
27. DiMauro, S., Mendell, J.R., Sahenk, Z. *et al.* (1980) Fatal infantile mitochondrial myopathy and renal dysfunction due to cytochrome-c-oxidase deficiency. *Neurology*, **30**, 795–804
28. Heiman Patterson, T.D., Bonilla, E., DiMauro, S., Foreman, J. and Schotland, D.L. (1982) Cytochrome-c-oxidase deficiency in a floppy infant. *Neurology*, **32**, 898–901
29. Minchom, P.E., Dormer, R.L., Hughes, I.A. *et al.* (1983) Fatal infantile mitochondrial myopathy due to cytochrome c oxidase deficiency. *Journal of Neurological Sciences*, **60**, 453–463
30. Muller-Hocker, J., Pongratz, D., Deufel, Th., Trijbels, J.M.F., Endres, W. and Hubner, G. (1983) Fatal lipid storage myopathy with deficiency of cytochrome c oxidase and carnitine. *Virchows Archives A*, **399**, 11–23
31. Bresolin, N., Zeviani, M., Bonilla, E. *et al.* (1985) Fatal infantile cytochrome c oxidase deficiency: decrease of immunologically detectable enzyme in muscle. *Neurology*, **35**, 802–812
32. Zeviani, M., Nonaka, I., Bonilla, E. *et al.* (1985) Fatal infantile mitochondrial myopathy and renal dysfunction caused by cytochrome c oxidase deficiency: immunological studies in a new patient. *Annals of Neurology*, **17**, 414–417
33. Nonaka, I., Koga, Y., Shikura, K. *et al.* (1988) Muscle pathology in cytochrome c oxidase deficiency. *Acta Neuropathologica*, **77**, 152–160
34. Tritschler, H.J., Bonilla, E., Lombes, A. *et al.* (1991) Differential diagnosis of fatal and benign cytochrome c oxidase-deficient myopathies of infancy: an immunohistochemical approach. *Neurology*, **41**, 300–305
35. Letellier, T., Malgat, M., Coquet, M., Moretto, B., Parrot-Roulaud, F. and Mazat, J.P. (1992) Mitochondrial myopathy studies on permeabilized muscle fibers. *Pediatric Research*, **32**, 17–22

36. Nonaka, I., Koga, Y., Ohtaki, E. and Yamamoto, M. (1989) Tissue specificity in cytochrome c oxidase deficient myopathy. *Journal of Neurological Sciences*, **92**, 193–203
37. Nonaka, I., Koga, Y., Okino, E., Kikuchi, A., Fujisawa, K. and Miyabayashi, S. (1988) Defects in muscle fiber growth in fatal infantile cytochrome c oxidase deficiency. *Brain Development*, **10**, 223–230
38. Romero, N., Marsac, C., Fardeau, M., Droste, M., Schneyder, B. and Kadenbach, B. (1990) Immunohistochemical demonstration of fibre type-specific isozymes of cytochrome c oxidase in human skeletal muscle. *Histochemistry*, **94**, 211–215
39. Muller-Hocker, J., Droste, M., Kadenbach, B., Pongratz, D. and Hubner, G. (1989) Fatal mitochondrial myopathy with cytochrome-c-oxidase deficiency and subunit-restricted reduction on enzyme protein in two siblings. *Human Pathology*, **20**, 666–672
40. Rimoldi, M., Bottacchi, E., Rossi, L., Cornelio, F., Uziel, G. and DiDonato, S. (1982) Cytochrome-c-oxidase deficiency in muscles of a floppy infant without mitochondrial myopathy. *Journal of Neurology*, **227**, 201–207
41. Sengers, R.C., Trijbels, J.M., Bakkeren, J.A. *et al.* (1984) Deficiency of cytochromes b and aa₃ in muscle from a floppy infant with cytochrome oxidase deficiency. *European Journal of Pediatrics*, **141**, 178–180
42. Zeviani, M., Van Dyke, D.H., Servidei, S. *et al.* (1986) Myopathy and fatal cardiopathy due to cytochrome c oxidase deficiency. *Archives of Neurology*, **43**, 1198–1202
43. Muller-Hocker, J., Ibel, H., Paetzke, I. *et al.* (1991) Fatal infantile mitochondrial cardiomyopathy and myopathy with heterogeneous tissue expression of combined respiratory chain deficiencies. *Virchows Archives A, Pathology and Anatomy*, **419**, 355–362
44. Munnich, A., Rustin, P., Rotig, A. *et al.* (1992) Clinical aspects of mitochondrial disorders. *Journal of Inherited Metabolic Diseases*, **15**, 448–455
45. Tanaka, M., Ino, H., Ohno, K. *et al.* (1990) Mitochondrial mutation in fatal infantile cardiomyopathy. *Lancet*, **2**, 1452
46. Zeviani, M., Gellera, C., Antozzi, C. *et al.* (1991) Maternally inherited myopathy and cardiomyopathy: association with mutation in mitochondrial DNA tRNA(Leu)(UUR). *Lancet*, **338**, 143–147
47. Silvestri, G., Shanske, S.B., Whitley, C.B.A., Schimmenti, L.A., Smith, S.A. and DiMauro, S. (1993) A new mtDNA mutation in the tRNA(LeuUUR) gene associated with cardiomyopathy and ragged-red fibers. *Human Mutations*, in press
48. DiMauro, S., Nicholson, J.F., Hays, A.P., Eastwood, A.B., Koenigsberger, R. and DiVivo, D.C. (1981) Benign infantile mitochondrial myopathy due to reversible cytochrome c oxidase deficieny. *Transactions of the American Neurological Association*, **106**, 205–207
49. Zeviani, M., Peterson, P., Servidei, S., Bonilla, E. and DiMauro, S. (1987) Benign reversible muscle cytochrome c oxidase deficiency: a second case. *Neurology*, **37**, 64–67
50. Servidei, S., Bertini, E., Dionisi-Vici, C. *et al.* (1988) Benign infantile mitochondrial myopathy due to reversible cytochrome c oxidase deficiency: a third case. *Clinical Neuropathology*, **7**, 209–210
51. Salo, M.K., Rapola, J., Somer, H. *et al.* (1992) Reversible mitochondrial myopathy with cytochrome c oxidase deficiency. *Archives of Diseases in Children*, **67**, 1033–1035
52. Jerusalem, F., Angelini, C., Engel, A.G. and Groover, R.V. (1973) Mitochondrial-lipid-glycogen (MLG) disease of muscle. *Archives of Neurology*, **29**, 162–169
53. Miranda, A.F., Shanske, S. and DiMauro, S. (1982) Developmentally regulated isozyme transitions in normal and diseased human muscle. In *Muscle Development: Molecular and Cellular Control* (eds M.L. Pearson and H.F. Epstein), Cold Spring Harbor Laboratory, Cold Spring Harbor, NY, pp. 515–525
54. Haller, R.G., Lewis, S.F., Estabrook, R.W., DiMauro, S., Servidei, S. and Foster, D.W. (1980) Exercise intolerance, lactic acidosis, and abnormal cardiopulmonary regulation in exercise associated with adult skeletal muscle cytochrome c oxidase deficiency. *Journal of Clinical Investigation*, **84**, 155–161
55. Cros, D., Palliyath, S., DiMauro, S., Ramirez, C., Shamsnia, M. and Wizer, B. (1992) Respiratory failure revealing mitochondrial myopathy in adults. *Chest*, **101**, 824–828
56. Servidei, S., Lazaro, R.P., Bonilla, E., Barron, K.D., Zeviani, M. and DiMauro, S. (1987) Mitochondrial encephalomyopathy and partial cytochrome c oxidase deficiency. *Neurology*, **37**, 58–63
57. Jinnai, K., Yamada, H., Kanda, F. *et al.* (1990) A case of mitochondrial myopathy, encephalopathy and lactic acidosis due to cytochrome c oxidase deficiency with neurogenic muscular changes. *European Neurology*, **30**, 56–60
58. Zheng, X., Shoffner, J.M., Lott, M.T. *et al.* (1989) Evidence in a lethal infantile mitochondrial disease for a nuclear mutation affecting respiratory complexes I and IV 'see comments'. *Neurology*, **39**, 1203–1209

59. Roodhooft, A.M., Van Acker, K.J., Martin, J.J., Ceuterick, C., Scholte, H.R. and Luyt Houwen, I.E. (1986) Benign mitochondrial myopathy with deficiency of NADH–CoQ reductase and cytochrome c oxidase. *Neuropediatrics*, **17**, 221–226
60. Takamiya, S., Yanamura, W., Capaldi, R.A. *et al.* (1986) Mitochondrial myopathies involving the respiratory chain: a biochemical analysis. *Annals of the New York Academy of Sciences*, **488**, 33–43
61. Robinson, B.H. (1989) Lactic acidemia. In *The Metabolic Basis of Inherited Disease* (eds C.R. Scriver, A.L. Beaudet, W.S. Sly and D. Valle), McGraw-Hill, New York, pp. 869–888
62. Van Erven, P.M., Gabreels, F.J., Ruitenbeek, W., Renier, W.O. and Fischer, J.C. (1987) Mitochondrial encephalomyopathy: association with an NADH dehydrogenase deficiency. *Archives of Neurology*, **44**, 775–778
63. Tatuch, Y., Christodoulou, J., Feigenbaum, A. et al. (1992) Heteroplasmic mtDNA mutation (T→G) at 8993 can cause Leigh disease when the percentage of abnormal mtDNA is high. *American Journal of Human Genetics*, **50**, 852–858
64. Hammans, S.R., Sweeney, M.G., Brockington, M., Morgan-Hughes, J.A. and Harding, A.E. (1991) Mitochondrial encephalopathies: molecular genetic diagnosis from blood samples. *Lancet*, **337**, 1311–1313
65. Lombes, A., Nakase, H., Tritschler, H.J. *et al.* (1991) Biochemical and molecular analysis of cytochrome c oxidase deficiency in Leigh's syndrome. *Neurology*, **41**, 491–498
66. Johnson, M.A., Bindoff, L.A. and Turnbull, D.M. (1993) Cytochrome c oxidase activity in single muscle fibers: assay techniques and diagnostic applications. *Annals of Neurology*, **33**, 28–35
67. Willems, J.L., Monnens, L.A.H., Trijbels, J.M.F. *et al.* (1977) Leigh's encephalomyelopathy in a patient with cytochrome c oxidase deficiency in muscle tissue. *Pediatrics*, **60**, 850–857
68. DiRocco, M., Veneselli, E., Ciccone, M.O., Taccone, A., Stroppiano, M. and Cottafava, F. (1988) Cytochrome c oxidase deficiency in three patients with Leigh's disease. *Journal of Inherited Metabolic Diseases*, **11**(Suppl.2), 189–192
69. Ogier, H., Lombes, A., Scholte, H.R. *et al.* (1988) deToni–Fanconi–Debre syndrome with Leigh syndrome revealing severe muscle cytochrome c oxidase deficiency. *Journal of Pediatrics*, **112**, 734–739
70. Jacobs, J.M., Harding, B.N., Lake, B.D., Payan, J. and Wilson, J. (1990) Peripheral neuropathy in Leigh's disease. *Brain*, **113**, 447–462
71. Van Coster, R., Lombes, A., DeVivo, D.C. *et al.* (1991) Cytochrome c oxidase-associated Leigh syndrome: phenotypic features and pathogenetic speculations. *Journal of the Neurological Sciences*, **104**, 97–111
72. Hrebicek, M., Zeman, J., Petrak, B. *et al.* (1992) Unusual clinical presentation in two boys with cytochrome c oxidase deficiency. *Journal of Inherited Metabolic Diseases*, **15**, 320–322
73. Reichmann, H., Scheel, H., Bier, B., Ketelsen, U.P. and Zabransky, S. (1992) Cytochrome c oxidase deficiency and long-chain acyl Coenzyme A dehydrogenase deficiency with Leigh's subacute necrotizing encephalomyelopathy. *Annals of Neurology*, **31**, 107–109
74. DiMauro, S., Servidei, S., Zeviani, M. *et al.* (1987) Cytochrome c oxidase deficiency in Leigh syndrome. *Annals of Neurology*, **22**, 498–506
75. Glerumd, M., Robinson, B.H., Spratt, C., Wilson, J. and Patrick, D. (1987) Abnormal kinetic behavior of cytochrome c oxidase in a case of Leigh disease. *American Journal of Human Genetics*, **41**, 584–593
76. Glerum, D.M., Yanamura, W., Capaldi, R.A. and Robinson, B.H. (1988) Characterization of cytochrome c oxidase mutants in human fibroblasts. *FEBS Letters*, **236**, 100–104
77. Glerum, D.M., Robinson, B.H. and Capaldi, R. (1989) Fibroblasts and cytochrome c oxidase deficiency. In *Molecular Basis of Membrane-Associated Diseases* (eds A. Azzi, Z. Drahota and S. Papa), Springer-Verlag, Berlin, pp. 228–238
78. Sinjorgo, K.M.C., Muijsers, A.O., Scholte, H.R. *et al.* (1989) Human cytochrome c oxidase deficiencies: structural and functional aspects. In *Molecular Basis of Membrane-Associated Diseases* (eds A. Azzi, Z. Drahota and S. Papa), Springer-Verlag, Berlin, pp. 239–253
79. Miyabayashi, S., Narisawa, K., Tada, K., Sakai, K., Kobayaski, K. and Kobayashi, Y. (1983) Two siblings with cytochrome c oxidase deficiency. *Journal of Inherited Metabolic Diseases*, **6**, 121–122
80. Arts, W.F.M., Scholte, H.R., Loonen, M.C.B. *et al.* (1987) Cytochrome c oxidase deficiency in subacute necrotizing encephalomyelopathy. *Journal of the Neurological Sciences*, **77**, 103–115
81. Miranda, A.F., Ishii, S., DiMauro, S. and Shay, J.W. (1989) Cytochrome c oxidase deficiency in Leigh's syndrome: genetic evidence for a nuclear DNA-encoded mutation. *Neurology*, **39**, 697–702
82. Ruitenbeek, W., Sengers, R., Albani, M. *et al.* (1988) Prenatal diagnosis of cytochrome c oxidase deficiency by biopsy of chorionic villi. *New England Journal of Medicine*, **319**, 1095
83. Ruitenbeek, W., Sengers, R.C.A., Trijbels, J.M.F., Ianssen, A.J.M. and Bakkeren, J.A.J.M. (1992) The use of chorionic villi in prenatal diagnosis of mitochondriopathies. *Journal of Inherited Metabolic Diseases*, **15**, 303–306

84. Kloeckener-Gruissem, B., McEwen, J.E. and Poyton, R.O. (1987) Nuclear functions required for cytochrome c oxidase biogenesis in *Saccharomyces cerevisiae*: multiple trans-acting nuclear genes exert specific effects on expression of each of the cytochrome c oxidase subunits encoded on mitochondrial DNA. *Current Genetics*, **12**, 311–322

85. Prick, M.J.J., Gabreels, F.J.M., Trijbels, J.M.F. *et al.* (1983) Progressive poliodystrophy (Alpers disease) with a defect in cytochrome aa_3 in muscle: a report of two unrelated patients. *Clinical Neurosurgery*, **85**, 57–70

86. Parker, W.D., Filley, C.M. and Parks, J.K. (1990) Cytochrome oxidase deficiency in Alzheimer's disease. *Neurology*, **40**, 1302–1303

87. Bardosi, A., Creutzfeldt, W., DiMauro, S. *et al.* (1987) Myo-, neuro-, gastrointestinal encephalopathy (MNGIE syndrome) due to partial deficiency of cytochrome-c-oxidase. A new mitochondrial multisystem disorder. *Acta Neuropathologica (Berlin)*, **74**, 248–258

88. Hirano, M., Silvestri, G., Blake, D.M. *et al.* (1994) Mitochondrial neuro-gastrointestinal encephalomyopathy (MNGIE): clinical, biochemical and genetic features of an autosomal recessive mitochondrial disorders. *Neurology*, in press

89. Holt, I.J., Harding, A.E. and Morgan-Hughes, J.A. (1988) Deletions of muscle mitochondrial DNA in patients with mitochondrial myopathies. *Nature*, **331**, 717–719

90. Wallace, D.C. (1992) Diseases of the mitochondrial DNA. *Annual Reviews of Biochemistry*, **61**, 1175–1212

91. Multiple Authors (1992) Mitochondrial encephalomyopathies. *Brain Pathology*, **2**, 153–205

92. DiMauro, S. (1993) Mitrochondrial encephalomyopathies. In *The Metabolic and Genetic Basis of Neurological Disease* (eds R.N. Rosenberg, S.B. Prusiner, S. DiMauro, R.L. Barchi and L.M. Kunkel), Butterworth-Heinemann, Boston, pp. 665–694

93. Bonilla, E., Sciacco, M., Tanji, K., Sparaco, M., Petruzzella, V. and Moraes, C.T. (1992) New morphological approaches to the study of mitochondrial encephalomyopathies. *Brain Pathology*, **2**, 113–119

94. Johnson, M.A., Turnbull, D.M., Dick, D.J. and Sherratt, H.S. (1983) A partial deficiency of cytochrome c oxidase in chronic progressive external ophthalmoplegia. *Journal of the Neurological Sciences*, **60**, 31–53

95. Muller-Hocker, J., Pongratz, D. and Hubner, G. (1983) Focal deficiency of cytochrome c oxidase in skeletal muscle of patients with progressive external ophthalmoplegia. *Virchows Archives [A]*, **402**, 61–71

96. Byrne, E., Dennet, X. and Trounce, I. (1985) Partial cytochrome c oxidase deficiency in chronic progressive external ophthalmoplegia. *Journal of the Neurological Sciences*, **71**, 257–271

97. Yorifuji, S., Ogasahara, S., Takahashi, M. and Tarui, S. (1985) Decreased activities in mitochondrial inner membrane electron transport system in muscle from patients with Kearns–Sayre syndrome. *Journal of the Neurological Sciences*, **71**, 65–75

98. Muller Hocker, J., Johannes, A., Droste, M., Kadenbach, B., Pongratz, D. and Hubner, G. (1986) Fatal mitochondrial cardiomyopathy in Kearns–Sayre syndrome with deficiency of cytochrome c oxidase in cardiac and skeletal muscle. An enzyme histochemical–ultra-immunocytochemical–fine structural study in longterm frozen autopsy tissue. *Virchows Archives [B]*, **52**, 353–367

99. Bresolin, N., Moggio, M., Bet, L. *et al.* (1987) Progressive cytochrome c oxidase deficiency in a case of Kearns–Sayre syndrome: morphological, immunological, and biochemical studies in muscle biopsies and autopsy tissues. *Annals of Neurology*, **21**, 564–572

100. Reichmann, H. (1988) Enzyme activity measured in single muscle fibers in partial cytochrome c oxidase deficiency. *Neurology*, **38**, 244–249

101. Yamamoto, M. and Nonaka, I. (1988) Skeletal muscle pathology in chronic progressive external ophthalmoplegia with ragged-red fibers. *Acta Neuropathologica*, **76**, 558–563

102. Desnuelle, C., Pellissier, J.F., Serratrice, G., Pouget, J. and Turnbull, D.M. (1989) Le syndrome de Kearns et Sayre: encephalopathie mitochondriale par deficit de la chaine respiratoire. *Reviews in Neurology*, **145**, 842–850

103. Yamamoto, M., Koga, Y., Ohtaki, E. and Nonaka, I. (1989) Focal cytochrome c oxidase deficiency in various neuromuscular diseases. *Journal of the Neurological Sciences*, **91**, 207–213

104. Turnbull, D.M., Johnson, M.A., Dick, D.J., Cartlidge, N.E.F. and Sherratt, H.S.A. (1985) Partial cytochrome oxidase deficiency without subsarcolemmal accumulation of mitochondria in chronic progressive external ophthalmoplegia. *Journal of the Neurological Sciences*, **70**, 93–100

105. Sherratt, H.S.A., Johnson, M.A. and Turnbull, D.M. (1986) Defects of complex I and complex IV in skeletal muscle from patients with chronic progressive external ophthalmoplegia. *Annals of the New York Academy of Sciences*, **488**, 508–510

106. Martens, M.E., Peterson, P.L., Lee, C.P. *et al.* (1988) Kearns–Sayre syndrome: biochemical studies of mitochondrial metabolism. *Annals of Neurology*, **24**, 630–637

107. Nishikawa, Y., Takahashi, M., Yorifuji, S. *et al.* (1989) Long-term coenzyme Q10 therapy for a mitochondrial encephalomyopathy with cytochrome c oxidase deficiency: A ^{31}P NMR study. *Neurology*, **39**, 399–403
108. Johnson, M.A., Kadenbach, B., Droste, M., Old, S.L. and Turnbull, D.M. (1988) Immunocytochemical studies of cytochrome oxidase subunits in skeletal muscle of patients with partial cytochrome oxidase deficiency. *Journal of the Neurological Sciences*, **87**, 75–90
109. Zeviani, M., Moraes, C.T. and DiMauro, S. (1988) deletions of mitochondrial DNA in Kearns–Sayre syndrome. *Neurology*, **38**, 1339–1346
110. Moraes, C.T., DiMauro, S., Zeviani, M. *et al.* (1989) Mitochondrial DNA deletions in progressive external ophthalmoplegia and Kearns–Sayre syndrome. 'see comments'. *New England Journal of Medicine*, **320**, 1293–1299
111. Rotig, A., Colonna, M., Bonnefont, J.P. *et al.* (1989) Mitochondrial DNA deletion in Pearson's marrow/pancreas syndrome 'letter; comment'. *Lancet*, **1**, 902–903
112. Rotig, A., Cormier, V., Blanche, S. *et al.* (1990) Pearson's marrow-pancreas syndrome. A multisystem mitochondrial disorder in infancy. *Journal of Clinical Investigation*, **86**, 1601–1608
113. Shoffner, J.M., Lott, M.T., Voljavec, A.S., Soueidan, S.A., Costigan, D.A. and Wallace, D.C. (1989) Spontaneous Kearns–Sayre/chronic external ophthalmoplegia plus syndrome associated with a mtDNA deletion: a slip-replication model and metabolic therapy. *Proceedings of the National Academy of Sciences USA*, **86**, 7952–7956
114. Holt, I.J., Harding, A.E., Cooper, J.M. *et al.* (1989) Mitochondrial myopathies: clinical and biochemical features of 30 patients with major deletions of muscle mitochondrial DNA. *Annals of Neurology*, **26**, 699–708
115. Romero, N., Lestienne, P., Marsac, C. *et al.* (1989) Immunocytological and histochemical correlation in Kearns–Sayre syndrome with mtDNA deletion and partial cytochrome c oxidase deficiency in skeletal muscle. *Journal of the Neurological Sciences*, **93**, 297–309
116. Gerbitz, K.D., Obermaier-Kusser, B., Zierz, S., Pongratz, D., Muller-Hocker, J. and Lestienne, P. (1990) Mitochondrial myopathies: divergences of genetic deletions, biochemical defects and clinical syndromes. *Journal of Neurology*, **237**, 5–10
117. Trounce, I., Byrne, E., Marzuki, S. *et al.* (1991) Functional respiratory chain studies in subjects with chronic progressive external ophthalmoplegia and large heteroplasmic mitochondrial DNA deletions. *Journal of the Neurological Sciences*, **102**, 92–99
118. Santorelli, F., Shanske, S., Sciacco, M. *et al.* (1992) A new etiology for Leigh's syndrome: mitochondrial DNA mutation in the ATPase 6 gene. *Annals of Neurology*, **32**, 467–468 (abstract)
119. Yamamoto, M., Clemens, P.R. and Engel, A.G. (1991) Mitochondrial DNA deletions in mitochondrial cytopathies: observations in 19 patients. *Neurology*, **41**, 1822–1828
120. Nakase, H., Moraes, C.T., Rizzuto, R., Lombes, A., DiMauro, S. and Schon, E.A. (1990) Transcription and translation of deleted mitochondrial genomes in Kearns–Sayre syndrome: implications for pathogenesis. *American Journal of Human Genetics*, **46**, 418–427
121. Hammans, S.R., Sweeney, M.G., Holt, I.J. *et al.* (1992) Evidence for intramitochondrial complementation between deleted and normal mitochondrial DNA in some patients with mitochondrial myopathy. *Journal of the Neurological Sciences*, **107**, 87–92
122. Goto, Y., Koga, Y., Horai, S. and Nonaka, I. (1990) Chronic progressive external ophthalmoplegia: a correlative study of mitochondrial DNA deletions and their phenotypic expression in muscle biopsies. *Journal of the Neurological Sciences*, **100**, 63–69
123. Matsuoka, T., Goto, Y.I., Hasegawa, H. and Nonaka, I. (1992) Segmental cytochrome c oxidase deficiency in CPEO: teased muscle fiber analysis. *Muscle & Nerve*, **15**, 209–213
124. Degoul, F., Nelson, I., Lestienne, P. *et al.* (1991) Deletions of mitochondrial DNA in Kearns–Sayre syndrome and ocular myopathies: genetic, biochemical and morphological studies. *Journal of the Neurological Sciences*, **101**, 168–177
125. Mita, S., Schmidt, B., Schon, E.A., DiMauro, S. and Bonilla, E. (1989) Detection of 'deleted' mitochondrial genomes in cytochrome-c-oxidase-deficient muscle fibers of a patient with Kearns–Sayre syndrome. *Proceedings of the National Academy of Sciences USA*, **86**, 9509–9513
126. Shoubridge, E.A., Karpati, G. and Hastings, K.E. (1990) Deletion mutants are functionally dominant over wild-type mitochondrial genomes in skeletal muscle fiber segments in mitochondrial disease. *Cell*, **62**, 43–49
127. Collins, S., Rudduck, C., Marzuki, S., Dennett, X. and Byrne, E. (1991) Mitochondrial genome distribution in histochemically cytochrome c oxidase-negative muscle fibres in patients with a mixture of deleted and wild type mitochondrial DNA. *Biochimica et Biophysica Acta*, **1097**, 309–317
128. Hammans, S.R., Sweeney, M.G., Wicks, D.A.G., Morgan-Hughes, J.A. and Harding, A.E. (1992) A molecular genetic study of focal histochemical defects in mitochondrial encephalomyopathies. *Brain*, **115**, 343–365

129. Oldfors, A., Larsson, N.G., Holme, E., Tulinius, M., Kadenbach, B. and Droste, M. (1992) Mitochondrial DNA deletions and cytochrome c oxidase deficiency in muscle fibres. *Journal of the Neurological Sciences*, **110**, 169–177
130. Moraes, C.T., Ricci, E., Petruzzella, V. *et al.* (1992) Molecular analysis of the muscle pathology associated with mitochondrial DNA deletions. *Nature and Genetics*, **1**, 359–367
131. Collins, S., Dennett, X., Byrne, E. and Marzuki, S. (1991) Chronic progressive external ophthalmoplegia in patients with large heteroplasmic mitochondrial DNA deletions: an immunocytochemical study. *Acta Neuropathologica*, **82**, 185–192
132. Hayashi, J.I., Ohta, S., Kikuchi, A., Takemitsu, M., Goto, Y.I. and Nonaka, I. (1991) Introduction of disease-related mitochondrial DNA deletions into HeLa cells lacking mitochondrial DNA results in mitochondrial dysfunction. *Proceedings of the National Academy of Sciences USA*, **88**, 1371–1375
133. Sancho, S., Moraes, C.T., Tanji, K. and Miranda, A.F. (1992) Structural and functional mitochondrial abnormalities associated with high levels of partially deleted mitochondrial DNAs in somatic cell hybrids. *Somatic Cell Molecular Genetics*, **18**, 431–442
134. Brown, M.D., Voljavec, A.S., Lott, M.T., Torroni, A., Yang, C.C. and Wallace, D.C. (1992) Mitochondrial DNA complex I and II mutations associated with Leber's hereditary optic neuropathy. *Genetics*, **130**, 163–173
135. Holt, I.J., Harding, A.E., Petty, R.K. and Morgan-Hughes, J.A. (1990) A new mitochondrial disease associated with mitochondrial DNA heteroplasmy. *American Journal of Human Genetics*, **46**, 428–433
136. Larsson, N.G., Anderson, O., Holme, E., Oldfors, A. and Wahlstrom, J. (1991) Leber's hereditary optic neuropathy and complex I deficiency in muscle. *Annals of Neurology*, **30**, 701–708
137. Shoffner, J.M., Fernhoff, P.M., Krawiecki, N.S. *et al.* (1992) Subacute necrotizing encephalopathy: oxidative phosphorylation defects and the ATPase 6 point mutation. *Neurology*, **42**, 2168–2174
138. Sakuta, R., Goto, Y., Horai, S. *et al.* (1992) Mitochondrial DNA mutation and Leigh's syndrome. *Annals of Neurology*, **32**, 597–598
139. Shoffner, J.M., Lott, M.T., Lezza, A.M.S., Seibel, P., Ballinger, S.W. and Wallace, D.C. (1990) Myoclonic epilepsy and ragged-red fiber disease (MERRF) is associated with a mitochondrial RNA tRNA^{Lys} mutation. *Cell*, **61**, 931–937
140. Saski, H., Kuzuhara, S., Kanazawa, I., Nakanishi, T. and Ogata, T. (1983) Myoclonus, cerebellar disorder, neuropathy, mitochondrial myopathy, and ACTH deficiency. *Neurology*, **33**, 1288–1293
141. Lombes, A., Mendell, J.R., Nakase, H. *et al.* (1989) Myoclonic epilepsy and ragged-red fibers with cytochrome oxidase deficiency: neuropathology, biochemistry, and molecular genetics. *Annals of Neurology*, **26**, 20–33
142. Haginoya, K., Miyabayashi, S., Iinuma, K. and Tada, K. (1990) Mosaicism of mitochondria in mitochondrial myopathy: an electromicroscopic analysis of cytochrome c oxidase. *Acta Neuropathologica*, **80**, 642–648
143. Wallace, D.C., Zheng, X., Lott, M.T. *et al.* (1988) Familial mitochondrial encephalomyopathy (MERRF): genetic, pathophysicalogical and biochemical characterization of a mitochondrial DNA disease. *Cell*, **55**, 601–610
144. Byrne, E., Trounce, I., Marzuki, S. *et al.* (1991) Functional respiratory chain studies in mitochondrial cytopathies. Support for mitochondrial DNA heteroplasmy in myoclonus epilepsy and ragged red fibers (MERRF) syndrome. *Acta Neuropathologica (Berlin)*, **81**, 318–323
145. Larsson, N.G., Tulinius, M.H., Holme, E. *et al.* (1992) Segregation and manifestations of the mtDNA tRNA^{Lys} A→G(8344) mutation of myoclonus epilepsy and ragged-red fibers (MERFF) syndrome. *American Journal of Human Genetics*, **51**, 1201–1212
146. Silvestri, G., Ciafaloni, E., Santorelli, F. *et al.* (1993) Clinical features associated with the A→G transition at nucleotide 8344 of mtDNA ("MERRF" mutation). *Neurology*, **43**, 1200–1206
147. Chomyn, A., Meola, G., Bresolin, N., Lai, S.T., Scarlato, G. and Attardi, G. (1991) In vitro genetic transfer of protein synthesis and respiration defects to mitochondrial DNA-less cells with myopathy-patient mitochondria. *Molceular and Cellular Biology*, **11**, 2236–2244
148. Boulet, L., Karpati, G. and Shoubridge, E.A. (1992) Distribution and threshold expression of the tRNA^{Lys} mutation in skeletal muscle of patients with myoclonic epilepsy and ragged-red fibers (MERRF). *American Journal of Human Genetics*, **51**, 1187–1200
149. Goto, Y., Nonaka, I. and Horai, S. (1990) A mutation in the tRNA$^{(Leu)(UUR)}$ gene associated with the MELAS subgroup of mitochondrial encephalomyopathies. *Nature*, **348**, 651–653
150. Hirano, M., Ricci, E., Koenigsberger, M.R. *et al.* (1992) MELAS: an original case and clinical criteria for diagnosis. *Neuromuscular Disorder*, **2**, 125–135
151. Obermaier-Kusser, B., Paetzke-Brunner, I., Enter, C. *et al.* (1991) Respiratory chain activity in

tissue from patients (MELAS) with a point mutation of the mitochondrial genome [tRNA leu(UUR)]. *FEBS Letters*, **286**, 67–70

152. Ciafaloni, E., Ricci, E., Shanske, S. *et al.* (1992) MELAS: Clinical features, biochemistry, and molecular genetics. *Annals of Neurology*, **31**, 391–398

153. Moraes, C.T., Ricci, E., Bonilla, E., DiMauro, S. and Schon, E.A. (1992) The mitochondrial tRNA$^{Leu(UUR)}$ mutation in mitochondrial encephalomyopathy, lactic acidosis, and strokelike episodes (MELAS): genetic, biochemical, and morphological correlations in skeletal muscle. *American Journal of Human Genetics*, **50**, 934–949

154. Moraes, C.T., Ciacci, F., Silvestri, G. *et al.* (1993) Atypical clinical presentations associated with the MELAS mutation at position 3243 of human mitochondrial DNA. *Neuromuscular Disorder*, **3**, 43–50

155. Chomyn, A., Martinuzzi, A., Yoneda, M. *et al.* (1992) MELAS mutation in mtDNA binding site for transcription termination factor causes defects in protein synthesis and in respiration but no change in levels of upstream and downstream mature transcripts. *Proceedings of the National Academy of Sciences USA*, **89**, 4221–4225

156. King, M.P., Koga, Y., Davidson, M. and Schon, E.A. (1992) Defects in mitochondrial protein synthesis and respiratory chain activity segregate with the tRNA$^{Leu(UUR)}$ mutation associated with mitochondrial myopathy, encephalopathy, lactic acidosis, and stroke-like episodes. *Molecular and Cellular Biology*, **12**, 480–490

157. Zeviani, M., Servidei, S., Gellera, C., Bertini, E., DiMauro, S. and DiDonato, S. (1989) An autosomal dominant disorder with multiple deletions of mitochondrial DNA starting at the D-loop region. *Nature*, **339**, 309–311

158. Zeviani, M. (1992) Nucleus-driven mutations of human mitochondrial DNA. *Journal of Inherited Metabolic Disorders*, **15**, 456–471

159. Servidei, S., Zeviani, M., Manfredi, G. *et al.* (1991) Dominantly inherited mitochondrial myopathy with multiple deletions of mitochondrial DNA: clinical, morphologic, and biochemical studies. *Neurology*, **41**, 1053–1059

160. Suomalainen, A., Majander, A., Haltia, M. *et al.* (1992) Multiple deletions of mitochondrial DNA in several tissues of a patient with severe retarded depression and familial progressive external ophthalmoplegia. *Journal of Clinical Investigation*, **90**, 61–66

161. Moraes, C.T., Shanske, S., Tritschler, H.-J. *et al.* (1991) MtDNA depletion with variable tissue expression: a novel genetic abnormality in mitochondrial diseases. *American Journal of Human Genetics*, **48**, 492–501

162. Boustany, R.N., Aprille, J.R., Halperin, J., Levy, H. and DeLong, G.R. (1983) Mitochondrial cytochrome deficiency presenting as a myopathy with hypotonia, external ophthalmoplegia, and lactic acidosis in an infant and as fatal hepatopathy in a second cousin. *Annals of Neurology*, **14**, 462–470

163. Aprille, J.R. (1985) Tissue specific cytochrome deficiencies in human infants. In *Achievements and Perspectives of Mitochondrial Research* (eds E. Quagliariello, E.C. Slater, F. Palmieri, C. Saccone and A.M. Kroon), Elsevier, Amsterdam pp. 465–476

164. Tritschler, H.J., Andreetta, F., Moraes, C.T. *et al.* (1992) Mitochondrial myopathy of childhood associated with depletion of mitochondrial DNA. *Neurology*, **42**, 209–217

165. Ricci, E., Moraes, C.T., Servidei, S., Tonali, P., Bonilla, E. and DiMauro, S. (1992) Disorders associated with depletion of mitochondrial DNA. *Brain Pathology*, **2**, 141–147

166. Figarella-Branger, D., Pellissier, J.F., Scheiner, C., Wernert, F. and Desnuelle, C. (1992) Defects of the mitochondrial respiratory chain complexes in three pediatric cases with hypotonia and cardiac involvement. *Journal of the Neurological Sciences*, **108**, 105–113

167. Dalakas, M., Illa, I., Pezeshkpour, G.H., Laukaitis, J.P., Cohen, B. and Griffin, J.L. (1990) Mitochondrial myopathy caused by long-term zidovudine therapy. *New England Journal of Medicine*, **322**, 1098–1105

168. Mhiri, C., Baudrimont, M., Bonne, G. *et al.* (1991) Zidovudine myopathy: a distinctive disorder associated with mitochondrial dysfunction. *Annals of Neurology*, **29**, 606–614

169. Arnaudo, E., Dalakas, M., Shanske, S., Moraes, C.T., DiMauro, S. and Schon, E.A. (1991) Depletion of muscle mitochondrial DNA in AIDS patients with zidovudine-induced myopathy. *Lancet*, **337**, 508–510

170. Charlot, P. and Gherardi, R. (1991) Partial cytochrome c oxidase deficiency and cytoplasmic bodies in patients with zidovudine myopathy. *Neuromuscular Disorders*, **1**, 357–363

171. Lewis, W., Gonzalez, B., Chomyn, A. and Papoian, T. (1992) Zidovudine induces molecular, biochemical, and ultrastructural changes in rat skeletal muscle mitochondria. *Journal of Clinical Investigation*, **89**, 1354–1360

172. Herzberg, N.H., Zorn, I., Zwart, R., Portegies, P. and Bolhuis, P.A. (1992) Major growth reduc-

tion and minor decrease in mitochondrial enzyme activity in cultured human muscle cells after exposure to zidovudine. *Muscle & Nerve*, **15**, 706–710

173. Menkes, J.H. (1993) Disorders of copper metabolism. In *The Molecular and Genetic Basis of Neurological Disease* (eds R.N. Rosenberg, S.B. Prusiner, S. DiMauro, R.L. Barchi and L.M. Kunkel), Butterworth-Heinemann, Boston, pp. 325–341

174. Nooijen, J.L., DeGroot, C.J., van der Hamer, C.J.A., Monnens, L.A.H., Willemse, J. and Niermeijer, M.F. (1981) Trace element studies in three patients and a fetus with Menke's disease. Effect of copper therapy. *Pediatric Research*, **15**, 284–289

175. French, J.H., Sherard, E.S., Lubell, H., Brotz, M. and Moore, C.L. (1972) Trichopoliodystrophy: I. Report of a case and biochemical studies. *Archives of Neurology*, **26**, 229–244

176. Seki, K., Sato, T., Ishigaki, Y., Nakamura, S., Ishihara, Y. and Ozawa, T. (1989) Decreased activity of cytochrome c oxidase in the macular mottled mouse: an immuno-electron microscopic study. *Acta Neuropathologica*, **77**, 465–471

177. Turnbull, D.M., Spoors, T.C., Cullen, M.J., Johnson, M.A. and Sherratt, H.S.A. (1983) The effect of copper deficiency on skeletal and cardiac muscle mitochondria. In *Mitochondrial Pathology in Muscle Diseases* (eds G. Scarlato and C. Cerri), Piccin, Padova, pp. 252–260

178. Bennett, M.C., Diamond, D.M., Stryker, S.L., Parks, J.K. and Parker, W.D. (1992) Cytochrome oxidase inhibition: a novel animal model of Alzheimer's disease. *Journal of Geriatric Psychiatry and Neurology*, **5**, 93–101

179. Huart, P., Brasseur, R., Goormaghtigh, E. and Ruysschaert, J.M. (1984) Antimitotics induce cardiolipin cluster formation. Possible role in mitochondrial enzyme inactivation. *Biochimica et Biophysica Acta*, **799**, 199–202

180. Arai, M., Gordon, E.R. and Lieber, C.S. (1984) Decreased cytochrome oxidase activity in hepatic mitochondria after chronic ethanol consumption and the possible role of decreased cytochrome aa_3 content and changes in phospholipids. *Biochimica et Biophysica Acta*, **797**, 320–327

181. Cardellach, F., Galofre, J., Grau, J.M., Casademont, J., Hoek, J.B., Rubin, E. and Urbano-Marquez, A. (1992) Oxidative metabolism in muscle mitochondria from patients with chronic alcoholism. *Annals of Neurology*, **31**, 515–518

6
Mitochondrial encephalomyopathies: lumping, splitting and melding

Lewis P. Rowland

INTRODUCTION

Nineteen ninety-one was a banner year in the history of mitochondrial diseases because friendly foes in nomenclature found common ground. For the first time, we inveterate splitters had to admit that both molecular genetics and the overlapping clinical features of syndromes generated too many inconsistencies for precise classification. We presented the evidence for this view at an international meeting in Venice in November 1991, and I here update the story to the end of 1992.

Also, for the very first time, decades after the eponym had been introduced, investigators at Queen Square, London, finally designated the Kearns–Sayre syndrome (KSS) as the diagnosis in some of their cases [1].

All agreed that deciphering the molecular biology of mitochondrial DNA (mtDNA) had created a wondrous new world, one that could not have been imagined a decade, or even five years, before. But these advances, as fantastic as they might be, had not resolved the problems of lumpers and splitters. In fact, analysis of mtDNA in these syndromes had actually created some new problems, and new challenges. But classification was only the most recent of several controversies in the history of these diseases. The mitochondrial diseases have generated debate for three decades, ever since they were first recognized. Even today, there is no consensus about the meaning of the popular phrases we use to describe them: 'mitochondrial myopathy', 'mitochondrial encephalomyopathy', 'mitochondrial cytopathy' or 'mitochondrial disease'. The uncertainty is a measure of the different criteria used to select the syndromes, which were first clinical and biochemical; then morphological definitions, moving from ultrastructure to histochemistry, were used, followed by human genetics and, finally, molecular genetics – or combinations of the above. The attention directed to these rare conditions and the emotional heat of the debates tell us something about the intellectual challenge the conditions have posed. (This discussion centres on disorders of the mitochondrial genome, not metabolic conditions caused by abnormalities of nuclear DNA that affect mitochondrial biochemical function [2,3].)

CONTROVERIAL QUESTION NO. 1:
Do structurally abnormal mitochondria define a specific group of diseases?

ANSWER:
Not as the sole criterion. There ought to be a characteristic clinical syndrome, a biochemical abnormality of mitochondrial function, a mutation in mitochondrial DNA (mtDNA) or a mutation in nuclear DNA that secondarily affects mtDNA

The notion that there are mitochondrial diseases was probably conceived by Milton Shy. With Nicholas Gonatas, Shy [4] pioneered the use of histochemistry and electron microscopy to study human muscle diseases. They were impressed with the diversity of diseases in which they found abnormal organelles, including conditions that were characterized by abnormal number, size, or morphology of intracellular structures. They found conditions in which there were too many mitochondria ('pleoconial myopathies') and some in which the mitochondria were too large ('megaconial'); often, the two abnormalities were seen together, and often there were 'paracrystalline' structures in some mitochondria.

They also knew that these changes might accompany a disorder of mitochondrial function, because Luft *et al.* [5], in Sweden, had already described a syndrome of hypermetabolism without hyperthyroidism in a woman with these structural changes. That patient's disorder defined a condition in which there were both functional and anatomical changes in mitochondria. Remarkably, there has been only one other patient with these characteristics, a Jordanian woman described by Haydar *et al.* [6] in 1971 and later studied by DiMauro *et al.* [7]; no new cases have been recognized anywhere in the world for the last 20 years.

Electron microscopy was still a recondite science in those days, and not widely available for diagnostic purposes. Diagnosis was then facilitated and made practical by Engel and Cunningham [8] when they described a simple histochemical method (the modified Gomori trichrome stain) that identified collections of abnormal mitochondria with the light microscope. In addition to the method, they provided the catchy name, 'ragged red fibres' (RRF). Before very long, this finding became the hallmark of a new class of disease, the mitochondrial myopathies.

However, there were doubters, including me. There seemed to be too many different clinical syndromes in which RRF were seen. They were described in diverse well-recognized disorders, including myotonic muscular dystrophy, Duchenne dystrophy, polymyositis and also neurogenic disorders, too many for comfort of classification [3,9]. Moreover, it was often unclear how many RRF were needed to be deemed pathological.

This debate was never clearly articulated, but there now seems to be a reconciliation. The aforementioned conditions are not properly regarded as mitochondrial disorders because RRF are not characteristic but exceptional in those conditions, and there are neither biochemical nor genetic abnormalities of mitochondria. On the other hand, there are diseases in which RRF are seen in almost all cases; in these conditions, the RRF are not the only hallmark of mitochondrial dysfunction; there are also biochemical abnormalities, mutations of mtDNA, or both. These are true mitochondrial myopathies. Even so, deletions of mtDNA may be seen in some patients who lack RRF [10].

There is yet another group of disorders that comprises conditions in which there are neither typical RRF nor abnormal mtDNA, but there is a documented deficiency of a mitochondrial enzyme. However, because of the demonstrable biochemical abnormality, there has not been much debate about the mitochondrial nature of these conditions and, anyway, they have little in common clinically with conditions that show RRF. Lack of carnitine palmitoyltransferase is an example of this class.

Sometimes, only the combination of specific symptoms and signs with RRF is available to designate the disorder as a mitochondrial disease [11].

There follow some (non-controversial?) definitions. A disease is a **mitochondrial myopathy** if there is a characteristic clinical syndrome, and if RRF are found in muscle of the great majority of cases, if there is a consistent abnormality of a mitochondrial enzyme, or if there is a consistent abnormality of mtDNA. If there is a maternal inheritance, with or without RRF, a mitochondrial disorder can reasonably be suspected. [Leber hereditary optic neuropathy (LHON) was in that category until the first point mutation in mtDNA was identified.]

Even though it is cumbersome, **mitochondrial encephalomyopathy** has persisted in the literature to describe multisystem disorders that involve the brain, as most of the major RRF diseases do. **Mitochondrial disease** is equivalent. **Mitochondrial cytopathy** has no clear distinction.

CONTROVERSIAL QUESTION NO. 2:
Is KSS a specific disorder?

ANSWER:
This question has been the subject of intense debate for about 15 years. It is a recognizable clinical disorder and it is only rarely familial, but that is only part of the story; molecular genetics is discussed later

Kearns and Sayre described their eponymic disease in 1958 [12]. There was not much controversy until 1975 when I was writing a review of progressive external ophthalmoplegia (PEO) [13] and noted that almost all of the 25 then-published cases of this disorder were simplex (sporadic), with no other cases in the family. Moreover, there was a sufficiently consistent pattern to conclude that the disorder was a unique and specific disorder.

In 1977, we analysed our own cases and the 35 other reports of the time [14]. The findings led to diagnostic criteria that served well for the next 15 years: the three invariable manifestations were onset before age 20, PEO and pigmentary retinopathy. Additionally, there should be at least one of the following triad: heart block, CSF protein content around 100 mg/dl or cerebellar syndrome.

Both of these contentions were challenged immediately. Many thought that the diverse manifestations were seen in other multisystem diseases and there was too much overlap to consider the disorder unique. David Drachman [15] introduced another catchy phrase that became popular and is still used, 'ophthalmoplegia-plus', describing the combination of ophthalmoplegia with retinopathy, dementia, cerebellar syndromes, peripheral neuropathy and other symptoms. Ironically, Drachman [15] used this term because he thought it premature to designate specific syndromes but, before long, others were using the term as though it were a specific diagnosis.

The sporadic nature of the disorder was also challenged. Bastianssen [16] and others reported familial cases, but most of those reports were deficient in one way or another. Often, the propositus did not have the complete KSS. To this day, there are reports of only four families in which one person had KSS and a first-degree relative also had a complete or partial syndrome: two families in which siblings were affected [17], one pair of homozygous twins [18] and one father–son pair [19A]. In the last family, it is strange that the father was affected rather than the mother (if the disease were mitochondrially inherited). In another family [19B], the propositus had an incomplete and atypical KSS; his brother and a maternal uncle had ophthalmoplegia and the parents were cosanguineous.

In 1982, the identity of KSS was bolstered indirectly at the first international meeting on mitochondrial diseases because two new RRF conditions were recognized. They differed clinically from KSS, and they were frequently familial. Fukuhara *et al.* [20,21] emphasized the maternal inheritance of a syndrome they called myoclonus epilepsy with ragged red fibres (MERRF), and Rowland *et al.* [9] described what they called mitochondrial myopathy and lactic acidosis (MELA), which was later modified by Pavlakis *et al.* [22] to add the characteristic features of dementia and stroke, so that name became mitochondrial encephalomyopathy with lactic acidosis and stroke (MELAS).

In addition to the distinct clinical characteristics that separated these syndromes from KSS and from each other, these new conditions differed because they were often familial and, remarkably, in a pattern consistent with maternal inheritance. The splitters thought they were justified by the recognition of these apparently specific syndromes.

But the debates continued. The investigative team at Queen Square, London, was especially resistant to the concept that KSS is a unique disorder because they found so many cases in which the elements of one condition merged with the others [23]. This seemed especially true of MERRF and MELAS because myoclonus, seizures and dementia were found frequently in both conditions. Also, there seemed to be manifestations that were common in all three syndromes, such as short stature, deafness, diabetes mellitus and calcification of the basal ganglia. Furthermore, they were concerned because features of KSS were sometimes seen with manifestations of MELAS and MERRF. They were therefore content with 'mitochondrial cytopathy' to indicate the multisystem nature of the conditions.

CONTROVERSIAL QUESTION NO. 3:
Does molecular genetics resolve the debate?

ANSWER:
Sometimes, not always

A new chapter opened in 1988, when Holt *et al.* [24] found deletions of mtDNA in patients with these diseases; the Queen Square investigators did not identify the patients by clinical syndrome but, almost immediately, Zeviani *et al.* [25] and then Moraes *et al.* [26] found that almost all patients with KSS had such a deletion, and so did many patients with sporadic forms of PEO. However, patients with familial PEO did not show deletions, and neither did patients with MERRF or MELAS. Once again, the splitters felt vindicated.

The debates continued for several reasons: there was no consistency in the site or size of the deletion and the clinical manifestations; the identical deletion might be found in patients with full KSS or isolated PEO. Moreover, there was no consistent relationship of the site or size of the deletion and mitochondrial biochemistry.

Also in 1988, Wallace *et al.* [27] found a point mutation in Leber hereditary optic neuropathy (LHON) and that was followed in short order by the 1990 reports of different point mutations in MELAS [28] and MERRF [29]. And the Queen Square group identified a point mutation in yet another syndrome – neuropathy, ataxia and retinitis pigmentosa (NARP) [30]. Another maternally inherited RRF disease was recognized, but there was no identifiable abnormality of mtDNA in most cases (presumably an unrecognized point mutation); the few cases have been reported under several different names, including MNGIE for myoneurogastrointestinal encephalopathy [31]. That name, however, does not include the essential feature of ophthalmoplegia or the constant finding of RRF, so Rowland [32] proposed MEPOP, for mitochondrial encephalomyopathy, polyneuropathy, ophthalmoplegia and pseudo-obstruction. Hirano *et al.* [31] prefer to keep the acronym MNGIE but would change the full title to *m*itochondrial *n*eurogastro*i*ntestinal *e*ncephalomyopathy.

Also, autosomal dominant PEO in some families was associated with multiple deletions of mtDNA, attributed to an abnormality of nuclear DNA that made mtDNA more susceptible to deletions [33,34] (Table 6.2). This was considered an anomalous interaction between nuclear and mitochondrial genomes, under the control of nuclear DNA and therefore inherited in an autosomal dominant fashion. Another disorder of the interaction of nuclear and mtDNA is 'depletion' of mtDNA, which was first recognized in childhood myopathies but has been seen with other syndromes, including zidovudine treatment for AIDS, a non-genetic and acquired form of mitochondrial disease [3].

The variety of clinical disorders and the separation of sporadic from maternally inherited syndromes in conformity with deletions or point mutations all seemed to vindicate the splitters. Furthermore, it seemed that we would soon have a rational genetic classification. For instance, two patients met all criteria for the diagnosis of KSS but they had strokes, a seeming 'overlap' with MELAS; the question was raised whether DNA analysis could resolve the diagnosis, and it did; both patients had mtDNA deletions as seen in KSS; they did not have the point mutation of MELAS, so confirming the clinical view that this was an atypical clinical manifestation in patients with KSS [35].

Again, the splitters felt vindicated but their exhilaration was once again short-lived because of the arrival of controversy no. 4.

CONTROVERSIAL QUESTION NO. 4:
Can mitochondrial diseases be identified by mutations in mtDNA?

ANSWER:
Often, not always. The problem is bidirectional: not all patients with the same clinical syndrome show the same mutation in mtDNA and, conversely, a single mutation can be associated with totally different clinical syndromes

The lack of specificity of DNA diagnosis of mitochondrial disease became apparent with the first identification of deletions in KSS, because these abnormalities were found in 12 of 15 patients, not all of them [25,26]. At first, it was not clear what had happened in the other three cases; it was thought that some had deletions too small to be identified by Southern blot, or that there might be point mutations. The first possibility seemed reasonable; the second proved to be correct.

Diverse mtDNA mutations with the same clinical syndrome

The lack of specificity was also apparent because deletions were found in about 50% of patients with non-familial PEO who lacked the other manifestations of

Table 6.1 mtDNA deletions and mutations in PEO and KSS

	No.	*Percentage*
Large deletions	74	53
KSS	21	15
Partial KSS	7	5
PEO alone	46	33
Multiple deletions	7	5
MELAS 3243 mutation	16	12
MERRF 8344 mutation	1	0.7
No mtDNA abnormality	40	29

Total no. of patients 138. Data from Moraes *et al.* [38].

KSS [26]; no deletions were found among familial cases. later, Zeviani *et al.* [36] found multiple deletions in some patients with PEO, and Lombes *et al.* [37] found depletion of mtDNA in KSS.

In further analysing the clinical correlations of mtDNA deletions, Moraes *et al.* [38] found deletions in 53% of 138 patients with different forms of PEO (Table 6.1). Some 20% had KSS and 33% had PEO alone (none familial). Multiple deletions were manifest in 5% of the patients, some in an autosomal dominant pattern of inheritance. However, they also found the MELAS 3243 mutation in 12% and there was even one patient with the MERRF 8344 mutation. 29% still had no identified mutation.

Similar inconsistencies were found in studies of MELAS. Goto *et al.* [28] identified a point mutation at nt 3243 in the gene for tRNA$^{Leu(UUR)}$ and found it in 26 of 31 patients with the clinical syndrome [39]. Ciafaloni *et al.* [40] found the same mutation in 21 of 23 patients. Nonaka [41] reported that 10% of cases were due to another mutation in the same gene but at nt 3721 [42]. To complicate matters even further, Lertrit *et al.* [43] described a mutation at position 11084, not in a tRNA but in a structural gene, the ND4 subunit of Complex I. Nevertheless, there are still some mutations that have not yet been identified in MELAS.

The situation is similar in MERRF. A point mutation in mtDNA was identified by Shoffner *et al.* [29] at nt 8344 in the gene for tRNA^{Lys}. It is present in most patients with the syndrome, but not all. Contrariwise, in both MELAS and MERRF, there are some asymptomatic family members who show the mutation but have no symptoms; it is not known whether these people are ultimately destined to develop symptoms.

Diverse clinical syndromes with the same mtDNA mutation

Deletions of mtDNA of similar extent and map location have been found in KSS, PEO, maternally inherited diabetes mellitus [44] and the Pearson syndrome (a sideroblastic anaemia of infancy). The few survivors of the Pearson syndrome have developed KSS. Multiple small deletions have been found not only in autosomal dominant PEO and two brothers with recurrent myoglobinuria, but also in the syndrome of mitochondrial encephalomyopathy with peripheral neuropathy, ophthalmoplegia and gastrointestinal pseudo-obstruction (MEPOP or MNGIE), which seems to be an autosomal recessive disease [31].

Table 6.2 Diseases associated with multiple deletions of mtDNA

PEO (some with cataracts, optic atrophy, ataxia, peripheral neuropathy, mental retardation)
Recurrent myoglobinuria
KSS
MEPOP (also MNGIE) (mitochondrial encephalomyopathy with peripheral neuropathy,
    ophthalmoplegia, and pseudo-obstruction)
DAD (diabetes mellitus and deafness)
Limb-girdle myopathy
Dilating cardiomyopathy

Table 6.3 Clinical syndromes of 91 patients with RRF but not MELAS

	Number with MELAS 3243 mutation
Total patients	91
With mutation	21
PEO	16 (76%)*
Ptosis, LA, myopathy	1
Myopathy, PR, hearing loss	2
Myopathy, LA, seizures	2

*16/21 patients (76%).
LA, lactic acidosis; PR, pigmentary retinopathy. Data from Moraes *et al.* [38].

The diversity of clinical syndromes with the same mutation is perhaps best illustrated by the MELAS mutation. Moraes *et al.* [38] found this mutation in 15 of 59 (25%) of those patients with ophthalmoplegia who did not have KSS or a deletion (Table 6.3). One of their patients with the MELAS mutation seemed to meet criteria for KSS [onset in adolescence, ophthalmoplegia, retinopathy (undescribed), CSF protein content of 100 mg/dl, RRF, lactic acidosis and ataxia]. However, there was never heart block or need for a pacemaker even though the patient was aged 60 at the last examination [38] (Table 6.4). Another KSS patients with the MELAS mutation was mentioned by Johns and Hurko [45] but was not described. Other syndromes with the MELAS 3243 mutation were encountered by Moraes *et al.* [38]: ptosis, lactic acidosis and myopathy (one patient); myopathy, with pigmentary retinopathy and hearing loss (two patients); and myopathy with lactic acidosis and seizures (two patients). Inui *et al.* [46] found the mutation in a family with cardiomyopathy and dementia but no strokes.

Among the most mysterious challenges of molecular genetics is the association of allelic mutations with different clinical syndromes. The parallel in studies of mtDNA is the finding of three different point mutations in the same gene for tRNA$^{Leu(UUR)}$. In addition to the MELAS mutation at nt 3243, two other mutations have been found in that gene, one at nt 3271 in a patient with MELAS [47] and one at nt 3260 in a maternally inherited infantile cardiomyopathy [48].

Similar divergences have been seen with the MERRF mutation (Table 6.5). Silvestri *et al.* [49] found the typical mutation in 11 of 15 patients with the full syndrome, as well as in relatives who had partial syndromes that would not have been diagnosed as MERRF had there not been a typical case in the family. Among other syndromes associated with this mutation were one patient with autopsy-proven Leigh syndrome, one with PEO, and one with limb-girdle myopathy (Table 6.5).

Table 6.4 Mitochondrial DNA mutations in ophthalmoplegic syndromes

	KSS	*Incomplete KSS*	*PEO*
Number of patients	22	7	109
Single mtDNA deletion	21	7	46 (42%)
Multiple mtDNA deletions	0	0	7 (6%)
mtDNA MELAS 3243 mutation	1*	0	15 (14%)
mtDNA MERRF 8344 mutation	0	0	40 (37%)

*This patient was atypical in late age at onset (after age 30) but had ophthalmoplegia, pigmentary retinopathy and CSF protein content of 100 mg/dl. There was no cardiopathy. The patient was studied by Dr S. Servidei and her colleagues in Rome, Italy.
Data from Moraes *et al.* [38].

Table 6.5 Syndromes associated with the MERRF mutation (nt 8344)

	No. of patients	*A→G nt 8344*
MERRF	15	11
Myoclonic epilepsy	44	0
Other mitochondrial diseases	23	2
PEO with RRF, no deletion	40	1
Limb-girdle myopathy	5	2
Normal controls	15	0

Data from Silvestri *et al.* [50].

Also in parallel with MELAS, Silvestri *et al.* [50] have found a second MERRF mutation in the same gene.

Both sets of variations – multiple clinical syndromes with the same mutation, and the converse, multiple mutations with the same clinical syndrome – make it impossible to define these diseases solely by criteria based on analysis of DNA.

CONTROVERSIAL QUESTION NO. 5:
Can the mitochondrial diseases be classified on the basis of biochemical abnormality?

ANSWER:
Not yet by biochemical criteria alone

The cardinal precept of molecular biology has been modified through the years from one gene – one enzyme to one gene – one protein, to one gene – one polypeptide. However, every inherited disease must have a key gene product. Now, the specific gene product in these human diseases can be a tRNA, but there may not be a specific abnormality of a specific enzyme or structural protein. In mitochondrial diseases more than one gene product and more than one enzyme may be affected.

MELAS provides a convenient example. In muscle biopsies from 16 patients, Ciafaloni *et al.* [51] found low levels of Complexes I, II, III and IV in various combinations. This was most striking for Complex IV (cytochrome oxidase) which was depressed in all but two patients. In contrast, they found normal activity for succinate dehydrogenase, an enzyme coded by nuclear DNA. This pattern

Table 6.6 Biochemistry in 42 MELAS patients

	Number of patients
Complex I deficiency alone	17
+ IV	7
+ III + IV	2
+ II + III + IV	4
Complex IV deficiency alone	3
(+ I, II, III as above)	(13)
Normal	9

Data from Hirano *et al.* [52].

suggested a generalized depression in the translation or transcription of mtDNA. There was no relationship to the proportion of mutant mtDNA in the muscle. Others, however, have found that Complex I was most likely to be affected. For instance, in reviewing the literature, Hirano *et al.* [52] found that biochemical studies had been done in 42 MELAS patients. Complex I deficiency was seen alone in 17 patients; Complex IV deficiency alone in three patients; or Complex IV with I, and II or III in 13 (Table 6.6).

DiMauro *et al.* [53] attribute the lack of consistent biochemical abnormality in these diseases to two characteristic phenomena: (1) Some of the point mutations result in a generalized impairment in synthesis of proteins encoded in mtDNA, so that more than one enzyme is affected. (2) The mutations of mtDNA are usually heteroplasmic, with both normal and mutant mtDNAs in the same cell. As a result, any enzyme deficiency resulting from the mutants would be compensated for by the normal mtDNA, and any enzyme deficiencies would be partial. The proportion of mutant and normal mtDNA in different tissues might also affect biochemical analysis; for instance, the proportion of mutant mtDNA might be high in retina and ocular muscles of a person with KSS, but a lower proportion in skeletal muscle might lead to normal biochemical activities in that tissue.

Nevertheless, it is diagnostically appropriate to continue to correlate biochemical functions with clinical and mtDNA data because abnormal mitochondrial function is sometimes identified in this way, even though no DNA abnormality is found with currently available methods [54,55].

Moreover, biochemical studies are needed for studies of pathogenesis. For instance, DiMauro *et al.* [53] have summarized their studies of cytochrome *c* oxidase (COX) deficiency which has been found in a fatal infantile myopathy that is inherited in an autosomal recessive pattern and shows an isolated defect of subunit VIIa, one of only two tissue-specific subunits of human COX. In KSS, they found decreased activity of all mtDNA-encoded subunits of COX as well as impaired protein synthesis.

CONTROVERSIAL QUESTION NO. 6:
Is there any sense in trying to classify syndromes by clinical criteria?
ANSWER:
Yes, but not by clinical criteria alone

All of the evidence cited above weakens the case for the classification of mitochondrial diseases according to clinical syndrome. Yet nothing else suffices alone either.

The diseases cannot be separated solely on the basis of morphology, pattern of inheritance, biochemical abnormality or mutation of mtDNA. As suggested by Moraes *et al.* [2] all have to be considered together.

The following arguments can be adduced for keeping eponyms and acronyms for the syndromes discussed.

(1) Recognition of KSS, MERRF and MELAS predicts separation into deletions and point mutations and, for practical purposes of genetic counselling, sporadic and maternally inherited diseases.

(2) Recognition of clinical syndromes identifies the scientific challenge of elucidating the pathogenesis of the syndrome, the path from mutation of mtDNA to clinical expression.

(3) Despite the overlaps and exceptions, the correspondence between different clinical syndromes and different mutations implies that there is something specific in the clinical identification.

(4) Failure to recognize clinical differences may obscure important biological or genetic differences. The ancient argument about the identity of KSS has taken a new form. Because PEO may occur with or without other features of KSS and because deletions of mtDNA are found in both KSS and PEO alone, Nonaka [41] and others consider PEO to be a partial expression of KSS, assuming that both have the same pathogenesis. This may yet prove to be true, but we know that there are differences. For instance, if the two conditions were due to the same pathogenesis, there ought to be many intermediate cases, many cases of 'incomplete KSS', but there are in fact very few. In our 1989 series [56], there were 15 cases of PEO, five with KSS, and only one with incomplete KSS. Also, non-familial PEO is not always due to deletion, which is found in about 50% of cases. In contrast, deletions are found in 90% of KSS cases. Among the non-deleted non-familial cases of PEO, the MELAS mutation accounts for about a third of the cases, but this mutation is rarely seen in KSS. Only two cases of the MELAS mutation have been reported in patients with KSS [38,45], neither described in detail; in contrast, Hirano *et al.* [52] found the elements of KSS in some of 69 people with MELAS, often enough to consider the syndrome 'overlapping' but actually only a few; 11% each had PEO or retinopathy, but both together were found in only 3%, and there were none with PEO and heart block, and none with the full KSS syndrome.

(5) These are complex clinical syndromes and it has been necessary to make lists of symptoms and signs that would qualify for diagnosis. 'Partial' syndromes are often seen in the relatives of probands or in individual patients. The need to set some kind of criteria is exemplified by the suspicion that some cases classed as progressive myoclonus epilepsy have a limited form of MERRF, but this has not been proven in molecular terms, or that the Ramsay–Hunt syndrome is really synonymous with MERRF, a possibility that has been excluded by molecular analysis.

(6) Munnich *et al.* [57] denoted the diversity of clinical manifestations that raise the possibility of a mitochondrial disorder in infants and children: ketoacidotic coma, hepatic failure, anaemia, cardiomyopathy, failure to thrive, diarrhoea, renal tubulopathy, diabetes, dwarfism, hypoparathyroidism, Leigh syndrome, ataxia, optic atrophy, epilepsy or myopathy as well as the manifestations of MELAS, MERRF and KSS. The diversity of clinical syndromes with RRF encompasses much of paediatrics, internal medicine, neurology, ophthalmology

Table 6.7 Mitochondrial diseases associated with point mutations

Disease	Nucleotide	Gene
LHON*	11778	ND4
	3460	ND1
	4160	ND1
	15257	CYB
	7444	CO1
	3394	ND1
MERRF	8344	tRNA^{Lys}
	8356	tRNA^{Lys}
MELAS	3243,3271	tRNA^{Leu}
	11084	ND4
Diabetes mellitus, maternally inherited	3243	tRNA^{Leu}
PEO, sporadic	3243	tRNA^{Leu}
NARP	8993	ATPase 6
Leigh disease	8993	ATPase 6
Cardiomyopathy, limb myopathy	3260	tRNA^{Leu}

*In addition, there are several 'synergistic' mtDNA mutations in LHON [58].

and psychiatry. Symptoms that were considered incidental to the main disorders are now linked to point mutations (Table 6.7). This has happened for diabetes and deafness; it may yet happen for other manifestations. Targets include short stature, complicated migraine, stroke before age 40, dementia in adolescence or young adults, or any symptom inherited in maternal fashion.

Investigators need a language to communicate with each other. Although there is a general impatience about the use of eponyms and acronyms, this shorthand is more informative than a blanket term such as mitochondrial encephalomyopathy, which is needed to encompass the field (and the title of this paper) but covers such a diversity of syndromes that it is almost meaningless. Besides, the term and its alter egos are almost unpronounceable.

CONTROVERSIAL QUESTION NO. 8: It would be inappropriate to leave our argumentative field totally devoid of controversy, which has been so stimulating and productive in the past. So here are some suggestions that can be targets for future debate

(1) Among syndromes with RRF (morphological criteria), we should continue to separate clinical syndromes (clinical criteria).
(2) If there is a characteristic abnormality of mtDNA in the disease in question and it is not found in an individual case or family with the syndrome, or if some other mtDNA mutation is found, we recognize that there is genetic heterogeneity. The same clinical syndrome can be found with different mutations.
(3) If, in a syndrome that is usually associated with deletions, no deletion is found, a point mutation should be considered. Similarly, if no mutation is found in a

syndrome that is usually characterized by a particular point mutation, some other point mutation should be considered.

(4) If the clinical syndrome is blurred by manifestations of two of the major disorders in the same patient, and if mtDNA shows the mutation of one syndrome, the diagnosis shall be determined by the mtDNA mutation.

(5) If, in such a clinically indeterminate case, neither of the characteristic mutations is found, then the syndrome may be delineated as, for instance, 'MELAS–MERRF syndrome, aetiology undetermined'.

(6) Even if we discover the affected gene products in these diseases, the problem of nomenclature may still not be settled, but we would have a more rational basis of classification. We might then have hope for understanding pathogenesis and perhaps even a rational therapy.

ACKNOWLEDGEMENTS

The author is totally dependent on a team of remarkable investigators which is led by S. DiMauro and E.A. Schon and included (in alphabetical order) the following luminaries: E. Bonilla, F. Ciacci, M. Davidson, D.C. De Vivo, A.P. Hays, M. Hirano, M.P. King, Y. Koga, J. Masucci, A.F. Miranda, C.T. Moraes, M. Sciacco, and S. Shanske, A. Suomalainen, and F. Santorelli. Research Fellows who have been involved in the work cited include, E. Arnaudo, E. Ricci, E. Ciafaloni and G. Silvestri. Other Fellows who now lead their own laboratories include M. Zeviani, R. Rizzuto, S. Mita, H. Nakase, S. Servidei and N. Bresolin. They all supplied the data and tempered my interpretations; the errors are my own. This work was supported by Clinical Center Grants from the Muscular Dystrophy Association and the National Institute of Neurological Diseases and Stroke (GRRC RR00645; NS 11766).

REFERENCES

1. Harding, A.E. (1991) Neurological disease and mitochondrial genes. *Trends in Neurological Sciences*, **14**, 132–138
2. Moraes, C.T., Schon, E.A. and DiMauro, S. (1991) Mitochondrial diseases: toward a more rational classification. In *Current Neurology* (ed. S.H. Appel), Mosby Year Book, St. Louis, Vol. 11, pp. 83–112
3. DiMauro, S., Tonin, P. and Servidei, S. (1992) Metabolic myopathies. *Handbook of Clinical Neurology*, **62**, 479–526
4. Shy, G.M., Gonatas, N.K. and Perez, M. (1966) Childhood myopathies with abnormal mitochondria. I. Megaconial myopathy – pleoconial myopathy. *Brain*, **89**, 133–158
5. Luft, R., Ikkos, D., Palmieri, G. *et al.* (1962) Severe hypermetabolism of nonthyroid origin with a defect in the maintence of mitochondrial respiratory control. *Journal of Clinical Investigation*, **41**, 1776–1804
6. Haydar, N.A., Conn, H.L., Afifi, A. *et al.* (1971) Severe hypermetabolism with primary abnormality of skeletal muscle mitochondria. *Annals of International Medicine*, **74**, 548–558
7. DiMauro, S., Bonilla, E., Lee, C.P. *et al.* (1976) Luft's disease; further biochemical and ultrastructural studies of skeletal muscle in the second case. *Journal of Neurological Sciences*, **27**, 217–232
8. Engel, W.K. and Cunningham, C.G. (1963) Rapid examination of muscle tissue: an improved trichrome stain method for fresh-frozen biopsy specimens. *Neurology*, **13**, 919–923
9. Rowland, L.P., Hays, A.P., DiMauro, S., De Vivo, D.C. and Behrens, M. (1983) Diverse clinical disorders associated with morphological abnormalities of mitochondria. In *Mitochondrial Pathology in Muscle Diseases* (eds. C. Cerri and G. Scarlato), Piccin Editore, Padua, pp. 141–158

10. Hammans, S.R., Sweeney, M.G., Brockington, M., Morgan-Hughes, J.A. and Harding, A.E. (1991) Mitochondrial encephalopathies: molecular genetic diagnosis from blood samples. *Lancet*, **357**, 1311–1313

11. Vaamonde, J., Muruzabal, J., Tunon, T. *et al.* (1992) Abnormal muscle and skin mitochondria in family with myoclonus, ataxia, and deafness (May and White syndrome). *Journal of Neurology, Neurosurgery and Psychiatry*, **55**, 128–132

12. Kearns, T.P. and Sayre, G. (1958) Retinitis pigmentosa, external ophthalmoplegia and complete heart block. *Archives of Ophthalmology*, **60**, 280–289

13. Rowland, L.P. (1975) Progressive external ophthalmoplegia. *Handbook of Clinical Neurology*, **22**, 177–202

14. Berenberg, R.A., Pellock, J.M., DiMauro, S. *et al.* (1977) Lumping or splitting? 'Ophthalmoplegia-plus" or Kearns–Sayre syndrome? *Annals of Neurology*, **1**, 37–54

15. Drachman, D.A. (1975) Ophthalmoplegia-plus; classification of disorders associated with progressive external ophthalmoplegia. *Handbook of Clinical Neurology*, **22**, 203–216

16. Bastienssen, L.A.K., Joosten, E.N.G. and DeRooj, J.A.M. (1978) Ophthalmoplegia-plus; a real nosological entity. *Acta Neurologica Scandinavica*, **58**, 9–34

17. Schnitzler, E.R. and Robertson, E.C. (1979) Familial Kearns–Sayre syndrome. *Neurology*, **29**, 1172–1174

18. Rowland, L.P., Hausmanowa-Petrusewicz, I., Warburton, D., Barduska, B., Niebroj-Dobosz, I., DiMauro, S. and Johnson, W.G. (1988) Kearns–Sayre syndrome in twins; lethal dominant mutation or acquired disease? *Neurology*, **38**, 1399–1402

19A. Jankowitz, E., Berger, E., Kurasz, S., Winogrodzka, W. and Elasz, L. (1977) Familial external ophthalmoplegia with abnormal muscle mitochondria. *European Neurology*, **15**, 318–324

19B. Desnvelle, C., Pellisier, J.F., Serratrice, G. *et al.* (1989) Le syndrome de Kearns et Sayres encéphalomyopathie mitochondriale par défecit de la chaine respiratoire. *Revue Neurologique (Paris)*, **145**, 842–850

20. Fukuhara, N., Tokiguchi, S., Shirakawa, K. and Tsubaki, T. (1980) Myoclonus epilepsy associated with ragged red fibers (mitochondrial abnormalities): disease entity or syndrome? *Journal of Neurological Sciences*, **47**, 117–133

21. Fukuhara, N. (1983) Myoclonus epilepsy and mitochondrial myopathy. In *Mitochondrial Pathology in Muscle Diseases*, (eds C. Cerri and G. Scarlato), Piccin Editore, Padua, pp. 87–111

22. Pavlakis, S.G., Phillips, P.C., DiMauro, S., De Vivo, D.C. and Rowland, L.P. (1984) Mitochondrial myopathy, encephalopathy, lactic acidosis, and stroke-like episodes; a distinctive clinical syndrome. *Annals of Neurology*, **16**, 481–487

23. Petty, R.K.H., Harding, A.E. and Morgan-Hughes, J.A. (1986) The clinical features of mitochondrial myopathy. *Brain*, **109**, 915–938

24. Holt, I.J., Harding, A.E. and Morgan-Hughes, J.A. (1988) Deletions of muscle mitochondrial DNA in patients with mitochondrial myopathy. *Nature*, **331**, 717–718

25. Zeviani, M., Moraes, C.T., DiMauro, S., Nakase, H., Bonilla, E., Schon, E.A. and Rowland, L.P. (1988) Deletions of mitochondrial DNA in Kearns–Sayre syndrome. *Neurology*, **38**, 1339–1346

26. Moraes, C.T., DiMauro, S., Zeviani, M. *et al.* (1989) Mitochondrial DNA deletions in progressive external ophthalmoplegia and Kearns–Sayre syndrome. *New England Journal of Medicine*, **320**, 1293–1299

27. Wallace, D.C., Singh, G., Lott, M.T. *et al.* (1988) Mitochondrial DNA mutation associated with Leber's hereditary optic atrophy. *Science*, **242**, 1427–1430

28. Goto, Y.I., Nonaka, I. and Horai, S. (1990) A mutation in the tRNA[Leu(UUR)] gene associated with the MELAS subgroup of mitochondrial encephalopathies. *Nature*, **348**, 651–653

29. Shoffner, J.M., Lott, M.T., Lezza, A.M.S. *et al.* (1990) Myoclonic epilepsy and ragged red fiber disease (MERRF) is associated with a mitochondrial DNA tRNA[Lys] mutation. *Cell*, **61**, 931–937

30. Holt, I.J., Harding, A.E., Petty, R.K.H. and Morgan-Hughes, J.A. (1990) A new mitochondrial disease associated with mitochondrial DNA heteroplasmy. *American Journal of Human Genetics*, **46**, 428–483

31. Hirano, M., Silvestri, G., Blake, D.M. *et al.* (1993) Mitochondrial neuro-gastrointestinal encephalomyopathy (MNGIE): clinical, biochemical and genetic features of an autosomal recessive disorder. *Neurology*, in press

32. Rowland, L.P. (1992) Progressive external ophthalmoplegia. *Handbook of Clinical Neurology*, **18**, 287–329

33. Zeviani, M. and Antozzi, C. (1992) Defects of mitochondrial DNA. *Brain Pathology*, **2**, 121–132

34. Haltia, M., Suomalainen, A., Majander, A. and Somer, H. (1992) Disorders associated with multiple deletions of mitochondrial DNA. *Brain Pathology*, **2**, 133–139

35. Zupanc, M., Moraes, C.T., Shanske, S. *et al.* (1991) Mitochondrial DNA deletion in patients with combined features of Kearns–Sayre and MELAS syndromes. *Neurology*, **41**, 680–683

36. Zeviani, M., Servidei, S., Gellera, C. *et al.* (1989) Autosomal dominant disorder with multiple deletions of mitochondrial DNA starting at the D-loop region. *Nature,* **154**, 1240–1247
37. Lombes, A. *et al.* (1993) Kearns–Sayre syndrome with depletion of mtDNA. *Lancet,* in press
38. Moraes, C.R., Ciacci, F., Silvestri, G., Shanske, S., Sciacco, M., Hirano, M., Schon, E.A., Bonilla, E. and DiMauro, S. (1993) Atypical clinical presentation with the MELAS mutation at position 3243 of human mitochondrial DNA. *Neuromuscular Disorders,* **3**, 43–50
39. Goto, Y.I., Horai, S., Matsuoka, T. *et al.* (1992) MELAS: correlative study of clinical features and mitochondrial DNA mutation. *Neurology,* **42**, 545–549
40. Ciafaloni, R., Ricci, E., Shanske, S. *et al.* (1992) MELAS: clinical features, biochemistry and molecular genetics. *Annals of Neurology,* **31**, 391–398
41. Nonaka, I. (1992) Mitochondrial diseases. *Current Opinions in Neurology and Neurosurgery,* **5**, 622–631
42. Goto, Y.I., Nonaka, I. and Horai, S. (1990) A mutation in the tRNA$^{Leu(UUR)}$ gene associated with the MELAS subgroup of mitochondrial encephalopathies. *Nature,* **348**, 651–653
43. Lertrit, P., Noer, A.S., Jean-Francois, M.J.B. *et al.* (1992) A new disease-related mutation for MELAS affects the ND4 subunit of the respiratory complex I. *American Journal of Human Genetics,* **51**, 457–468
44. Ballinger, S.W., Shoffner, J.M., Hedaya, E.V. *et al.* (1992) Maternally transmitted diabetes and deafness associated with a 10.4 kb mitochondrial DNA deletion. *Nature and Genetics,* **1**, 11–15
45. Johns, D.R. and Hurko, O. (1991) Mitochondrial leucine tRNA mutation in neurological diseases. *Lancet,* **337**, 927–928
46. Inui, K., Fukushima, H., Tsukamoto, H. *et al.* (1992) Mitochondrial encephalomyopathies with the mutation of the mitochondrial tRNA$^{Leu(UUR)}$ gene associated with MELAS. *Molecular and Cellular Biology,* **12**, 480–490
47. Goto, Y.I., Tojo, M., Tohyama, J., Horai, S. and Nonaka, I. (1992) A novel point mutation in the mitochondrial tRNA$^{Leu(UUR)}$ gene in a family with mitochondrial myopathy. *Annals of Neurology,* **31**, 672–675
48. Zeviani, M., Gellera, C., Antozzi, C. *et al.* (1991) Maternally inherited myopathy and cardiomyopathy associated with mutation in mitochondrial DNA tRNA$^{Leu(UUR)}$. *Lancet,* **338**, 143–147
49. Silvestri, G., Ciafaloni, E., Santorelli, F.M. *et al.* (1993) Clinical features associated with the A→G transition at nucleotide 8344 of mtDNA ('MERRF mutation'). *Neurology,* **43**, 1200–1206
50. Silvestri, G., Moraes, C.T., Shanske, S., Oh, S.J. and DiMauro, S. (1992) A new mtDNA mutation in the tRNA^{Lys} gene associated with MERRF. *American Journal of Human Genetics,* **51**, 1213–1217
51. Ciafaloni, E., Ricci, E., Shanske, S. *et al.* (1992) MELAS: clinical features, biochemistry and molecular genetics. *Annals of Neurology,* **31**, 391–398
52. Hirano, M., Ricci, E., Koenigsberger, M.R. *et al.* (1992) MELAS: an original case and clinical criteria for diagnosis. *Neuromuscular Disorder,* **2**, 125–135
53. DiMauro, S., Moraes, C.T., Shanske, S. *et al.* (1991) Mitochondrial encephalomyopathies: biochemical approach. *Revue Neurologique (Paris),* **147**, 443–449
54. Bundey, S., Poulton, K., Whitwell, H., Curtis, E., Brown, I.A.R. and Fielder, A.R. (1992) Mitochondrial abnormalities in the DIDMOAD (Wolfram) syndrome. *Journal of Inherited Metabolic Diseases,* **15**, 315–319
55. Van Hove, J.L.K., Shanske, S., Cicacci, F. *et al.* (1994) Mitochondrial myopathy with anemia, cardiomyopathy and lactic acidosis; a distinct late-onset mitochondrial disease. *Journal of Neurological Science,* in press
56. Rowland, L.P., Blake, D.M., Hirano, M. *et al.* (1991) Clinical syndromes associated with ragged red fibers. *Revue Neurologique (Paris),* **290**, 457–465
57. Munnich, A., Rustin, P., Rotig, A. *et al.* (1992) Clinical aspects of mitochondrial disorders. *Journal of Inherited Metabolic Diseases,* **15**, 448–455
58. Shoffner, J.M. and Wallace, D.C. (1992) Mitochondrial genetics: principles and practice. *American Journal of Human Genetics,* **51**, 1179–1186

7
Intermediary Metabolism

Darryl C. De Vivo

INTRODUCTION

The mitochondrial environment is central to the process of intermediary metabolism. The glycolytic pathway converts glucose to pyruvate in the cytoplasm. Pyruvate is in equilibrium with lactate, and is translocated across the inner mitochondrial membrane by a specific monocarboxylic carrier mechanism. Pyruvate has two principal intramitochondrial fates. The first involves oxidative decarboxylation and activation to acetyl-CoA; the second involves carboxylation to oxaloacetate. These two reaction products, acetyl-CoA and oxaloacetate, can condense to form citrate, a 6-carbon molecule; and citrate is decarboxylated in the tricarboxylic acid cycle yielding carbon dioxide and reducing equivalents. The reducing equivalents (NADH, $FADH_2$) are reoxidized by Complexes I and II of the respiratory chain. The hydrogen ion reacts with molecular oxygen to form water. The liberated thermic energy is trapped in the high-energy phosphate bond of adenosine triphosphate (ATP) by the coupling of oxidation to phosphorylation. The net reaction of glycolysis and oxidative metabolism for 1 mol of glucose ($C_6H_{12}O_6$) is shown in equation 1:

$$C_6H_{12}O_6 + 6O_2 + 36ADP + 36P_i \rightarrow 6CO_2 + 6H_2O + 36ATP \tag{1}$$

This equation assumes that 100% of the pyruvate formed by glycolysis enters the mitochondrial compartment. Evidence suggests that some small fraction (perhaps 10–15% in brain) is converted to lactate. This percentage is influenced by the tissue energy demand as reflected by the cellular energy potential. Numerous inborn metabolic errors interfere with the process of intermediary metabolism. The classic metabolic signature of a biochemical defect involving oxidative metabolism is the elevation of tissue pyruvate and lactate concentrations. As stated earlier, these two organic acids are in equilibrium, and the ratio is influenced by the tissue hydrogen pressure, or the cellular oxidation–reduction potential (redox state):

$$\text{Pyruvate} + \text{NADH} + \text{H}^+ \leftrightarrow \text{lactate} + \text{NAD}^+ \tag{2}$$

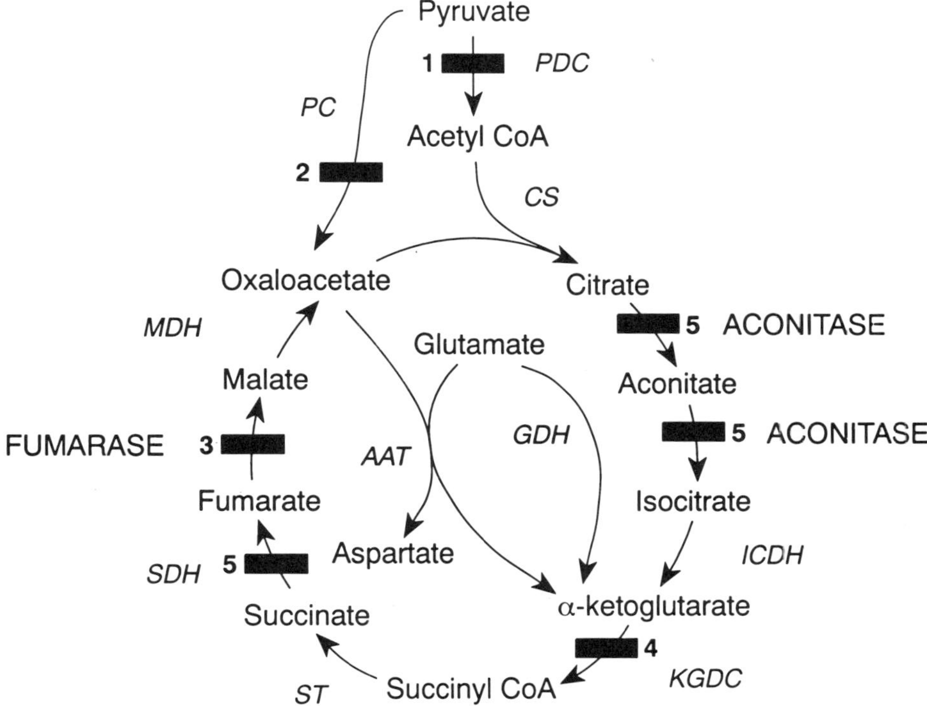

Figure 7.1 Schematic representation of intermediary metabolism. The reported defects are indicated by the numbered solid bars (see the text for details). Abbrevations: PDC, pyruvate dehydrogenase complex; PC, pyruvate carboxylase; CS, citrate synthase; ICDH, isocitrate dehydrogenase; KGDC, α-ketoglutarate dehydrogenase complex; ST, succinate thiokinase; SDH, succinate dehydrogenase; MDH, malate dehydrogenase; GDH, glutamate dehydrogenase; AAT, aspartate aminotransferase

The lactate/pyruvate ratio is normally approximately 10. Ratios exceeding 20 are abnormal, indicating a shift of the cellular redox state to a more reduced potential. Lack of oxygen (e.g. asphyxia) is a classic clinical condition associated with an elevated lactate/pyruvate ratio. Similarly, biochemical defects involving the respiratory chain (see Chapter 5) cause accumulation of reducing equivalents and an elevation of the lactate/pyruvate ratio. Biochemical defects involving pyruvate metabolism directly, or indirectly in the tricarboxylic acid cycle, should not perturb the cellular redox state. Therefore, the lactate/pyruvate ratio should remain normal even though the organic acids accumulate in the tissue.

Biochemical defects involving intermediary metabolism also affect the tissue concentrations of acetyl-CoA. Acetyl-CoA is the central metabolite in intermediary metabolism, formed mainly from the decarboxylation of pyruvate, β-oxidation of fatty acids and the oxidation of ketone bodies in extrahepatic tissues.

Expansion of the acetyl-CoA intracellular pool, as might occur with decreased synthesis of oxaloacetate, a block in the tricarboxylic acid cycle or a respiratory chain defect, results in the increased synthesis of ketone bodies. Acetyl-CoA is diverted to the β-hydroxy-β-methylglutaryl-CoA cycle under these conditions. A

Table 7.1 Molecular genetics of the pyruvate dehydrogenase complex

Components	Subunits	Chromosome
Pyruvate dehydrogenase E_1 ($\alpha_2 \beta_2$)	$E_1 \alpha$	X, 4
	$E_1 \beta$	3
Dihydrolipoyl transacetylase E_2 (α_{60})	$E_2 \alpha$	
Dihydrolipoyl dehydrogenase E_3 (α_2)	$E_3 \alpha$	7
Protein X	X	
E_1-Kinase ($\alpha \beta$)	α	
	β	
PhosphoE$_1$-phosphatase ($\alpha \beta$)	α	
	β	

mixed metabolic acidosis with elevations of lactate, pyruvate, β-hydroxybutyrate and acetoacetate often accompanies deficiencies of pyruvate carboxylase, tricarboxylase, tricarboxylic acid cycle enzymes and respiratory chain enzymes. The ratio of lactate/pyruvate reflects the cytoplasmic redox state, and the ratio of β-hydroxybutyrate/acetoacetate reflects the mitochondrial redox state.

These metabolic profiles are helpful in anticipating the site of the enzyme defect in intermediary metabolism (Figure 7.1).

PYRUVATE DEHYDROGENASE COMPLEX DEFICIENCY

The biochemistry and molecular genetics of the pyruvate dehydrogenase multienzyme complex (PDC) are complicated [1]. This complexity is transmitted to the clinical domain, and to the understanding of the relationship between phenotypic expression and the genotypic abnormality. The complex is composed of five enzyme reactions and at least nine different proteins as shown in Table 7.1.

The three catalytic proteins (E_1, E_2 and E_3) facilitate the conversion of pyruvate to acetyl-CoA as shown in equation 3:

$$\text{Pyruvate} + \text{CoASH} + \text{NAD}^+ \rightarrow \text{acetyl-CoA} + \text{NADH} + \text{H}^+ + \text{CO}_2 \qquad (3)$$

The E_1 component forms a 2-(1-hydroxyethylidene)thiamine pyrophosphate complex and carbon dioxide. The E_2 component transfers the acetate group to CoASH forming acetyl-CoA. The E_3 component is a flavin-requiring enzyme that reduces NAD^+ to NADH and H^+. Lipoyl groups are covalently linked to the E_2 component and transfer the acetyl group and hydrogen to the respective components as schematically represented in Figure 7.2.

Four genes have been cloned for the PDC as shown in Table 7.1. Two genes, one located on the X-chromosome, encode the $E_1 \alpha$ subunit. This subunit is pivotal in the catalytic activity of the enzyme complex (Figure 7.2). The E_1 component is tightly regulated by a phosphorylation (inactivated)–dephosphorylation (activated) mechanism. Phosphorylation involves three serine residues on the $E_1 \alpha$ subunit, and is controlled by the E_1-kinase and the phosphoE$_1$-phosphatase. A mutation of the $E_1 \alpha$ subunit gene may affect the catalytic efficiency of the complex directly, or indirectly through the kinetic alteration of the phosphorylation–dephosphorylation regulation sequence.

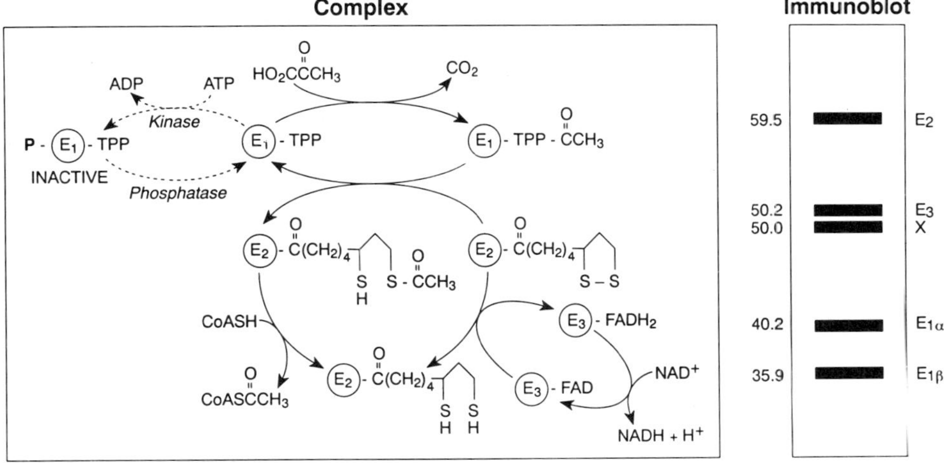

Figure 7.2 Schematic representation of the pyruvate dehydrogenase complex (PDC) and an immunoblot of the PDC components. E_1, E_2 and E_3 represent the catalytic components of the complex. The E_1 component contains an α and β subunit. The complex is regulated by phosphorylation–dephosphorylation of the E_1 α subunit under the catalytic regulation of the E_1-kinase and the phosphoE$_1$-phosphatase

More than 100 patients with PDC defects have been described since the pioneering observations of Blass and co-workers in 1970 [2]. Most patients with PDC deficiency have a molecular defect involving the E_1 α subunit [3–7]. Robinson and colleagues, in 1987, [4] reviewed 30 patients from their experience with an E_1 defect, and culled an additional 24 cases from the literature. These 54 cases were broadly grouped in four categories depending on the age at clinical presentation and death. Thirteen patients died between birth and age 6 months; 12 patients died between ages 7 and 36 months; 23 patients were alive at the time of the review and had evidence of chronic psychomotor retardation; and six patients had a recurrent syndrome manifested primarily by ataxia. These patients were neurologically well between attacks. The degree of lactic acidosis correlated with this clinical grouping, as has been noted by others [5,8]. The residual enzyme activity in cultured fibroblasts correlated less well except for the most severely affected infants presenting in the newborn period. These patients had the lowest residual enzyme activity and the most severe lactic acidosis. Our review of the literature has led us to similar conclusions regarding the clinical presentations (R.C. van Coster and D.C. De Vivo, unpublished observations). We also noted that the majority of patients had a defect of the E_1 component, including a few patients with presumed abnormality of the phosphoE$_1$-phosphatase. These patients were subdivided into three presentations: neonatal, infantile–early childhood and benign.

The neonatal presentation was manifested by hypotonia, episodic apnea, convulsions, weak suck, dysmorphic features, lethargy, low birth weight, failure to thrive and coma [4,7,9–21]. There was a male predominance in the neonatal group reflecting the molecular genetics of the E_1 α subunit. The dysmorphic features included broad nasal bridge, upturned nose, micrognathia, low-set and posteriorly rotated ears, short fingers, short arms, Simian creases, hypospadias and anteriorly placed anus. Patients presenting in the newborn period usually died before age 8 months. The neuropathological abnormalities included cystic degeneration of the subcortical white matter, basal ganglia and brainstem. Associated abnormalities included dysmyelination, optic atrophy, hydrocephalus, agenesis of the corpus callosum, ectopic olivary nuclei, reactive gliosis, spongy degeneration of the neuropil, neuronal loss and vascular proliferation. Some of the histopathological features, although not distinctive, were reminiscent of the light microscopic features seen in Leigh syndrome.

The majority of patients displayed an infantile phenotype with onset of symptoms between 3 and 6 months of age [4,7,9,11,12,18,20,22–45]. These symptoms included psychomotor retardation, hypotonia, convulsions, episodic apnoea, ataxia, pyramidal tract signs, lethargy, dysmorphic features, deceleration of head growth with acquired microcephaly, ophthalmoplegia, optic atrophy, peripheral neuropathy, ptosis, dysphagia, deceleration in somatic growth, extrapryramidal signs and cranial nerve palsies. Autopsy results were reported in 21 cases, and 83% of these patients had neuropathological abnormalities consistent with Leigh syndrome.

Seven male children had the benign phenotype with episodic symptomatology which included post-exercise fatigue, fluctuating ataxia, transient paraparesis and thiamine responsiveness [4,9,11,33,46,47]. These boys demonstrated normal mental and motor development between episodes.

One patient with an E_2 defect has been reported [48]. This girl, aged 3½ years at the time of the report, had psychomotor retardation, acquired microcephaly, moderate lactic acidosis and associated hyperammonaemia.

Two additional patients have been described with abnormalities of protein X [48]. The first patient was symptomatic in early infancy with hypotonia and poor head control. By age 4 years, he was severely delayed in psychomotor development, and unable to walk or speak. His parents were first cousins. The second patient was asymptomatic until age 3 years when he demonstrated gait instability, difficulty in running, and frequent falling. The remainder of the neurological findings were normal, although the CT scan revealed bilateral basal ganglia lucencies and abnormalities in the putamen. Protein X is a 50 kDa protein containing a lipoyl group attached to a lysine residue (Figure 7.2). This protein plays an important role in the organization and function of the PDC and participates in the shuttling of electrons and acetyl groups between the E_2 and E_3 components [1].

Several patients with a deficiency of the E_3 component have been described [49–52]. The first patient described with this condition died at age 7 months. Progressive neurological symptomatology included inspiratory stridor, laboured respirations, lethargy, hypotonia alternating with irritability, optic atrophy and hyperreflexia. A persistent metabolic acidosis was present. Autopsy findings included cavitation of the basal ganglia, thalami and brainstem, and loss of myelin. The organic acid profile is distinctive because the E_3 component is shared by pyruvate dehydrogenase, α-ketoglutarate dehydrogenase, and the α-ketoacid dehydrogenase. This patient had elevated lactate, pyruvate, alanine, α-ketoglutarate and branched-chain α-ketoacids. The E_3 component is encoded by a gene on chromosome 7 as shown in Table 7.1.

In summary, the clinical phenotypes associated with PDC defects include: a neonatal presentation with severe congenital lactic acidosis and dysmorphic features; an infantile presentation with psychomotor retardation, acquired microcephaly, and the clinical and neuropathological correlates of Leigh syndrome; and a benign condition with episodic ataxia and limb weakness. Other clinical presen-tations include sudden infant death syndrome [13], central hypoventilation syndrome [25] and limb weakness associated with a chronic motor neuropathy [45,53].

The region of the X chromosome containing the gene for the E_1 α subunit is highly conserved. Twenty-three mutations of the E_1 α subunit gene have been described [6,54,55]. These mutations show striking allelic heterogeneity with point mutations, microdeletions and insertions. Eighteen of the mutations are unique and there is a clustering of these mutations at the C-terminus of the gene involving exons 10 and 11. Most of the mutations appear to be sporadic, and the clinical expression of the genotypic abnormality in the female is influenced by the non-random pattern of X-chromosome inactivation [56].

PYRUVATE CARBOXYLASE DEFICIENCY

The direct carboxylation of pyruvate to oxaloacetate was documented convincingly in 1963 by Utter and Keech [57,58]. The reaction, as shown in equation 4, is catalysed by the enzyme pyruvate carboxylase, one of the four known biotin-containing proteins found in mammalian tissues:

$$\text{Pyruvate} + \text{ATP} + \text{HCO}_3 \rightarrow \text{oxaloacetate} + \text{ADP}^- + \text{P}_i \tag{4}$$

The enzyme is composed of four identical subunits, and is dependent on Mg^{2+} and acetyl-CoA for full activity. It is located exclusively in the mitochondrial matrix, and widely distributed throughout the body tissues. Highest concentrations are present in liver, kidney, adrenal gland, lactating mammary gland and newborn brown adipose tissue; measurable activities are present in brain, white adipose tissue, heart and testes; and very low activities are present in intestinal mucosa and skeletal muscle [59]. Although there is only one form of pyruvate carboxylase in body tissues, the roles of the enzymes are tissue-specific. The enzyme role in gluconeogenesis is important in liver and kidney; its role in lipogenesis is important in the adrenal gland, lactating mammary gland and adipose tissue. In all tissues, it subserves an important anapleurotic role in the synthesis of oxaloacetate, the rate-limiting intermediate in the Krebs cycle (Figure 7.1). A deficiency of pyruvate carboxylase therefore has profound implications in intermediary metabolism. Aspartic acid, in equilibrium with oxaloacetate through a transamination reaction, may be depleted in severe pyruvate carboxylase deficiency with resulting alterations in the cytoplasmic and mitochondrial redox states, and ureagenesis.

cDNA clones of human pyruvate carboxylase have been isolated and characterized, and the gene has been localized to the long arm of chromosome 11 [60]. Available evidence suggests that only one form of pyruvate carboxylase exists in human tissues [61,62]. Therefore, prenatal and postnatal diagnoses are possible by assaying the enzyme in cultured fibroblasts, amniocytes or lymphocytes.

Classical pyruvate carboxylase deficiency has been described in 34 cases, and may have two distinctive clinical presentations. These phenotypes are determined by the presence or absence of any residual enzyme activity [63].

The North American phenotype has been reported in 18 patients, and is associated with the presence of immunoreactive enzyme protein and residual enzyme activity [63–74]. These patients have severe psychomotor delay and death in infancy or early childhood. The metabolic features include lactic acidosis, ketosis, an increased lactate/pyruvate ratio and decreased β-hydroxybutyrate/acetoacetate ratio.

The French phenotype has been reported in 16 patients, and is associated with decreased or absent cross-reacting material and negligible residual enzyme activity [62,63,75–83]. As a result, there is no carboxylation of pyruvate, and the tissue oxaloacetate concentrations are depleted. Aspartic acid levels are markedly decreased and interfere with the conversion of citrulline to argininosuccinate, an important step in the urea cycle. Clinically, these patients are severely ill in the newborn period and die within the first 4 months of life. Metabolic features are similar to the findings in the North American phenotype and include, in addition, hyperammonaemia, citullinaemia and hyperlysinaemia. Intermittent hypercholesterolaemia has been observed but hypoglycaemia is less common, probably because alternative gluconeogenic precursors are available. Impaired Krebs cycle activity is probably the most devastating consequence of pyruvate carboxylase deficiency leading to cellular energy failure. Six reports have discussed the apparent association of pyruvate carboxylase deficiency with Leigh syndrome, but the prevailing consensus suggests that pyruvate carboxylase deficiency is not causally related to this specific neuropathological entity [84–89].

A single patient has been described with a benign variant of pyruvate carboxylase deficiency associated with normal growth and development [90]. This 7-year-old girl was hospitalized repeatedly from age 7 months for vomiting and dehydration. Between attacks her psychomotor development was normal. The metabolic abnormalities during the crises were consistent with the North American phenotype. Residual enzyme activity in cultured skin fibroblasts was 1.8% of normal. Enzyme cross-reacting material was present in normal abundance and electrophoretic mobility. The molecular basis for this benign variant is not known.

The neuropathology associated with classical pyruvate carboxylase deficiency reveals abnormalities of myelin formation and neuronal migration [64]. There is severe depletion of cerebral cortical neurons, and the presence of ectopic neurons and glia in the white matter and subarachnoid space. Paraventricular cystic cavitation also is seen in the French phenotype. These findings suggest that the principal insults to brain associated with pyruvate carboxylase deficiency occur after mid-gestation and continue into the postnatal period. These conclusions are consistent with the lack of neuropathological abnormalities in a 20-week-old affected fetus [80].

The molecular basis for pyruvate carboxylase deficiency has not been studied in detail. It is presumed that all cases result from mutations of the gene located on chromosome 11q. The condition is transmitted as an autosomal recessive trait, and the heterozygote state can be documented by residual enzyme activity in cultured skin fibroblasts or lymphocytes [91].

FUMARASE DEFICIENCY

The fumarate hydratase gene is located on the long arm of chromosome 1. This single gene is responsible for six electrophoretically different fumarase isoforms, two located in the cytoplasm and four in the mitochondria [92]. These isoforms share a high degree of structural homology, and are not readily differentiated by immunological methods. The cytosolic isoenzyme is a homotetramer with subunit molecular mass of 45 000 Da. The mitochondrial isoenzyme is also a homotetramer. The subunit molecular mass is 50 000 Da before the 5000 Da leader sequence is cleaved by the mitochondrial endopeptidase following subcellular importation. The rat liver fumarase mRNA contains two separate messages. The two separate isoenzymes are synthesized by differential splicing of this single genetic transcript. The reaction catalysed by fumarase is shown in equation 5 (Figure 7.1):

$$\text{Fumarate} + \text{H}_2\text{O} \rightarrow \text{malate} \tag{5}$$

This reaction is central to Krebs cycle activity in the mitochondrion. In the cytoplasm, fumarase activity is important in the malate–aspartate shuttle system. Cytoplasmic fumarate is generated in the urea cycle and in purine nucleotide biosynthesis.

Seven patients have been described since 1986 with fumarase deficiency [92–97]. Two additional patients were described with fumaric aciduria, mental retardation and impaired speech [98]. These adult siblings were thought to have a defect in renal clearance, and fumarase activity was not determined.

Two phenotypes have emerged from this small series of cases, both present in infancy. Three patients had a fatal infantile encephalopathy with psychomotor deterioration, deceleration of head growth and hypotonia [93–95]. Death occurred between 6 and 8 months. Fumaric aciduria was the most important metabolic feature. Lactate values in blood and urine were more variable, normal on occasions and markedly elevated on other occasions particularly during metabolic crises. One autopsy revealed microcephaly, hypomyelination, heterotopias in cerebellar and posterior cerebral regions [95]. Neuroimaging revealed enlarged cerebral ventricles, and agenesis of the corpus callosum in one case, associated with dysmorphic features and hepatocellular damage [94]. Generalized seizures were noted in a second case [95]. Cytosolic and mitochondrial fumarase isoenzymes were deficient in two cases [93,95], and liver total fumarase activity was decreased in the third case [94].

The second phenotype is more protracted after presentation in infancy [92,96,97]. One child presented at 6 months with hypotonia, microcephaly and delayed development [92]. Parents were consanguineous. A second child had congenital hypotonia, and later was diagnosed as having cerebral palsy with psychomotor retardation and hypotonia [96]. At age 5 years, fumaric acid was detected in the urine of this patient. Two brothers had polyhydramnios and cerebral ventriculomegaly *in utero* [97]. After their birth, they had developmental delay, infantile spasms, severe cerebral atrophy and a hypsarrhythmic EEG pattern. The older boy died at age 64 months, and the younger sibling was alive at age 34 months. Fumarase deficiency was documented in cultured fibroblasts (both isoenzymes) and lymphocytes. In the other patients, total fumarase deficiency was found in fibroblast homogenates in one case [96], and a 20% residual activity was found in the mitochondrial isoenzyme in the other case [92].

These seven cases indicate that fumarase deficiency may be associated with a rapidly progressive fatal infantile encephalopathy, or a relatively non-progressive neurological syndrome and, like PDC deficiency, may be associated with brain malformations.

α-KETOGLUTARATE DEHYDROGENASE DEFICIENCY

α-Ketoglutarate dehydrogenase is a multienzyme complex that catalyses the oxidative decarboxylation of α-ketoglutarate to succinyl-CoA (Figure 7.1) as shown in equation 6:

$$\alpha\text{-Ketoglutarate} + CoASH + NAD^+ \rightarrow Succinyl\text{-CoA} + CO_2 + NADH + H^+ \quad (6)$$

This complex resembles the pyruvate dehydrogenase complex and the branched-chain α-ketoacid dehydrogenase complex. Each complex shares lipoamide dehydrogenase as the E_3 component. α-Ketoglutarate dehydrogenase, unlike the other two complexes, is not regulated by a phosphorylation–dephosphorylation mechanism. α-Ketoglutarate decarboxylase (E_1) is a dimer of identical subunits, and lipoate succinyl transferase (E_2) is a monomeric peptide chain and serves as the structural core of the multienzyme complex. The gene for the E_3 component has been assigned to chromosome 7 (Table 7.1).

Deficiencies of the α-ketoglutarate dehydrogenase complex have been described in isolation [99,100], or in combination with the other two multienzyme complexes as the result of a deficiency of the shared E_3 component as discussed previously [49–52,101–104].

Kohlschuetter and colleagues [99] reported a familial progressive neurodegenerative disease with α-ketoglutarate aciduria, and proposed that the enzyme defect involved the E_2 component. Two siblings, a boy and a girl, were affected with a slowly progressive neurodegenerative disease following normal development during the first year of life. The initial clinical features were dominated by extrapyramidal symptoms, and, later, pyramidal tract findings also became evident. By age 5 years, the patients were severely disabled and bedridden with less severe involvement of mentation. The siblings were born to a consanguineous Tunisian couple, suggesting an autosomal recessive pattern of inheritance.

Bonnefont and colleagues [100] reported three male siblings born to a consanguineous Algerian couple. These children were immediately symptomatic after birth with hypotonia, metabolic acidosis and elevated blood lactate values. The first son suffered generalized seizures, and pyramidal tract dysfunction initially, with superimposed extrapyramidal signs by age 10 months. He died suddenly at age 32 months. The second son followed a similar course and died suddenly at age 30 months. The third son, alive at 20 months, had deteriorated and developed oculogyric crises, painful stiffening and oral dyskinesias. The plasma lactate concentrations and the lactate/pyruvate ratio were elevated in each case. The blood ketone values were mildly increased, but the β-hydroxybutyrate/acetoacetate ratio was low. This profile is compatible with an enzyme defect involving Krebs cycle activity. Enzyme activities in cultured fibroblasts revealed decreased activity of the α-ketoglutarate dehydrogenase in two siblings, and normal E_3 activity and PDC activity in one sibling. The primary locus of the enzyme defect was not determined, but the circumstantial evidence suggested involvement of either E_1 or E_2 components.

The prominence of the extrapyramidal signs in these two families is noteworthy. Similar findings have been seen in glutaric aciduria type I [105] and Leigh syndrome [106].

Combined deficiencies of the three multienzyme complexes were first recognized in 1976 when Haworth and colleagues [101] described three siblings who had lactic acidosis, increased pyruvic and α-ketoglutaric acids in the urine, and partial deficiencies of the pyruvate and α-ketoglutarate dehydrogenase complexes. The parents were probably consanguineous. Robinson and colleagues described a male child who died at age 7 months with progressive neurological deterioration and persistent metabolic acidosis. These investigators documented elevations of blood pyruvate, lactate, α-ketoglutarate and the branched-chain amino acids. Dihydrolipoyl dehydrogenase (E_3) was decreased in liver, brain, kidney, skeletal muscle and heart.

This clinical phenotype is now well established, and affected patients can be identified by the distinctive urinary metabolic profile implicating combined involvement of the three multienzyme complexes. A favourable response to oral lipoic acid treatment has been reported in one case [103]. The outcome, in general, is otherwise poor with death in early childhood following a rapidly progressive neurological syndrome.

COMBINED SUCCINATE DEHYDROGENASE AND ACONITASE DEFICIENCIES

A combined deficiency of skeletal muscle succinate dehydrogenase and aconitase has been described recently [107]. This patient was a 22-year-old man with exercise intolerance, and episodes of increased muscle fatigue and weakness associated with muscle pain and pigmenturia. The patient also suffered from cardiac palpitations and dyspnoea, and prominent calf hypertrophy. Polarographic studies of isolated skeletal muscle mitochondria implicated a defect of Complex II. Biochemical studies documented succinate dehydrogenase deficiency with decreased amounts of the 30 kDa and 13.5 kDa proteins. Decreased mitochondrial aconitase activity was also documented. The investigators hypothesized that a common and unidentified defect related to the iron-sulphur centres in both enzymes could explain the combined deficiency. The authors also propose that the combined defect impaired muscle intermediary metabolism by limiting the rate of NADH production by the Krebs cycle. This defect could lead to progressive cellular energy failure during prolonged exercise.

CONCLUSIONS

In summary, approximately 200 patients with primary enzyme defects involving intermediary metabolism have been described since 1970. More than half of these cases involve a primary defect of the pyruvate dehydrogenase multienzyme complex or a combined defect of the three multienzyme complexes that share lipoamide dehydrogenase as the E_3 component. These patients usually are symptomatic at birth or in infancy with a fatal encephalopathy associated with lactic acidosis. The most common mutation involves the E_1 α subunit of the pyruvate dehydrogenase complex. This X-linked condition causes a devastating illness in males during early infancy, and a chronic neurological syndrome in affected females. Pyruvate carboxylase deficiency has been reported in 35 cases. In general, this enzyme deficiency is associated with death in infancy or early childhood. Only one patient has had normal development. Dysmorphic facial features and brain malformations have been described in patients with deficiencies of pyruvate dehydrogenase and fumarase, and extrapyramidal signs were prominent in the five cases of α-ketoglutarate dehydrogenase deficiency. One patient has presented with a muscle-tissue-specific syndrome of lifelong exercise intolerance and episodic limb weakness, muscle swelling and pain, and pigmenturia associated with a combined defect of succinate dehydrogenase and aconitase.

Lactic acidosis is a prominent metabolic abnormality in patients with defects of the pyruvate dehydrogenase complex and pyruvate carboxylase. Lactic acidosis is less striking and inconstant in patients with defect of the Krebs cycle. Ketonaemia is absent in patients with pyruvate dehydrogenase deficiency, and present in patients with pyruvate carboxylase deficiency and defects of the Krebs cycle. Ketonaemia, when present, is characteristically associated with a low β-hydroxybutyrate/acetoacetate ratio in contrast with patients with defects of the respiratory chain. These conditions generally are inherited as autosomal recessive traits except for the X-linked form of pyruvate dehydrogenase deficiency.

ACKNOWLEDGEMENTS

The author thanks Ms Alice Marti and Mrs Christine Wade for their valued assistance and skill in the preparation of this chapter. The studies have been supported by the Colleen Giblin Foundation for Pediatric Neurology Research.

REFERENCES

1. Patel, M.S. and Roche, T.E. (1990) Molecular biology and biochemistry of pyruvate dehydrogenase complexes. *FASEB Journal*, **4**, 3224–3233
2. Blass, J.P., Avigan, J. and Uhlendorf, B.W. (1970) A defect in pyruvate decarboxylase in a child with an intermittent movement disorder. *Journal of Clinical Investigation*, **49**, 423–432
3. Schneffner, D. and Wille, L. (1973) Acute infantile hemiplegia due to obstruction of intracranial arterial vessels. *Neuropaediatrie*, **4**, 7–19
4. Robinson, B.H., MacMillan, H., Petrova-Benedict, R. *et al.* (1987) Variable clinical presentation in patients with defective E1 component of pyruvate dehydrogenase complex. *Journal of Pediatrics*, **111**, 525–533
5. Brown, G.K., Brown, R.M., Scholem, R.D. *et al.* (1989) The clinical and biochemical spectrum of human pyruvate dehydrogenase complex deficiency. *Annals of the New York Academy of Science*, **573**, 360–368
6. Dahl, H.-H.M., Brown, G.K., Brown, R.M. *et al.* (1992) Mutations and polymorphisms in the pyruvate dehydrogenase E1 alpha gene. *Human Mutation*, **1**, 97–102
7. Wexler, I.D., Kerr, D.S., Ho, L. *et al.* (1988) Heterogeneous expression of protein and mRNA in pyruvate dehydrogenase deficiency. *Proceedings of the National Academy of Sciences USA*, **85**, 7336–7340
8. Stansbie, D., Wallace, S.J. and Marsac, C. (1986) Disorders of the pyruvate dehydrogenase complex. *Journal of Inherited Metabolic Disease*, **9**, 105–119
9. Wicking, C.A., Scholem, R.D., Hunt, S.M. *et al.* (1986) Immunochemical analysis of normal and mutant forms of human pyruvate dehydrogenase. *Biochemical Journal*, **239**, 89–96
10. Dahl, H.-H.M., Hunt, S.M., Hutchinson, W.M. *et al.* (1987) The human pyruvate dehydrogenase complex. Isolation of cDNA clones for the E1 alpha subunit, sequence analysis, and characterization of the mRNA. *Journal of Biological Chemistry*, **262**, 7398–7403
11. Old, S.E. and De Vivo, D.C. (1989) Pyruvate dehydrogenase complex deficiency: biochemical and immunoblot analysis of cultured skin fibroblasts. *Annals of Neurology*, **26**, 746–751
12. Kitano, A., Endo, F., Matsuda, I. *et al.* (1989) Mutation of the E1 alpha subunit of the pyruvate dehydrogenase complex, in relation to heterogeneity. *Journal of Inherited Metabolic Disease*, **12**, 97–107
13. Sperl, W., Ruitenbeek, W., Kerkhof, C.M. *et al.* (1990) Deficiency of the alpha and beta subunits

of pyruvate dehydrogenase in a patient with lactic acidosis and unexpected sudden death. *European Journal of Pediatrics*, **149**, 487–492

14. Dahl, H.-H.M., Maragos, C., Brown, R.M. *et al.* (1990) Pyruvate dehydrogenase deficiency caused by deletion of a 7-bp repeat sequence in the E1 alpha gene. *American Journal of Human Genetics*, **47**, 286–293

15. Matsuo, M., Ookita, K., Takemine, H. *et al.* (1985) Fatal case of pyruvate dehydrogenase deficiency. *Acta Paediatrica Scandinavica*, **74**, 140–142

16. Stromme, J.H., Borud, O. and Moe, P.J. (1976) Fatal lactic acidosis in a newborn attributable to a congenital defect of pyruvate dehydrogenase. *Pediatric Research*, **10**, 62–66

17. Aleck, K.A., Kaplan, A.M., Sherwood, W.G. *et al.* (1988) In utero central nervous system damage in pyruvate dehydrogenase deficiency. *Archives of Neurology*, **45**, 987–989

18. Wick, H., Schweitzer, K. and Baumgartner, R. (1977) Thiamine dependency in a patient with congenital lacticacidaemia due to pyruvate dehydrogenase deficiency. *Agents and Actions*, **7**, 405–410

19. Farrell, D.F., Clark, A.F., Scott, C.R. *et al.* (1975) Absence of pyruvate decarboxylase activity in man: a cause of congenital lactic acidosis. *Science*, **187**, 1082–1084

20. McKay, N., Petrova-Benedict, R., Thoene, J. *et al.* (1986) Lacticacidemia due to pyruvate dehydrogenase deficiency with evidence of protein polymorphism in the alpha-subunit of the enzyme. *European Journal of Pediatrics*, **144**, 445–450

21. Papanastasiou, D., Lehnert, W., Schuchmann, L. *et al.* (1980) Chronische laktataziose bei einem Saugling. *Helvetica Pediatrica Acta*, **35**, 253–260

22. Huq, A.H., Ito, M., Naito, E. *et al.* (1991) Demonstration of an unstable variant of pyruvate dehydrogenase protein (E1) in cultured fibroblasts from a patient with congenital lactic acidemia. *Pediatric Research*, **30**, 11–14

23. Birch-Machin, M.A., Shepherd, I.M., Solomon, M. *et al.* (1988) Fatal lactic acidosis due to deficiency of E1 component of the pyruvate dehydrogenase complex. *Journal of Inherited Metabolic Disease*, **11**, 207–217

24. Olson, S., Song, B.J., Huh, T.L. *et al.* (1990) Three genes for enzymes of the pyruvate dehydrogenase complex map to human chromosomes 3, 7, and X. *American Journal of Human Genetics*, **46**, 340–349

25. Johnston, K., Newth, C.J., Sheu, K.F. *et al.* (1984) Central hypoventilation syndrome in pyruvate dehydrogenase complex deficiency. *Pediatrics*, **74**, 1034–1040

26. Kretzschmar, H.A., DeArmond, S.J., Koch, T.K. *et al.* (1987) Pyruvate dehydrogenase complex deficiency as a cause of subacute necrotizing encephalopathy (Leigh disease). *Pediatrics*, **79**, 370–373

27. Kerr, D.S., Ho, L., Berlin, C.M. *et al.* (1987) Systemic deficiency of the first component of the pyruvate dehydrogenase complex. *Pediatric Research*, **22**, 312–318

28. Kerr, D.S., Berry, S.A., Lusk, M.M. *et al.* (1988) A deficiency of both subunits of pyruvate dehydrogenase which is not expressed in fibroblasts. *Pediatric Research*, **24**, 95–100

29. De Vivo, D.C., Haymond, M.W., Obert, K.A. *et al.* (1979) Defective activation of the pyruvate dehydrogenase complex in subacute necrotizing encephalomyelopathy (Leigh disease). *Annals of Neurology*, **6**, 483–494

30. Hansen, T.L., Christensen, E. and Brandt, N.J. (1982) Studies on pyruvate carboxylase, pyruvate deacrboxylase and lipoamide dehydrogenase in subacute necrotizing encephalomyelopathy. *Acta Paediatrica Scandinavica*, **71**, 263–267

31. Evans, O.B. (1981) Pyruvate decarboxylase deficiency in subacute necrotizing encephalomyelopathy. *Archives of Neurology*, **38**, 515–519

32. Toshima, K., Kuroda, Y., Hashimoto, T. *et al.* (1982) Enzymologic studies and therapy of Leigh's disease associated with pyruvate decarboxylase deficiency. *Pediatric Research*, **16**, 430–435

33. Miyabayashi, S., Ito, T., Narasawa, K. *et al.* (1985) Biochemical studies in 28 children with lactic acidosis in relation to Leigh's encephalomylopathy. *European Journal of Pediatrics*, **143**, 278–283

34. Naito, E., Kuroda, Y., Takeda, E. *et al.* (1988) Detection of pyruvate metabolism disorders by culture of skin fibroblasts with dichloroacetate. *Pediatric Research*, **23**, 561–564

35. Kohlschuetter, A., Kraus-Ruppert, R., Rohrer, T. *et al.* (1978) Myelin studies in a case of subacute necrotizing encephalomyelopathy. *Journal of Neuropathology and Experimental Neurology*, **37**, 155–164

36. Ohtake, M., Takada, G., Miyabayashi, S. *et al.* (1982) Pyruvate decarboxylase deficiency in a patient with Leigh's encephalomyelopathy. *Tohoku Journal of Experimental Medicine*, **137**, 379–386

37. Evans, O.B. (1984) Episodic weakness in pyruvate decarboxylase deficiency. *Journal of Pediatrics*, **105**, 961–963

38. Falk, R.E., Cederbaum, S.D., Blass, J.P. *et al.* (1976) Ketonic diet in the management of pyruvate dehydrogenase deficiency. *Pediatrics*, **58**, 713–721

39. Cederbaum, S.D., Blass, J.P., Minkoff, N. *et al.* (1976) Sensitivity to carbohydrate in a patient with familial intermittent lactic acidosis and pyruvate dehydrogenase deficiency. *Pediatric Research*, **10**, 713–720
40. Farmer, T.W., Veath, L., Miller, A.L. *et al.* (1973) Pyruvate decarboxylase activity in familial intermittent cerebellar ataxia. *Transcripts of the American Neurological Association*, **98**, 260–262
41. Prick, M., Gabreels, F.J., Renier, W. *et al.* (1981) Pyruvate dehydrogenase deficiency restricted to brain. *Neurology*, **31**, 398–404
42. Willems, J.L., Monnens, L.A., Trijbels, J.M. *et al.* (1974) Letter: Pyruvate decarboxylase deficiency in liver. *New England Journal of Medicine*, **290**, 406–407
43. Blass, J.P., Cederbaum, S.D. and Dunn, H.G. (1976) Letter: Biochemical abnormalities in Leigh's disease. *Lancet*, **i**, 1237–1238
44. Kitano, A., Endo, F., Kuroda, Y. *et al.* (1989) Biochemical nature of pyruvate dehydrogenase complex in the patient with primary lactic acidaemia. *Journal of Inherited Metabolic Disease*, **12**, 379–385
45. Federico, A., Dotti, M.T., Fabrizi, G.M. *et al.* (1990) Congenital lactic acidosis due to a defect of pyruvate dehydrogenase complex (E1). Clinical, biochemical, nerve biopsy study and effect of therapy. *European Neurology*, **30**, 123–127
46. Blass, J.P., Avigan, J. and Uhlendorf, B.W. (1970) A defect in pyruvate decarboxylase in a child with an intermittent movement disorder. *Journal of Clinical Investigation*, **49**, 423–432
47. Kodama, S., Yagi, R., Ninomiya, M. *et al.* (1983) The effect of a high fat diet of pyruvate decarboxylase deficiency without central nervous system involvement. *Brain and Development*, **5**, 381–389
48. Robinson, B.H., MacKay, N., Petrova-Benedict, R. *et al.* (1990) Defects in the E2 lipoyl transacetylase and the X-lipoyl containing component of the pyruvate dehydrogenase complex in patients with lactic acidemia. *Journal of Clinical Investigation*, **85**, 1821–1824
49. Robinson, B.H., Taylor, J. and Sherwood, W.G. (1977) Deficiency of dihydrolipoyl dehydrogenase (a component of pyruvate and alpha-ketoglutarate dehydrogenase complexes): a cause of congenital chronic lactic acidosis in infancy. *Pediatric Research*, **11**, 1198–1202
50. Robinson, B.H., Taylor, J., Kahler, S.G. *et al.* (1981) Lactic acidemia, neurologic deterioration and carbohydrate dependence in a girl with dihydrolipoyl dehydrogenase deficiency. *European Journal of Pediatrics*, **136**, 35–39
51. Munnich, A., Saudubray, J.M., Taylor, J. *et al.* (1982) Congenital lectic acidosis, alpha-ketoglutaric aciduria and variant form of maple syrup urine disease due to a single enzyme defect: dihydrolipoyl dehydrogenase deficiency. *Acta Paediatrica Scandinavica*, **71**, 167–171
52. Yoshida, I., Sweetman, L., Kulovich, S. *et al.* (1990) Effect of lipoic acid in a patient with defective activity of pyruvate dehydrogenase, 2-oxoglutarate dehydrogenase, and branched-chain keto acid dehydrogenase. *Pediatric Research*, **27**, 75–79
53. Bonne, G., Benelli, C., De Meirleir, L. *et al.* (1993) E1 pyruvate dehydrogenase deficiency in a child with motor neuropathy. *Pediatric Research*, **33**, 284–288
54. Matthews, P.M., Marchington, D.R., Squier, M. *et al.* (1993) Molecular genetic characterization of an X-linked form of Leigh's syndrome. *Annals of Neurology*, **33**, 652–655
55. Fujii, T., van Coster, R.N., Old, S.E. *et al.* (1993) Pyruvate dehydrogenase complex deficiency due to a novel point mutation in exon 4 of the E1-alpha subunit gene. *Annals of Neurology*, **34**, 463
56. Brown, R.M. and Brown, G.K. (1993) X chromosome inactivation and the diagnosis of X linked disease in females. *Journal of Medical Genetics*, **30**, 177–184
57. Utter, M.F. and Keech, D.B. (1963) Pyruvate carboxylase: I. The nature of the reaction. *Journal of Biological Chemistry*, **238**, 2603–2614
58. Keech, D.B. and Utter, M.F. (1963) Pyruvate carboxylase II: properties. *Journal of Biological Chemistry*, **238**, 2609–2614
59. Ballard, F.J., Hanson, R.W. and Reshef, L. (1970) Immunochemical studies with soluble and mitochondrial pyruvate carboxylase activities from rat tissues. *Biochemical Journal*, **119**, 735–742
60. Freytag, S.O. and Collier, K.J. (1984) Molecular cloning of a cDNA for human pyruvate carboxylase. Structural realtionship to other biotin-containing carboxylase and regulation of mRNA content in differentiating preadipocytes. *Journal of Biological Chemistry*, **259**, 12831–12837
61. Renis, M., Cantatore, P., Polosa, P.L. *et al.* (1989) Content of mitochondrial DNA and of three mitochondrial RNAs in developing and adult rat cerebellum. *Journal of Neurochemistry*, **52**, 750–754
62. Vidailhet, M., Lebevre, E., Beley, G. *et al.* (1981) Neonatal lactic acidosis with pyruvate carboxylase inactivity. *Journal of Inherited Metabolic Disease*, **4**, 131–132
63. Robinson, B.H., Oei, J., Sherwood, W.G. *et al.* (1984) The molecular basis for the two different clinical presentations of classical pyruvate carboxylase deficiency. *American Journal of Human Genetics*, **36**, 283–294

64. Atkin, B.M., Buist, N.R., Utter, M.F. *et al.* (1979) Pyruvate carboxylase deficiency and lactic acidosis in a retarded child without Leigh's disease. *Pediatric Research*, **13**, 109–116
65. De Vivo, D.C., Haymond, M.W., Leckie, M.P. *et al.* (1977) The clinical and biochemical implications of pyruvate carboxylase deficiency. *Journal of Clinical Endocrinology and Metabolism*, **45**, 1281–1296
66. Brunette, M.G., Delvin, E., Hazel, B. *et al.* (1972) Thiamine-responsive lactic acidosis in a patient with deficient pyruvate carboxylase activity in liver. *Pediatrics*, **50**, 702–711
67. Haworth, J.C., Robinson, B.H. and Perry, T.L. (1981) Lactic acidosis due to pyruvate carboxylase deficiency. *Journal of Inherited Metabolic Disease*, **4**, 57–58
68. Maesaka, H., Komiya, K., Misugi, K. *et al.* (1976) Hyperalaninemia hyperpyruvicemia and lactic acidosis due to pyruvate carboxylase deficiency of the liver; treatment with thiamine and lipoic acid. *European Journal of Pediatrics*, **122**, 159–168
69. Oizumi, J., Shaw, K.N., Guidici, T.A. *et al.* (1983) Neonatal pyruvate carboxylase deficiency with renal tubular acidosis and cystinuria. *Journal of Inherited Metabolic Disease*, **6**, 89–94
70. Perry, T.L., Haworth, J.C. and Robinson, B.H. (1985) Brain amino acid abnormalities in pyruvate carboxylase deficiency. *Journal of Inherited Metabolic Disease*, **8**, 63–66
71. Tada, K., Takada, G., Omura, K. *et al.* (1978) Congenital lactic acidosis due to pyruvate carboxylase deficiency: absence of an inhibitor of TPP-ATP phosphoryl transferase. *European Journal of Pediatrics*, **127**, 141–147
72. Tsuchiyama, A., Oyanagi, K., Hirano, S. *et al.* (1983) A case of pyruvate carboxylase deficiency with later prenatal diagnosis of an unaffected sibling. *Journal of Inherited Metabolic Disease*, **6**, 85–88
73. van Biervliet, J.P., Bruinvis, L., van der Heiden, C. *et al.* (1977) Report of a patient with severe, chronic lactic acidaemia and pyruvate carboxylase deficiency. *Developmental Medicine and Child Neurology*, **19**, 392–401
74. Robinson, B.H., Taylor, J. and Sherwood, W.G. (1980) The genetic herteogeneity of lactic acidosis: occurrence of recognizable inborn errors of metabolism in pediatric population with lactic acidosis. *Pediatric Research*, **14**, 956–962
75. Bartlett, K., Ghneim, H.K., Stirk, J.H. *et al.* (1984) Pyruvate carboxylase deficiency. *Journal of Inherited Metabolic Disease*, **7** (Suppl. 1), 74–78
76. Coude, F.X., Ogier, H., Marsac, C. *et al.* (1981) Secondary citrullinemia with hyperammonemia in four neonatal cases of pyruvate carboxylase deficiency [letter]. *Pediatrics*, **68**, 914
77. Greter, J., Gustafsson, J. and Holme, E. (1985) Pyruvate-carboxylase deficiency with urea cycle impairment. *Acta Paediatrica Scandinavica*, **74**, 982–986
78. Merinero Cortes, B., Del Valle Martinez, J., Perez-Cerda Silvestre, C. *et al.* (1988) Acidosis lactica neonatal debida a deficiencia aislada de piruvato carboxilasa. *Anales Espania Pediatrica*, **29**, 57–60
79. Pollock, M.A., Cumberbatch, M., Bennett, M.J. *et al.* (1986) Pyruvate carboxylase deficiency in twins. *Journal of Inherited Metabolic Disease*, **9**, 29–30
80. Robinson, B.H., Toone, J.R., Benedict, R.P. *et al.* (1985) Prenatal diagnosis of pyruvate carboxylase deficiency. *Prenatal Diagnosis*, **5**, 67–71
81. Rutledge, S.L., Snead, O.C., 3d., Kelly, D.R. *et al.* (1989) Pyruvate carboxylase deficiency: acute exacerbation after ACTH treatment of infantile spasms. *Pediatric Neurology*, **5**, 249–252
82. Saudubray, J.M., Marsac, C., Cathelineau, C.L. *et al.* (1976) Neonatal congenital lactic acidosis with pyruvate carboxylase deficiency in two siblings. *Acta Paediatrica Scandinavica*, **65**, 717–724
83. Wong, L.T., Davidson, A.G., Applegarth, D.A. *et al.* (1986) Biochemical and histologic pathology in an infant with cross-reacting material (negative) pyruvate carboxylase deficiency. *Pediatric Research*, **20**, 274–279
84. Hommes, F.A., Polman, H.A. and Reerink, J.D. (1968) Leigh's encephalomyelopathy: an inborn error of gluconeogenesis. *Archives of Diseases of Childhood*, **43**, 423–426
85. Tang, T.T., Good, T.A., Dyken, P.R. *et al.* (1972) Pathogenesis of Leigh's encephalomyelopathy. *Journal of Pediatrics*, **81**, 189–190
86. Grover, W.D., Auerbach, V.H. and Patel, M.S. (1972) Biochemical studies and therapy in subacute necrotizing encephalomyelopathy (Leigh's syndrome). *Journal of Pediatrics*, **81**, 39–44
87. Grobe, H., Bassewitz, D.B., Dominick, H.C. *et al.* (1975) Subacute necrotizing encephalomyelopathy. Clinical, ultrastructural, biochemical and therpeutic studies in an infant. *Acta Paediatrica Scandinavica*, **64**, 755–762
88. van Biervleit, J.P., Duran, M. and Wadman, S.K. (1979) Leigh's disease with decreased activities of pyruvate carboxylase and pyruvate decarboxylase. *Journal of Inherited Metabolic Disease*, **2**, 15–18
89. Gilbert, E.F., Arya, S. and Chun, R. (1983) Leigh's necrotizing encephalomyelopathy with pyruvate carboxylase deficiency. *Archives of Pathology and Laboratory Medicine*, **107**, 162–166
90. van Coster, R.N., Fernhoff, P.M. and De Vivo, D.C. (1991) Pyruvate carboxylase deficiency: a benign variant with normal development. *Pediatric Research*, **30**, 1–4

91. Atkin, B.M., Utter, M.F. and Weinberg, M.B. (1979) Pyruvate carboxylase and phospho-enolpyruvate carboxykinase activity in leukocytes and fibroblasts from a patient with pyruvate carboxylase deficiency. *Pediatric Research*, **13**, 38–43
92. Petrova-Benedict, R., Robinson, B.H., Stacey, T.E. *et al.* (1987) Deficient fumarase activity in an infant with fumaricademia and its distribution between the different forms of the enzyme seen on isoelectric focusing. *American Journal of Human Genetics*, **40**, 257–266
93. Zinn, A.B., Kerr, D. and Hoppel, C.L. (1986) Fumarase deficiency: a new cause of mitochondrial encephalomyopathy. *New England Journal of Medicine*, **315**, 469–475
94. Walker, V., Mills, G.A., Hall, M.A. *et al.* (1989) A fourth case of fumarase deficiency. *Journal of Inherited Metabolic Disease*, **12**, 331–332
95. Gellera, C., Uziel, G., Rimoldi, M. *et al.* (1990) Fumarase deficiency is an autosomal recessive encephalopathy affecting both the mitochondrial and the cytosolic enzymes. *Neurology*, **40**, 495–499
96. Elpeleg, O.N., Amir, N. and Christensen, E. (1992) Variability of clinical presentation in fumarate hydratase deficiency. *Journal of Pediatrics*, **121**, 752–754
97. Remes, A.M., Rantala, H., Kalervo Hiltunen, J. *et al.* (1992) Fumarase deficiency: two siblings with enlarged cerebral ventricles and polyhydramnios in utero. *Pediatrics*, **89**, 730–734
98. Whelan, D.T., Hill, R.E. and McLorry, S. (1983) Fumaric aciduria: a new organic aciduria, associated with mental retardation and speech impairment. *Clinica Chimica Acta*, **132**, 301–308
99. Kohlschuetter, A., Behbehani, A.W., Langenbeck, U. *et al.* (1982) A familial progressive neuro-degenerative disease with 2-oxoglutaric aciduria. *European Journal of Pediatrics*, **138**, 32–37
100. Bonnefont, J.P., Chretien, D., Rustin, P. *et al.* (1992) Alpha-ketoglutarate dehydrogenase deficiency presenting as congenital lactic acidosis. *Journal of Pediatrics*, **121**, 255–258
101. Haworth, J.C., Perry, T.L., Blass, J.P. *et al.* (1976) Lactic acidosis in three sibs due to defects in both pyruvate dehydrogenase and alpha-ketoglutarate dehydrogenase complexes. *Pediatrics*, **58**, 564–572
102. Kuroda, Y., Kline, J.J., Sweetman, L. *et al.* (1979) Abnormal pyruvate and alpha-ketoglutarate dehydrogenase complexes in a patient with lactic acidosis. *Pediatric Research*, **13**, 928–931
103. Matalon, R., Stumpf, D.A., Michals, K. *et al.* (1984) Lipoamide dehydrogenase deficiency with primary lactic acidosis: favorable response to treatment with oral lipoic acid. *Journal of Pediatrics*, **104**, 65–69
104. Robinson, B.H. (1989) Lactic acidemia. In *The Metabolic Basis of Inherited Disease* (eds C.R. Scriver and A.L. Beaudet), McGraw-Hill, New York, pp. 869–888
105. Goodman, S.I., Markey, S.P., Moe, P.G. *et al.* (1975) Glutaric aciduria: a 'new' disorder of amino acid metabolism. *Biochemistry and Medicine*, **12**, 12–21
106. Macaya, A., Munell, F., Burke, R.E. *et al.* (1993) Disorders of movement in Leigh syndrome. *Neuropediatrics*, **24**, 60–67
107. Haller, R.G., Henriksson, K.G., Jorfeldt, L. *et al.* (1991) Deficiency of skeletal muscle succinate dehydrogenase and aconitase. Pathophysiology of exercise in a novel human muscle oxidative defect. *Journal of Clinical Investigation*, **88**, 1197–1206

8
Human defects of β-oxidation: clinical and molecular aspects

Stefano DiDonato

FATTY ACID OXIDATION

The source of chemical energy for the cell is the hydrolysis of ATP to ADP. The rephosphorylation of ADP to ATP requires the utilization of available fuels which are degraded in mitochondria. The main carbohydrate fuels are intracellular glycogen and blood glucose. The main lipid fuel is plasma free fatty acid, derived from adipose tissue, with a small contribution from intracellular lipid stores. In contrast with carbohydrate, which can be metabolized either aerobically or anaerobically, lipid can be metabolized only aerobically [1].

In order to be oxidized, long-chain acyl-CoA esters, synthesized from the corresponding long-chain fatty acids by the long-chain acyl-CoA synthetase of the outer mitochondrial membrane, have to cross the inner mitochondrial membrane. As the inner mitochondrial membrane is impermeable to acyl-CoA esters, the acyl groups are transferred into the mitochondria as acylcarnitine esters. L-Carnitine, two carnitine palmitoyltransferases located in the outer (CPT I) and in inner mitochondrial membrane (CPT II), and a carnitine–acylcarnitine translocase (CT), embedded in the inner mitochondrial membrane, are required in mammalian tissues to transfer long-chain acyl-CoAs across the inner membrane for β-oxidation in the matrix [2].

Fatty acyl-CoA esters then undergo β-oxidation to generate acetyl-CoA. Mitochondrial β-oxidation is a process characterized by repeated cycles of four concerted reactions. The first step is catalysed by a flavin-dependent dehydrogenase, the acyl-CoA dehydrogenase, which is present in mammalian mitochondria in three different molecular forms: the long-chain, the medium-chain and the short-chain acyl-CoA dehydrogenase (LCAD, MCAD and SCAD respectively) which act on acyl-CoA substrates of different chain length, and give rise to reduced flavin adenine dinucleotide (FAD) and enoyl-CoAs. Two different enoyl-CoA hydratases (short-chain and long-chain) catalyse the further step leading to 3-hydroxyacyl-CoA derivatives. At this point two different NAD$^+$-dependent dehydrogenases, i.e. the short-chain and the long-chain 3-hydroxyacyl-CoA dehydrogenase (HAD), catalyse the second dehydrogenation, leading to the 3-ketoacyl-CoA derivatives. The final step is the conversion of the 3-ketoacyl-CoA esters by two thiolases (an acetoacetyl-CoA-specific and a general 3-ketoacyl-CoA thiolase, again acting on substrates of different chain length) to form acetyl-CoA and a fatty acyl-CoA that is two carbon atoms shorter [3].

During fatty acid oxidation, the electrons are transferred to the respiratory chain. The electrons of the FAD-dependent acyl-CoA dehydrogenases are transferred to coenzyme Q through two flavoproteins: the electron-transferring flavoprotein (ETF) and the ETF–coenzyme Q oxidoreductase (ETF-QO). In contrast, the NAD^+-dependent 3-hydroxyacyl-CoA dehydrogenases transfer their electrons to the first complex of the respiratory chain [4].

HUMAN DISEASES CAUSED BY DEFECTIVE MITOCHONDRIAL β-OXIDATION

Mitochondrial fatty acid oxidation, in addition to its general role in cellular ATP synthesis which is important for most body cells, is crucial for the synthesis of ketone bodies in the liver and for energy production in those organs, such as skeletal and heart muscles, that perform conspicuous mechanical work. Indeed, most acquired and genetic defects of mitochondrial β-oxidation can cause either acute hepatic dysfunction or progressive lipid myopathy, or both. In addition, progressive cardiomyopathy and acute muscle energy failure with associated myoglobinuria can occur. These clinical considerations suggest that a classification of human diseases due to defects of β-oxidation should be based on aetiopathogenetic grounds, and identification of the fundamental biochemical defect, the clinical consequences of which may vary from tissue-specific to multiorgan dysfunction. Table 8.1 lists the recognized human genetic defects of β-oxidation.

Table 8.1. Human defects of β-oxidation

Enzyme/cofactor	Phenotype			
	Myopathy/ hypotonia	*Cardio- myopathy*	*Myoglobinuria*	*Hypoglycaemia/ hypoketonaemia*
Fatty acid transport				
Carnitine	++	++	–	++
Carnitine palmitoyltransferase	±	+	++	+
Carnitine–acylcarnitine translocase	+	+	–	+
β-Oxidation enzymes				
Long-chain acyl-CoA dehydrogenase	++	+	+	++
Medium-chain acyl-CoA dehydrogenase	±	±	–	++
Short-chain acyl-CoA dehydrogenase	+	±	–	+
3-Hydroxyacyl-CoA dehydrogenase	+	++	–	++
Combined defect of HAD and 3-ketothiolase	++	–	–	–
Transferring flavoproteins				
Electron-transfer flavoprotein (ETF)	±	±	–	++
ETF–coenzyme Q reductase (ETF-QO)	±	–	–	++
Riboflavin-sensitive forms	+	–	–	++

++ present; – absent; + sometimes present; ± rarely observed.

CARNITINE DEFICIENCY

Carnitine function and metabolism

L-Carnitine (β-hydroxy-γ-*N*-trimethylaminobutyrate) is recognized for its essential role in the transport of long-chain fatty acids into mitochondrial for β-oxidation. Furthermore, intramitochondrial carnitine and the matrix enzyme carnitine acetyltransferase can react with short- and medium-chain acyl-CoAs to produce acylcarnitines, which can be shuttled out of mitochondria [5]. Through this mechanism, carnitine is able to modulate the intracellular concentrations of free CoA and acetyl-CoA via reversible formation of acetylcarnitine [5,6]. Therefore, besides shuttling long-chain fatty acids into mitochondria, carnitine has additional functions; it facilitates the oxidation of pyruvate [7] and branched-chain amino acids [5], and it may also convey acyl moieties, shortened by the peroxisomal β-oxidation system, from peroxisomes to mitochondria for further oxidation [5,8].

Roughly 25% of the carnitine source for the body is endogenously synthesized from the immediate precursor γ-butyrobetaine, while the remaining 75% comes from the diet. Carnitine is present in tissues and biological fluids in free and esterfied forms. In humans, acylcarnitine esters account for about 25% of total carnitine in the serum and for about 15% of total carnitine in liver and skeletal muscle [8].

Total carnitine concentration in adult human tissues is higher in the heart and skeletal muscle [4.8±0.4 and 3.9±0.1 µmol/g respectively) than in the liver, kidney and brain (2.9±0.2, 1.0±0.1 and 0.3±0.5 µmol/g respectively). Carnitine in blood is 20 to 60 times less concentrated (30–60 nmol/ml) than in tissues [8,9]. Therefore carnitine must be actively concentrated from the blood into fatty acid-metabolizing organs. Cell receptors with high affinity for carnitine (K_m 2–6µM) have been identified in muscle [10], heart cells [11] and cultured fibroblasts [12]. The liver and the brain have low-affinity receptors [8], while the intestinal epithelial cells and the kidney [8,13] have intermediate-affinity receptors (K_m100–200 µM).

Identification of carnitine deficiency

Three independent pathological entities have been described as 'primary' carnitine deficiency (CD): muscle CD associated with progressive lipid myopathy was identified by Engel and Angelini in 1973 [14]; systemic CD associated with hepatic encephalopathy and myopathy was later reported by Karpati *et al.* in 1975 [15]; and systemic CD associated with progressive cardiomyopathy was described by Tripp *et al.* in 1981 [16]. The association of human diseases with critically low tissue carnitine suggested a therapeutic role for L-carnitine [17].

Since then, more than 100 patients with different forms of carnitine deficiency have been reported. However, the diffusion of diagnostic methods to identify CD made it clear that it was a relatively common finding in patients with a number of inherited and acquired diseases. Therefore, many authors started to distinguish primary CD from secondary CD [18,19].

MYOPATHIC AND SYSTEMIC CD
Since the first description of muscle CD [14], more than 20 patients with progressive muscular weakness and wasting, accumulation of lipid droplets in type I

muscle fibres, low carnitine in skeletal muscle, variable carnitine content in plasma, but normal carnitine in liver or heart, have been described as having muscular CD [20,21]. In addition to myopathic CD, several patients suffering in infancy or early childhood from hepatic encephalopathy, hypotonia, diffuse lipid storage in tissues, metabolic acidosis and decreased carnitine in plasma, muscle, liver and heart have been reported under the heading of systemic CD [20–22]. Some of these cases were familial and suggested autosomal recessive inheritance [23–25].

It should be noted that the very existence of these two forms of primary CD has been questioned. First, it is difficult to differentiate between myopathic and systemic CD because often extramuscular tissues, such as liver and heart, are not available to be tested [9]. Secondly, in patients that underwent reinvestigation, the diagnosis of primary CD was excluded, and most of the 'carnitine deficiencies' were found to be other diseases, mainly genetic defects of β-oxidation [9,18,19].

PRIMARY CD WITH CARDIOMYOPATHY

The first description of a disorder characterized by systemic CD, lipid myopathy and cardiomyopathy was reported by Morand *et al.* [26] in a girl who completely recovered from heart failure after carnitine supplementation. Subsequently, another young boy with systemic CD, cardiomyopathy, hypoketotic hypoglycaemia and dramatic response to carnitine therapy was described by Chapoy *et al.* [27]. Familial forms of the disease have been reported, suggesting autosomal recessive inheritance [16,28]. The disease was later identified as a carnitine membrane-transport defect [29–32]. The association of cardiomyopathy, hypoglycaemic attacks, systemic CD and, in some instances, defective carnitine transport has been reported in at least 20 patients [32,33].

The main symptom is progressive dilatative cardiomyopathy which, if untreated, leads to death. Attacks of hypoglycaemia with low ketones and generalized hypotonia or open myopathy are also frequently observed. Morphological features include lipid storage in skeletal muscle, heart and liver. Lipid accumulation in skeletal muscle is characterized by small and numerous lipid droplets in type I muscle fibres. Total and free carnitine in plasma is less than 10% of normal and carnitine esters are not increased. Total carnitine is 1–2% of the mean of controls in skeletal muscle and liver; it is also very low in the heart of the few patients in whom it could be measured.

Gas chromatographic–mass spectrometric (GC–MS) analysis of urine reveals that dicarboxylic aciduria is apparently not associated with CPT deficiency. The reason for this unexpected finding is still obscure. Different dicarboxylic acids are produced by shortening, via mitochondrial β-oxidation, of longer-chain dicarboxylic acids generated by microsomal ω-oxidation, and urinary excretion of increased amounts of dicarboxylic acids occurs in a variety of conditions such as ketosis, diabetes and various defects of the mitochondrial β-oxidation pathway. The absence of dicarboxylic aciduria in carnitine and CPT deficiencies might be a consequence of the inability of long-chain dicarboxylic fatty acids (in the form of dicarboxylcarnitine esters) to enter mitochondria, where they would be transformed into the shorter species and excreted in the urine.

The clinical response to carnitine supplementation (generally 2–6 g or oral L-carnitine per day) is dramatic. Normal heart function and muscle strength are soon recovered, and attacks of hypoglycaemia disappear. Chronically treated patients can live a normal life, whereas untreated patients die of cardiac failure [16]. In fact, some young patients were enlisted to receive heart transplants before the correct diagnosis of carnitine deficiency was available [33]. Excellent therapeutic results with carnitine have been reported in patients from different countries [32,34,35].

Although a recent overview suggests that most patients have a recognizable phenotype characterized by cardiomyopathy variably associated with hypoglycaemia [32], clinical heterogeneity has been suggested by the study of two patients from unrelated families [33]. One child had progressive cardiomyopathy and myopathy, while the other had recurrent metabolic attacks of hypoglycaemia and drowsiness, hepatomegaly and myopathy, but no cardiomyopathy. Furthermore, family studies showed that the elder brother of the first patient had hypertrophic cardiomyopathy, whereas two brothers of the second patient died at 10 months of age from a disorder which, in one sibling, was characterized by hypertrophic cardiomyopathy and low carnitine content in the heart and liver [33].

Although it is generally accepted that primary CD is due to defective carnitine transport, it is still not clear whether the low tissue carnitine content is secondary to the fall in blood carnitine, caused by defective intestinal and renal handling of carnitine, or whether it involves a more generalized defect of carnitine transport. Low intestinal uptake and renal reabsorption of carnitine have been demonstrated in patients with CD [28,36], suggesting that cardiac involvement might be secondary to reduced concentrations of carnitine in the plasma. However, the defect in carnitine transport might involve not only the gut and the kidney, but also the heart, the muscle and other organs.

Unfortunately, carnitine transport has only been explored in fibroblasts [29–33]. These studies show that patients have negligible carnitine uptake, so that K_m and V_{max} cannot be calculated. Obligate heterozygotes have a K_m for carnitine transport close to those of controls; the V_{max}, however, is reduced to approximately 50% of control. These data suggest autosomal recessive inheritance and imply that homozygotes lack functionally active high-affinity carnitine receptors, whereas heterozygotes retain approximately 50% of the receptor pool [32,33].

It is noteworthy that the same rates of carnitine uptake are detected in cultured fibroblasts from patients and normal controls when carnitine is at supraphysiological concentrations in the culture medium, suggesting that carnitine transport at high carnitine concentrations is normal in primary CD. Whether this effect is mediated by low-affinity receptors or simply reflects passive diffusion through the cell membrane is not known [33].

Secondary carnitine deficiencies

A description of these disorders is beyond the scope of this chapter. However, it is worth mentioning that these syndromes include a variety of genetic and acquired diseases, including the organic acidurias, those associated with primary defects of β-oxidation (see below), defects of the mitochondrial respiratory chain, chronic treatment with valproate and other drugs, and haemodialysis [17].

CARNITINE PALMITOYLTRANSFERASE DEFICIENCY

L-carnitine, CPT I, the carnitine–acylcarnitine translocase and CPT II provide the mechanisms whereby long-chain fatty acyl-CoAs are transferred from the cytosolic compartment to the matrix to undergo β-oxidation.

The relevance of CPT in intermediary metabolism has been stressed by the important work of McGarry and Foster [37], who proposed that carnitine and malonyl-CoA (i.e. a potent inhibitor of CPT I) might exert a reciprocal control of hepatic fatty acid oxidation and biosynthesis in ketotic and normal states. With carbohydrate feeding, when the plasma ratio of glucagon to insulin is low, malonyl-CoA concentration rises with concomitantly enhanced fatty acid synthesis and suppression of fatty acid oxidation. Conversely, in the fasting state or uncontrolled diabetes, where the plasma glucagon/insulin ratio is high, malonyl-CoA levels fall and carnitine levels increase in the liver, fatty acid synthesis is diminished and CPT is depressed favouring fatty acid oxidation and ketogenesis [2]. Although there is general agreement on this scheme of metabolic regulation proposed by McGarry *et al.* [2], it is still uncertain whether the two CPT activities represent two enzymes or one enzyme in two locations. Indeed, there is some evidence from the literature that a malonyl-CoA-binding protein is present at the outer mitochondrial membrane, but this protein seems to lack CPT activity [38]. Furthermore, there is some evidence that regulation of CPT activity might include covalent phosphorylation of CPT [39]. The recent molecular cloning of a cDNA encoding the entire sequence of rat [40] and human [41] CPT should help in addressing these ambiguities.

CPT deficiency occurs in humans with two different phenotypes: adult-onset 'muscular' CPT deficiency and infantile 'hepatic' CPT deficiency, which have been attributed to a defect of CPT II and CPT I respectively [42]. More recently, a 'hepatomuscular' form of the disease, associated with CPT II deficiency, has also been reported [43]. It is not yet clear whether clinical heterogeneity is the consequence of distinct molecular lesions affecting either CPT I or CPT II, as we lack information on the physical properties of the CPT enzymes. In particular, CPT I has not yet been purified to homogeneity and no molecular information on the gene encoding this protein is available.

Muscular CPT deficiency

Since the first description in 1973 by DiMauro and Melis DiMauro [44] of a young adult with exercise-induced myoglobinuria due to CPT deficiency in muscle, more than 50 patients complaining of muscle pain, rhabdomyolysis and myoglobinuria after prolonged exercise have been reported. The disease is inherited in an autosomal recessive fashion, but is most frequently seen in young adult males, probably because environmental and hormonal factors influence the expression of the defect. In typical patients, the attacks are triggered by prolonged exercise in fasting conditions. However, cold, fever or other stress conditions may induce the metabolic crisis. The morphology of muscle biopsy is generally normal, but may show evidence of transient fibre necrosis and lipid storage after acute episodes. Lipid storage in myocardium has been reported in a few patients with a peculiar phenotype characterized by congestive cardiomyopathy [45].

Biochemically, CPT activity in muscles of patients ranges from 'not detectable' to 30% of normal. Although the enzyme defect has been demonstrated in several tissues such as leukocytes, fibroblasts, platelets and liver, clinical expression of the disease is generally limited to muscle. However, the almost monomorphous clinical phenotype contrasts with some degree of biochemical heterogeneity, as some patients with muscle CPT deficiency exhibit defective CPT activity in the liver, as deduced by impaired ketogenesis on fasting or after lipid loading tests [46,47], while other patients do not [43].

As two CPT activities are present in mitochondria, much work has been carried out to try to understand whether in patients with muscular CPT deficiency the mutation affects CPT I or CPT II. Early studies with malonyl-CoA, a specific inhibitor of CPT I, suggested that patients with the adult muscular form have a deficiency of the malonyl-CoA-insensitive CPT II [47]. Other studies, however, suggested that CPT II deficiency might not be due to a 'catalytic' defect; rather, it could be the consequence of a mutation(s) giving rise to an enzyme with altered regulatory properties, which would be most vulnerable to inhibition by substrate and/or product. The enzyme defect would consequently be expressed only in conditions of stressed lipid metabolism [48].

More recently, analysis of cultured fibroblasts from patients with adult muscular CPT deficiency has shown normal malonyl-sensitive and detergent-labile CPT activity, whereas the malonyl-insensitive and detergent-extractable enzyme (i.e. CPT II) was decreased [42]. In agreement, subsequent immunological studies showed that the amount of CPT II antigen was either absent or markedly reduced in fibroblasts from patients with muscular CPT deficiency [49].

Hepatic CPT deficiency

A few infants with recurrent attacks of hypoketotic hypoglycaemia in the fasting state, leading to brain damage and death, have been described [50]. CPT activity was barely detectable in liver specimens from these infants; in their cultured fibroblasts the malonyl-sensitive CPT and detergent-labile enzyme, but not the malonyl-insensitive activity, was lowered pointing to a deficiency in CPT I. As mentioned above, immunological studies with antibodies raised against purified CPT II antigen [49] showed that the cells from patients with the infantile form retain a normal amount of CPT II antigen. Interestingly, these patients with the hepatic infantile phenotype and CPT I deficiency in liver and fibroblasts have normal CPT I and CPT II activities in skeletal muscle [51].

As mentioned previously, in addition to these phenotypes, an 'intermediate' form of CPT deficiency has been reported in an infant who suffered attacks of hypoketotic hypoglycaemia leading to coma and death (i.e. a hepatic phenotype); cultured fibroblasts from this patient displayed normal activity of the malonyl-CoA-sensitive activity and a profound deficiency of the malonyl-CoA-insensitive enzyme (i.e. a biochemical phenotype of CPT II deficiency). Cultured cells from this patient lacked CPT II antigen, similarly to cells from patients with the adult muscular phenotype [43]. Furthermore, recent contributions describing new patients with either neonatal muscular [52] or infantile non-muscular forms [53] suggest that the clinical spectrum of CPT deficiency may be wider than previously supposed.

Molecular studies of CPT deficiency

CPT has been purified to homogeneity from human liver, and the partial amino acid sequences from several tryptic peptides and the N-terminal region have been determined [54]. On the basis of these data, the cDNA encoding the entire CPT was cloned and its chromosomal localization determined (human chromosome 1, band 1p13.p11)[41]. This cDNA contains an open reading frame of 1974 bp which encodes a protein of 658 amino acid residues, including 25 residues of an N-terminal leader peptide. The human cDNA has been used for Northern blot analysis of normal cells and cells from CPT-deficient patients. It recognizes a message of 3.1 kb both in control and CPT-deficient fibroblasts. Isolation, reverse transcription and sequencing of mRNAs from cells of patients with different forms of CPT deficiency, including a new patient with a 'hepatomuscular' phenotype, allowed the recognition of point mutations associated with human CPT deficiency. This work suggests that discrete mutations at the same gene encoding CPT II could have different consequences on CPT function. Specifically, the following molecular lesions have now been identified [55].

(1) A missense mutation has been identified [55] in a patient with the *early-onset form* of CPT II deficiency presenting with hypoketotic hypoglycaemia and cardiomyopathy. cDNA and genomic DNA analysis demonstrated that the patient was homozygous for a mutant CPT II allele, which carried a C→T transition, predicting an Arg→Cys substitution (R631C). The normal and mutated CPT II cDNAs were expressed in COS cells. The R631C substitution drastically depressed the catalytic activity of CPT II, thus confirming that this is the crucial mutation. Biochemical characterization of mutant CPT II showed that this allele is associated with severe reduction of V_{max} (=90%), normal apparent K_m values and decreased protein stability.

(2) A different missense mutation is frequently found in patients with the classical adult muscular form characterized by recurrent exercise-induced myoglobinuria: a C→T transition, resulting in a non-conservative amino acid substitution Ser→Leu. Among 17 unrelated patients of different European ancestry, the mutation occurred on 20 of the 34 mutant alleles (59%). Thus, the Ser→Leu mutation appears to be the most frequent mutation in CPT II-deficient patients [56].

CARNITINE–ACYLCARNITINE TRANSLOCASE (CT) DEFICIENCY

The possibility that CT might be implicated in human disease was suspected, but never proven [57], in patients with evidence of defective fatty acid oxidation and low free carnitine but high long-chain acylcarnitine levels in plasma.

Recently, however, Stanley and co-workers [58] reported a 2½-year-old boy with a congenital defect in fatty acid oxidation presenting, in the neonatal period, with vomiting, lethargy and coma. Clinically, he had muscle weakness, cardiomyopathy and hypoglycaemia. Increased excretion of medium-chain dicarboxylic acids was present, in the absence of ketonuria. Plasma free carnitine concentrations were extremely low both before and during treatment with oral carnitine. In contrast, the level of long-chain acylcarnitine was markedly increased. Under L-carnitine treatment, nearly all the plasma carnitine remained esterfied. Oleate oxidation by the patient's cultured fibroblasts was negligible, but CPT activity was normal. Measurement of CT in the patients' cells showed a reduction in activity of less than 5% of the control. Translocase activity in the parents' cells was intermediate between the patient's and the control activities, suggesting heterozygosity [58].

DEFECTS OF THE β-OXIDATION ENZYME COMPLEX ACTING ON STRAIGHT-CHAIN FATTY ACIDS

These disorders are relatively common and, in some instances, can mimic primary CD [18,19]. Among these disorders, deficiencies in LCAD, MCAD and SCAD (i.e. the primary acyl-CoA dehydrogenase, which catalyse the initial step in β-oxidation) appear to be the most frequent. LCAD is active on substrates of more than 12 carbon atoms, MCAD acts on fatty acids 4–14 carbon atoms long, and SCAD acts on those with 4–6 carbon atoms. All enzymes have been purified to homogeneity from human liver. They are homotetramers with molecular masses of 160 and 180 kDa, the masses of mature subunits of LCAD, MCAD and SCAD being 45, 45 and 41 kDa respectively [59]. Each subunit contains 1 molecule of FAD. Molecular clones of cDNAs encoding human LCAD, MCAD and SCAD have been obtained; the genes corresponding to LCAD and MCAD have been mapped to human chromosomes 2 and 1 respectively [60–62].

Recently two new enzymes of β-oxidation have been identified in mitochondria: a very long-chain acyl-CoA dehydrogenase and a multifunctional enzyme having three β-oxidation activities [i.e. enoyl-CoA-hydratase, 3-hydroxyacyl-CoA dehydrogenase (HAD) and 3-ketoacyl-CoA thiolase activity] [63].

LCAD deficiency

Hale *et al.* [64] first reported on three children, from unrelated families, with non-ketotic hypoglycaemia and LCAD deficiency. All patients were infants and exhibited failure to thrive, hepatomegaly, cardiomegaly, hypotonia and metabolic crises characterized by hypoglycaemia and low ketones. GC–MS analysis of the urine showed marked C_6–C_{16} dicarboxylic aciduria. The clinical picture was therefore reminiscent of the more common MCAD deficiency, yet patients with LCAD deficiency tend to be more severely affected [65]. Hypoglycaemic attacks are more frequent and develop at an earlier age; proximal myopathy, muscle pain and myoglobinuria may be present; cardiac involvement is prominent and can be easily documented by ECG and echocardiography; cardiorespiratory arrest may occur during hypoglycaemic attacks [65,66]. Six more patients with documented LCAD deficiency with variable clinical presentations including both Reye-like encephalopathy, hypoglycaemia and sudden death, and milder courses with myopathy and cardiomyopathy, have been described [66–69].

Muscle weakness and hypotonia may be striking in LCAD deficiency. For instance, the two infants described by Hale *et al.* [64] who survived the first metabolic attacks, later presented with chronic muscle weakness and episodes of muscle pain and fatigue associated with high serum creatine kinase (CK); on one occasion myoglobinuria was noticed. Obvious hypoglycaemia was associated with muscle symptoms.

A biphasic clinical course was also found in two patients described later [66]. Typical episodes of hypoketotic hypoglycaemia and dicarboxylic aciduria occurred in early infancy, followed by a milder symptomatology in the subsequent years. Thus, myopathic signs and symptoms seem to be characteristic of patients who reach childhood or adulthood, whereas acute metabolic attacks are confined to infancy.

Pathologically, LCAD deficiency is characterized by liver steatosis [65]. Biochemically, total and free carnitine are lowered in plasma, liver and muscle, but long-chain carnitine esters tend to be increased [64,66]. The diagnosis is made by measuring LCAD activity in cultured fibroblasts, by a fluorimetric assay that follows the reduction of purified ETF in the presence of palmitoyl-CoA. Enzyme activity is less than 10% of control values in patients' fibroblasts, and it is also markedly reduced in leukocytes and liver homogenates. Carnitine treatment can improve the cardiac hypertrophy and prevent metabolic attacks, but fails to affect the hypotonia and weakness [66].

A cDNA for a full-length message of LCAD has been identified recently and characterized: it encodes a 430-amino acid polypeptide, including a leader peptide of 30 amino acids. The corresponding gene has been mapped to human chromosome 2, band q34–q35 [60].

MCAD deficiency

MCAD deficiency is, together with CPT deficiency, the most frequent disease of β-oxidation. More than 85 patients have been described in the literature since the first report of Kolvraa *et al.* in 1982 [70]. It is considered one of the more common causes of the sudden infant death (SID) syndrome [71]. Typical symptoms include fasting intolerance, nausea, vomiting, hypoketotic hypoglycaemia, lethargy and coma [66,72,73]. Hepatomegaly with fatty degeneration of the liver is characteristic. However, the clinical presentation of this disorder is variable. Some individuals may be asymptomatic and have been recognized only during studies on families of clinically symptomatic subjects [66].

The patients have dicarboxylic aciduria. Adipic, suberic and sebacic acids are detected in the urine [66,74–77], but C_{12}–C_{14} dicarboxylic acids, the hallmarks of LCAD deficiency, are absent. Carnitine deficiency has been reported in liver, muscle and plasma [73]. MCAD deficiency has been demonstrated in cultured fibroblasts, lymphocytes and liver from patients [66,72,73,77]. Enzyme activity is between 2 and 10% of normal. Cultured cells are unable to oxidize medium-chain fatty acid, but long- and short-chain substrates are oxidized normally.

Diagnosis *in vivo* is greatly facilitated by fast atom bombardment mass spectrometric (FAB-MS) data showing that in MCAD deficiency increased urinary levels of the most common dicarboxylic acids are associated with the excretion of hexanoylglycine, suberylglycine and phenylpropionylglycine; detection of these metabolites was considered pathognomonic of the disease [76]. However, the results of other FAB-MS studies indicate that the major urinary metabolites in MCAD deficiency are octanoylcarnitine and other medium-chain acylcarnitines; these substances are probably the most useful diagnostic markers of the disease, being detectable in high amount also between metabolic attacks [66,75].

A few patients with MCAD deficiency, may present a myopathic phenotype. In a review of 15 patients aged 2–26 years and diagnosed as 'systemic CD', Zierz *et al.* [78] showed that seven of these patients, who later developed a mild lipid myopathy, had MCAD deficiency.

MOLECULAR GENETICS OF MCAD DEFICIENCY

Immunoprecipitation assays in cultured fibroblasts from 13 patients with MCAD deficiency showed that the size of the defective MCAD protein was indistinguishable from the normal human enzyme, although the activity was 6–13% of normal. These results suggested that the genetic abnormality could be due to point mutations at the MCAD locus [79]. Other immunological studies with anti-MCAD antibodies and Northern blot analysis of cellular mRNA led to different conclusions. Cells from different affected individuals showed either reduced synthesis of a normal-sized MCAD precursor or the synthesis of an MCAD enzyme of abnormal size, suggesting that both point mutations and aberrant splicing of primary transcription products may cause MCAD deficiency in man [80].

The molecular cloning of a full-length cDNA encoding the human MCAD enzyme [61] and the assignment of the gene to the short arm of chromosome 1, band p31, allowed molecular studies in numerous independent patients. These studies proved that most patients with MCAD deficiency carry a point mutation at nucleotide 985 of the coding region [81–84]. This mutation is an A→G transition which changes a highly conserved lysine at codon 329 into glutamate. Most patients inherit this mutation in the homozygous configuration, suggesting a high frequency of the G985 mutation in the general population [84]. The mutation introduces a new *Nco*I restriction site, making the identification of the mutated allele(s) easy. On this basis and using the possibility of studying this mutation by polymerase chain reaction and *Nco*I restriction analysis in post-mortem fixed tissues, an interesting retrospective study of the incidence of MCAD deficiency in victims of SID syndrome was made. Seven families, in which a total of nine infants had died suddenly of unexplained causes, were studied. The diagnosis of MCAD deficiency had been established in subsequent live siblings on the basis of urine analysis or enzyme assay in fibroblasts. In all post-mortem samples from the dead infants, mutational analysis revealed the A→G transition at nucleotide 985 in the homozygous form, while the parents were heterozygous for the same mutation [85].

SCAD deficiency

This disease has been documented in only a few patients. One patient, the first to be described, was an adult with a progressive myopathy and massive lipid storage in type I muscle fibres [86]. Two were unrelated infants with failure to thrive and non-ketotic hypoglycaemia, who died at age 20 and 30 months respectively [87]. Another infant also exhibited failure to thrive, vomiting and severe myopathy, with associated lipid storage and decreased carnitine levels in skeletal muscle [88]. A second adult had limb-girdle myopathy [89]. In addition, an infant with failure to thrive, metabolic acidosis and excessive ethylmalonic acid in the urine had defective oxidation of short-chain fatty acids in cultured fibroblasts; SCAD activity, however, was not determined [90]. All SCAD-deficient patients excreted in the urine increased amounts of ethylmalonic and methylsuccinic acids and, in some instances, butyrylglycine.

As in LCAD deficiency, myopathy may be a prominent expression in SCAD deficiency. The 53-year-old woman reported by Turnbull *et al.* [86] had weakness and wasting of proximal limb muscles, low carnitine in skeletal muscle and lipid storage prevalent in type I muscle fibres. She oxidized fatty acids normally in the liver, as deduced by normal ketogenesis with fasting. In muscle, oxidation of short-chain fatty acids was impaired, whereas oxidation of octanoate and palmitate was similar to that in controls. SCAD activity was low, but not abolished, in her muscle mitochondria. Another adult patient, a 33-year-old male, who suffered episodes of nausea and lethargy in childhood, developed at age 31 progressive myopathy of the limbs and cervical muscles [89]. Lipid storage was present in muscle fibres. He produced normal ketones on fasting. SCAD activity in muscle mitochondria was 7% of that of controls. Muscle pathology was also a prominent feature in the baby described by Coates *et al.* [88], who suffered since infancy from progressive proximal muscle weakness, hypotonia and developmental delay. Lipid accumulation in her muscle biopsy was minimal and restricted to scattered type I muscle fibres. Carnitine content was reduced in skeletal muscle.

The latter patient, and the two infants with metabolic acidosis described by Amendt *et al.* [87], had marked deficiency of SCAD activity in cultured fibroblasts. In contrast, the myopathic patient described by Turnbull had normal enzyme activity in fibroblasts: as no SCAD-related CRM could be demonstrated in her skeletal muscle [91], it was suggested that she could be suffering from a SCAD deficiency restricted to skeletal muscle. It is still controversial, however, whether muscle-specific SCAD deficiency is due to an abnormality of a tissue-specific SCAD isoenzyme, or whether it is secondary to other enzyme defects [92].

A murine model of SCAD deficiency has also been reported. Similarities to human pathology include ethylmalonic and methylsuccinic aciduria, hypoglycaemia and fatty degeneration of the liver; however, muscle pathology and systemic CD are not seen in mice [93].

MOLECULAR STUDIES OF SCAD DEFICIENCY

Early studies with anti-SCAD antibodies in three unrelated infants with SCAD deficiency provided evidence for molecular heterogeneity of the enzyme defect. The enzyme was labile in fibroblasts from one patient, whereas it had normal stability in cell lines from the other two patients [94]. More information on the molecular pathology of SCAD deficiency came after the identification of a cDNA encoding the precursor of human SCAD. The 1852 bases of the full-length clone encoded for a 412-amino acid protein, including the 24 amino acids of the leader peptide [62]. The SCAD clone was utilized for studies of human SCAD-deficient cell lines. On Northern blot analysis, no differences in the sizes of transcripts were observed between normal and SCAD-deficient cell lines. Moreover, Southern blots showed no difference in the restriction pattern after digestion with four different endonucleases of genomic DNA. These data suggest that the defects in SCAD in the cell lines studied are caused by a point mutation [62].

3-Hydroxyacyl-CoA dehydrogenase (HAD) deficiency

HAD is the third enzyme of the β-oxidation cycle. It dehydrogenates 3-hydroxyacyl-CoAs to their corresponding 3-ketoacyl-CoA esters, and requires NAD^+ as an electron acceptor. Mammalian mitochondria contain two HAD enzymes [95]. One enzyme, active with long-chain 3-hydroxyacyl-CoA substrates, is bound to the inner membrane, whereas the other, active with short-chain 3-hydroxyacyl-CoA substrates, is located in the matrix. A third and a fourth HAD activity are also present in peroxisomes and mitochondria as part of the bifunctional protein in the peroxisomal β-oxidation system [95] and the trifunctional protein in mitochondria respectively [63].

Six patients with long-chain HAD deficiency have been reported [96–100]. Clinical features in these infants include Reye's syndrome-like episodes, hypoketotic hypoglycaemia, myopathy and/or cardiomyopathy and sudden infant death. In deceased patients, moderate lipid infiltration has been reported in post-mortem samples from liver, heart and skeletal muscle. In contrast with other patients with defects of β-oxidation, lipid infiltration is minimal in muscle biopsies of HAD-deficient patients [98]. One of the six patients reported had a peculiar phenotype characterized by progressive sensory-motor peripheral neuropathy and pigmental retinopathy [100]. Biochemically, all patients had a marked deficiency of the long-chain HAD in cultured fibroblasts, with normal activity of the enzyme active on short-chain substrates. The other enzymes of β-oxidation were all normal, except for a partial reduction in 3-ketoacyl-CoA thiolase observed in three patients [99,100].

3-Ketoacyl-CoA thiolase deficiency

This disease is a rare inherited metabolic disorder characterized clinically by ketoacidotic attacks, with headache, vomiting and coma, which presents most frequently as a late-onset disease of childhood. Ketoacidosis is associated with normal or high blood glucose levels (ketotic hyperglycaemia) [101,102]. There is a characteristic accumulation of 2-methyl-3-hydroxybutyric acid and 2-methylacetoacetic acid and tiglylglycine in the urine, a consequence of the metabolic block in isoleucine catabolism due to 3-ketoacyl-CoA thiolase deficiency. However, there are at least four different thiolase activities in mammalian cells: cytosolic acetoacetyl-CoA thiolase, mitochondrial acetoacetyl-CoA thiolase, mitochondrial 3-ketoacyl-CoA thiolase and peroxisomal 3-ketoacyl-CoA thiolase [103]. The disease is due to a deficiency in the mitochondrial K^+-dependent short-chain thiolase which plays a major role in isoleucine and ketone body metabolism [104] and strictly speaking is not a disease of fatty acid β-oxidation. The muscle pathology of this disease has never been reported.

Combined deficiency of β-oxidation enzymes

Quite recently, Jackson *et al.* [105] reported on a young girl who suffered progressive severe muscle weakness leading to death. The myopathy followed a brief prodromal illness characterized by nausea and anorexia. The activity of long-chain 3-hydroxyacyl-CoA dehydrogenase was severely decreased in her muscle, heart, liver and fibroblasts; also, long-chain, but not short-chain, 3-ketoacyl-CoA thiolase was markedly reduced in muscle, liver and heart, pointing to a combined defect of the last two enzymes of β-oxidation.

MULTIPLE ACYL-CoA DEHYDROGENASE DEFICIENCY OR GLUTARIC ACIDURIA TYPE II

Glutaric aciduria type II (GA II) is a genetic disorder which was first described by Przyrembel *et al.* [106]. It is characterized clinically by metabolic acidosis, hypoketotic hypoglycaemia and early death. Pathologically, fatty degeneration of several organs occurs, including the liver, the kidney, the heart and skeletal muscle [107]. The biochemical hallmark is excretion in the urine of massive amounts of numerous organic acids which derive from different acyl-CoA substrates [108]. Consequently, the disorder has also been classified as multiple acyl-CoA dehydrogenase deficiency

[107–109]. The disease is biochemically heterogeneous, as it can be associated with a deficiency in either ETF or ETF-QO [110,111]. A third type of GA II has no identified aetiology and, in contrast with the previous forms, is treatable with riboflavin [112].

ETF and ETF dehydrogenase deficiencies

ETF is a mitochondrial flavoprotein which transfers electrons from the reduced form of several acyl-CoA dehydrogenases to the respiratory chain via EFT-QO. ETF consists of an α subunit and a β subunit of 32 and 27 kDa respectively; both subunits are synthesized in the cytosol [113]. Whereas the α subunit is synthesized as a 35 kDa precursor and processed to the mature form in mitochondria, the β subunit is synthesized in its ultimate size [113]. Each molecule of β-ETF binds one molecule of FAD and interacts directly with acyl-CoA dehydrogenase [107,113]. ETF-QO is a 68 kDa protein, partially embedded in the inner mitochondrial membrane, which transfers electrons from ETF to ubiquinone. The protein monomer contains two different redox centres, i.e. a flavin site which interacts with ETF, and a Fe_4S_4 cluster which donates electrons to ubiquinone [107].

Typical patients with GA II exhibit neonatal acidosis and urinary excretion of glutaric, ethylmalonic, isovaleric, isobutyric, 2-methyl-butyric, saturated and unsaturated dicarboxylic acids and sarcosine. Oxidation of several metabolites that are oxidized via different specific acyl-CoA dehydrogenases (i.e. SCAD, MCAD and LCAD, isovaleryl-CoA dehydrogenase, 2-methyl branched-chain acyl-CoA dehydrogenase, glutaryl-CoA dehydrogenase, dimethylglycine dehydrogenase and sarcosine dehydrogenase) is decreased [108].

At least three groups of patients with GA II have been identified [107]: (1) patients with the neonatal form characterized by hypotonia, hepatomegaly, severe hypoglycaemia and metabolic acidosis, multiple congenital anomalies, typical 'sweaty feet' odour and early death; (2) infants without congenital anomalies and severe clinical course, similar to that of patients with the neonatal form, but with longer survival up to a few months of age; these patients may develop cardiomyopathy; (3) patients with later-onset and variable clinical presentation, frequently characterized by vomiting, hypoglycaemia, hepatomegaly and proximal myopathy. The latter group of patients is frequently characterized by the prevalent excretion in the urine of ethylmalonic and adipic acids (ethylmalonic–adipic aciduria) [114]. There is no evidence that these patients can be distinguished on a biochemical–genetic basis from patients with GA II [114]. Carnitine content in plasma, muscle and liver may be low both in the early-onset and late-onset cases [109,115].

Among the late-onset patients with a milder course, some present with, in addition to episodes of nausea, vomiting, lethargy and hypotonia, an overt progressive proximal myopathy [115,116].

MOLECULAR GENETICS OF GA II
Molecular cloning and complete sequencing of the cDNA encoding human α-ETF [117,118] and β-ETF polypeptides has been performed [119] and the corresponding genes mapped to human chromosomes 15 and 19 respectively. Recently the cDNA encoding the sequence of human liver ETF-QO has also been obtained. The open reading frame of this clone encodes a protein of 67 kDa, including a leader peptide of 34 amino acids [120].

Deficiencies of ETF α subunit and ETF-QO were first identified as causes of GA type II. In the case of ETF deficiency, studies with antibodies directed against ETF subunits showed that biosynthesis of the α subunit precursor was either virtually absent or gave rise to a polypeptide of abnormal size in cultured cells from three affected infants [113]. Biosynthesis of ETF β subunit was normal. These and other studies [103,113] suggested that GA II with ETF deficiency was predominantly associated with mutations of the gene encoding α-ETF. However, recent studies in 23 cell lines from infants with GA II showed clinical and biochemical heterogeneity [103]. Clinically, heterogeneity was exemplified by some patients presenting with a progressive extrapyramidal disorder with no organic aciduria. Biochemically, fibroblasts from all patients had severe deficiency of ETF or ETF-QO, but immunotitration in cultured cells gave variable results, showing either the absence or the presence of α- and β-ETF subunits, or ETF-QO CRM. In particular, one cell line with undetectable ETF enzyme activity had negligible β-ETF CRM [103]. In cultured cells from two additional infants with GA II and low ETF activity, β-ETF protein was also not detectable, whereas α-ETF was synthesized normally [121]. These are the first patients in whom a specific mutation of β-ETF gene is suspected [103,121].

Severe forms of GA II with congenital anomalies are more frequently associated with deficiency of ETF-QO. In the tissue and cultured cells from those infants that die early, there is an almost total deficiency of ETF-QO activity, which in most cases is associated with the absence of ETF-QO immunoreactive material [107]. Molecular studies in two unrelated infants with ETF-QO deficiency proved that cell lines from both patients carried a unique single-base deletion in a lysine codon close to the Fe_4S_4 cluster of ETF-QO. This frameshift mutation results in a protein that is not incorporated into mitochondria, but appears to be rapidly degraded [120].

Riboflavin-responsive GA II

This disorder has been described in both infants with acute metabolic attacks [108] and adult patients with progressive myopathy and lipid storage [122]. The patients exhibit a urinary pattern of organic acids compatible with either GA II or ethylmalonic adipic acidurea (EMA). Carnitine content in plasma and tissues is variably reduced. The clinical, morphological and biochemical responses to oral riboflavin supplementation are dramatic. In infants, metabolic attacks and hypotonia disappear [123,124]. In myopathic adults, muscle weakness and wasting improve over a matter of weeks [125,126]. In both groups of patients, organic acids in the urine normalize within a few days.

Biochemically, the disease is characterized by multiple deficiency of several flavin-dependent acyl-CoA dehydrogenases. Recent studies in muscle mitochondria isolated from these patients showed that the defect involves the primary dehydrogenases active on acyl-CoA substrates and is more marked at the level of SCAD and MCAD [92]. In a young girl, oral riboflavin was not only able to normalize the deficient SCAD and MCAD activities, but also restored to normal the amount of SCAD antigen in muscle mitochondria [126].

It is noteworthy that, in animal models of riboflavin deficiency such as weanling rats fed with a riboflavin-deficient diet, the activities of acyl-CoA dehydrogenases are consistently decreased in liver mitochondria, SCAD activity being depressed to the maximal extent [127,128].

ACKNOWLEDGEMENTS

This work was supported in part by a grant from ARIN (Associazione Italiana Promozione Ricerca Neurologica, Milano, Italy). The work of Barbara Bertagnolio, Barbara Garavaglia, Gaetano Finocchiaro, Marco Rimoldi and Franco Taroni, Divisione di Biochimica e Genetica del Sistema Nervoso, Istituto Nazionale Neurologico 'C. Besta', Milano, Italy is also acknowledged.

REFERENCES

1. Layzer, R.B. (1991) How muscles use fuel. *New England Journal of Medicine*, **324**, 411
2. McGarry, J.D., Woeltje, K.F., Kuwajima, M. and Foster, D.W. (1989) Regulation of ketogenesis and the renaissance of carnitine palmitolyltransferase. *Diabetes/Metabolism Reviews*, **5**, 271–284
3. Turnbull, D.M., Bartlett, K., Watmough, N.J. *et al.* (1987) Defects of fatty acid oxidation in skeletal muscle. *Journal of Inherited Metabolic Diseases*, **10**, 105–112
4. Crane, F.L. and Bennert, H. (1956) On the mechanisms of dehydrogenation of fatty acyl derivatives of coenzyme: II. The electron transferring flavoprotein. *Journal of Biological Chemistry*, **218**, 717–731
5. Bieber, L.L., Emans, R., Valkeur, K. and Farrel, S. (1982) Possible functions of short- and medium-chain carnitine acyltransferases. *Federation Proceedings*, **41**, 2858–2862
6. Bloisi, W., Colombo, I., Garavaglia, B. *et al.* (1990) Purification and properties of carnitine acetyltransferase from human liver. *European Journal of Biochemistry*, **189**, 539–546
7. Uziel, G., Garavaglia, B. and DiDonato, S. (1988) Carnitine stimulation of the pyruvate dehydrogenase complex (PDHC) in human muscle mitochondria. *Muscle & Nerve*, **11**, 720–724
8. Scholte, H.R. and de Jonge, P.C. (1987) Metabolism, function and transport of carnitine in health and disease. In *Carnitine in der Medizine* (eds R. Gitzelman, K. Baerlocker and B. Steinmann), Schattauer, Stuttgart and New York, pp. 22–59
9. DeVivo, D.C. and Tein, I. (1990) Primary and secondary disorders of carnitine metabolism. *International Pediatrics*, **5**, 134–140
10. Rebouche, C.J. (1977) Carnitine movement across muscle cell membranes. *Biochimica et Biophysica Acta*, **471**, 145–155
11. Bohmer, T., Eiklid, K. and Jonsen, J. (1977) Carnitine uptake into human heart cells in culture. *Biochimica et Biophysica Acta*, **465**, 627–634
12. Rebouche, C.J. and Engel, A.G. (1982) Carnitine transport in cultured muscle cells and skin fibroblasts from patients with primary carnitine deficiency. *In Vitro*, **18**, 495–502
13. Rebouche, C.J. and Mack, D.L. (1984) Sodium gradient stimulated transport of L-carnitine into renal brush border membrane vesicles: kinetics, specificity, and regulation by dietary carnitine. *Archives of Biochemistry and Biophysics*, **235**, 393–401
14. Engel, A.G. and Angelini, C. (1973) Carnitine deficiency of human skeletal muscle with associated lipid storage myopathy. A new syndrome. *Science*, **179**, 899–902
15. Karpati, G., Carpenter, S., Engel, A.G. *et al.* (1975) The syndrome of systemic carnitine deficiency. Clinical, morphological, biochemical and pathophysiological features. *Neurology*, **25**, 16–24
16. Tripp, M.E., Katcher, M.L., Peters, H.A. *et al.* (1981) Systemic carnitine deficiency presenting as familial endocardial fibroelastosis. A treatable cardiomyopathy. *New England Journal of Medicine*, **305**, 385–390
17. Editorial (1990) Carnitine deficiency. *Lancet*, **i**, 631–633
18. Stanley, C.A. (1987) New genetic defects in mitochondrial fatty acid oxidation and carnitine deficiency. *Advances in Pediatrics*, **34**, 59–88
19. DiDonato, S., Garavaglia, B., Bloisi, W. *et al.* (1989) Biochemical and molecular aspects of β-oxidation defects in skeletal muscle. *Advances in Myochemistry*, **2**, 151–163
20. Engel, A.G. (1986) Carnitine deficiency syndromes and lipid storage myopathies. In *Myology. Basic and Clinical* (eds A.G. Engel and B.Q. Banker), McGraw-Hill, New York, pp. 1663–1696
21. Cornelio, F. and DiDonato, S. (1985) Myopathies due to enzyme deficiencies. *Journal of Neurology*, **232**, 329–340
22. Cornelio, F., DiDonato, S., Peluchetti, D. *et al.* (1977) Fatal cases of lipid storage myopathy with carnitine deficiency. *Journal of Neurology, Neurosurgery and Psychiatry*, **40**, 170–177
23. Cruse, R.P., DiMauro, S., Towfighi, J. and Trevisan, C. (1984) Familial systemic carnitine deficiency. *Archives of Neurology*, **41**, 301

24. Hale, D.E., Cruse, R.P. and Engel, A.G. (1985) Familial systemic carnitine deficiency. *Archives of Neurology*, **42**, 1133

25. DiDonato, S., Peluchetti, D., Cornelio, F. *et al.* (1982) Evidence for autosomal recessive inheritance in systemic carnitine deficiency. *Annals of Neurology*, **11**, 190–192

26. Morand, P., Despert, F., Carrier, H.N. *et al.* (1979) Myopathie lipidique avec cardiomyopathie sévère par déficit généralisé en carnitine. *Archiv de Mal Coeur*, **5**, 536–544

27. Chapoy, P.F., Angelini, C., Brown, W.J. *et al.* (1980) Systemic carnitine deficiency: a treatable inherited lipid storage disease presenting as Reye's syndrome. *New England Journal of Medicine*, **303**, 1389

28. Waber, L.J., Valle, D., Neill, C. *et al.* (1982) Carnitine deficiency presenting as familial cardiomyopathy. A treatable defect in carnitine transport. *Journal of Pediatrics*, **101**, 700–705

29. Eriksson, B.O., Lindstedt, S. and Nordin, I. (1988) Hereditary defect in carnitine membrane transport is expressed in skin fibroblasts. *European Journal of Pediatrics*, **147**, 662–663

30. Eriksson, B.O., Gustafson, B., Lindsted, S. and Nordin, I. (1989) Transport of carnitine into cells in hereditary carnitine deficiency. *Journal of Inherited Metabolic Diseases*, **12**, 108–111

31. Treem, W.R., Stanley, C.A., Finegold, D.N. *et al.* (1988) Primary carnitine deficiency due to a failure of carnitine transport in kidney, muscle and fibroblasts. *New England Journal of Medicine*, **319**, 1331–1336

32. Tein, I., DeVivo, D.C., Bierman, F. *et al.* (1990) Impaired skin fibroblast carnitine uptake in primary systemic carnitine deficiency manifested by childhood carnitine-responsive cardiomyopathy. *Pediatric Research*, **28**, 247–255

33. Garavaglia, B., Uziel, G., Dworzak, F. *et al.* (1991) Primary carnitine deficiency: heterozygote and intrafamilial phenotypic variation. *Neurology*, **41**, 1691–1693

34. Taillard, F., Mundler, O. and Tillous-Borde, I. (1988) Value of radionuclide assessment with thallium 201 scintigraphy in carnitine deficiency cardiomyopathy. *European Heart Journal*, **9**, 811–818

35. Matsuishi, T., Hirata, K., Terasawa, K. *et al.* (1985) Successful carnitine treatment in two siblings having lipid storage myopathy with hypertrophic cardiomyopathy. *Neuropediatrics*, **16**, 6–12

36. Rodriguez-Pereira, R., Scholte, H.R., Luyt-Houven, I.E.M., Vaandrager-Verduin, M.H.M. (1988) Cardiomyopathy associated with carnitine loss in kidneys and small intestine. *European Journal of Pediatrics*, **148**, 193–197

37. McGarry, J.D. and Foster, D.W. (1979) In support of the roles of malonylCoA and carnitine acyltransferase I in the regulation of hepatic fatty acid oxidation and ketogenesis. *Journal of Biological Chemistry*, **254**, 8163–8168

38. Lund, H. and Woldegiorgis, G. (1986) Carnitine palmitoyltransferase: separation of enzyme activity and malonylCoA binding in rat liver mitochondria. *Biochimica et Biophysica Acta*, **878**, 243–249

39. Harano, Y., Kashiwagi, A., Kojima, H. *et al.* (1985) Phosphorylation of carnitine palmitoyltransferase and activation by glucagon in isolated rat hepatocytes. *FEBS Letters*, **188**, 267–272

40. Woeltie, K.F., Esser, V., Weis, B.C. *et al.* (1990) Cloning, sequencing, and expression of a cDNA encoding rat liver mitochondrial carnitine palmitoyltransferase II. *Journal of Biological Chemistry*, **265**, 10720–10725

41. Finocchiaro, G., Taroni, F., Rocchi, M. *et al.* (1991) cDNA cloning, sequence analysis, and chromosomal localization of the gene encoding human carnitine palmitoyltransferase. *Proceedings of the National Academy of Sciences USA*, **88**, 661–665

42. Demaugre, F., Bonnefont, J.P., Mitchell, G. *et al.* (1988) Hepatic and muscular presentations of carnitine palmitoyltransferase deficiency: two distinct entities. *Pediatric Research*, **24**, 308–311

43. Demaugre, F., Bonnefont, J.P., Colonna, M. *et al.* (1991) Infantile form of carnitine palmitoyltransferase II deficency with hepatomuscular symptoms and sudden death. *Journal of Clinical Investigation*, **87**, 859–864

44. DiMauro, S. and Melis DiMauro, P. (1973) Muscle carnitine palmitoyltransferase deficiency and myoglobinuria. *Science*, **182**, 929–931

45. Sacrez, A., Porte, A., Hindengland, C., Bieth, R. and Merain, B. (1982) Myocardiopathie avec surcharge lipidique et deficit en palmityl carnitine transferase (4PCT) leucocytaire. *Archiv de Mal Coeur*, **12**, 1371–1379

46. DiDonato, S., Castiglione, A., Rimoldi, M. *et al.* (1981) Heterogeneity of carnitine palmitoyltransferase deficiency. *Journal of Neurological Sciences*, **50**, 207–215

47. Trevisan, C.P., Angelini, C., Freddo, L., Isaya, G. and Martinuzzi, A. (1984) Myoglobinuria and carnitine palmitoyltransferase (CPT) deficiency/studies with malonylCoA suggest absence of only CPT II. *Neurology*, **34**, 353–356

48. Zierz, S. and Engel, A.G. (1985) Regulatory properties of a mutant carnitine palmitoyltransferase in human skeletal muscle. *European Journal of Biochemistry*, **149**, 207–214

49. Demaugre, F., Bonnefont, J.P., Cepanec, C. *et al.* (1990) Immunoquantitative analysis of human carnitine palmitoyltransferase I and II defects. *Pediatric Research*, **27**, 497–500
50. Bougnères, P.F., Saudubray, J.M., Marsac, C., Bernard, O., Odievre, M. and Girard, J. (1981) Fasting hypoglycemia resulting from hepatic carnitine palmitoyltransferase deficiency. *Journal of Pediatrics*, **98**, 742–746
51. Tein, I., Demaugre, F., Bonnefont, J.P. and Saudubray, J.M. (1989) Normal muscle CPT I and CPT II activities in hepatic presentation patients with CPT I deficiency in fibroblasts. *Journal of Neurological Sciences*, **92**, 229–245
52. Land, J., Mistry, S., Squier, W. *et al.* (1991) Neonatal CPT deficiency. In *Proceedings of the 2nd International Symposium on Clinical, Biochemical and Molecular Aspects of Fatty Acid Oxidation* (eds P.M. Coates and K. Tanaka), abstract o-32
53. Vianey-Saban, C., Mousson, B., Floret, D. *et al.* (1991) Carnitine palmitoyltransferase I deficiency presenting as a Reye-like syndrome without hypoglycemia. In *Proceedings of the 2nd International Symposium on Clinical, Biochemical amd Molecular Aspects of Fatty Acid Oxidation* (eds P.M. Coates and K. Tanaka), abstract p-33
54. Finocchiaro, G., Colombo, I. and DiDonato, S. (1990) Purification and partial amino acid sequence of human carnitine palmitoyltransferase. *FEBS Letters*, **274**, 163–166
55. Taroni, F., Verderio, E., Fiorucci, S. *et al.* (1993) Molecular characterization of inherited carnitine palmitoyltransferase II deficiency. *Proceedings of the National Academy of Sciences, USA*, **89**, 8429–8433
56. Verderio, E., Gellera, C., Cavadini, P., Foirucci, S., Finocchiaro, G., DiDonato, S. and Taroni, F. (1992) A missense mutation in the carnitine palmitoyltransferase gene causing carnitine palmitoyltransferase deficiency and myopathy. *Journal of Neurology*, **239**, suppl. 2, S58
57. Murty, M.S.R., Kamanna, V.S. and Pande, S.V. (1986) A carnitine/acylcarnitine translocase assay applicable to biopsied muscle specimens without requiring mitochondrial isolation. *Biochemical Journal*, **236**, 143–148
58. Stanley, C.A., Hale, D.E., Barry, G.T., Deleeuw, S., Boxer, J. and Bonnefont, J.P. (1992) A deficiency of carnitine-acylcarnitine translocase in the inner mitochondrial membrane. *New England Journal of Medicine*, **327**, 19–23
59. Finocchiaro, G., Ito, M. and Tanaka, K. (1988) Purification and properties of short-chain acylCoA, medium-chain acylCoA, and isovaleryl acylCoA dehydrogenases from human liver. *Journal of Biological Chemistry*, **262**, 798
60. Indo, Y., Yang-Feng, T., Glassberg, R. and Tanaka, K. (1991) Molecular cloning and nucleotide sequence of cDNAs encoding human long-chain acylCoA dehydrogenase (LCAD) and assignment of the location of its gene to chromosome 2. In *Proceedings of the 2nd International Symposium on Clinical, Biochemical and Molecular Aspects of Fatty Acid Oxidation* (eds P.M. Coates and K. Tanaka), abstract p-2
61. Matsubara, Y., Kraus, J.P., Yang-Feng, T.L. *et al.* (1986) Molecular cloning ofⓒDNAs encoding rat and human medium-chain acylCoA dehydrogenase and assignment of the gene to human chromosome 1. *Proceedings of the National Academy of Sciences USA*, **83**, 6543–6547
62. Naito, E., Ozasa, H., Ikeda, Y. and Tanaka, K. (1989) Molecular cloning and nucleotide sequence of complementary DNAs encoding human short-chain acylcoenzyme A dehydrogenase and the study of the molecular basis of human short-chain acylcoenzyme A dehydrogenase deficiency. *Journal of Clinical Investigation*, **83**, 1605–1613
63. Hashimoto, T. (1991) Mitochondrial and peroxysomal enzymes. In *Proceedings of the 2nd International Symposium on Clinical, Biochemical and Molecular Aspects of Fatty Acid Oxidation* (eds P.M. Coates and K. Tanaka), abstract o-2
64. Hale, D.E., Batshaw, M.L., Coates, P.M. *et al.* (1985) Long-chain acylCoA dehydrogenase deficiency. An inherited cause of non-ketotic hypoglycemia. *Pediatric Research*, **19**, 666–671
65. Treem, W.R., Witzeben, C.A., Piccoli, D.A. *et al.* (1986) Medium-chain and long-chain acylCoA dehydrogenase deficiency: clinical, pathologic and ultrastructural differentiation from Reye's syndrome. *Hepatology*, **67**, 1270–1278
66. Roe, C. and Coates, P. (1989) AcylCoA dehydrogenase deficiency. In *The Metabolic Basis of Inherited Disease* (eds C.R. Scriver, A.R. Beaudet, W.S. Sly and D. Valle), McGraw-Hill, New York, pp. 889–914
67. DiDonato, S., Gellera, C., Rimoldi, M. *et al.* (1988) Long-chain acylCoA dehydrogenase deficiency in muscle of an adult with lipid myopathy. *Neurology*, **38**, 269A
68. Chalmers, R.A., English, N., Hughes, E.A. *et al.* (1987) Biochemical studies on cultured fibroblasts from a baby with long-chain acylCoA dehydrogenase deficiency presenting as sudden neonatal death. *Journal of Inherited Metabolic Disease*, **10**, 260–262
69. Parini, R., Garavaglia, B., Saudubray, J.M. *et al.* (1991) Clinical diagnosis of long-chain

acylCoA dehydrogenase deficiency: use of stress and fat-loading test. *Journal of Pediatrics*, **119**, 77–80

70. Kolvraa, S., Gregersen, N., Christensen, E. *et al.* (1982) In vitro fibroblast studies in a patient with C_6-C_{10} dicarboxylic aciduria: evidence for a defect in general acylCoA dehydrogenase. *Clinica et Chimica Acta*, **126**, 53–67

71. Howat, A.J., Bennet, M.J., Variend, S. *et al.* (1985) Defects of metabolism of fatty acids in the sudden infant death syndrome. *British Medical Journal*, **290**, 1771–1773

72. Stanley, C.A., Hale, D.E., Coates, P.M. *et al.* (1983) Medium-chain acylCoA dehydrogenase deficiency in children with hypoketotic hypoglycemia and low carnitine levels. *Pediatric Research*, **17**, 877–884

73. Coates, P.M., Hale, D.E., Stanley, C.A. *et al.* (1985) Genetic deficiency of medium-chain acylcoenzyme A dehydrogenase: studies in cultured skin fibroblasts and peripheral mononuclear leukocytes. *Pediatric Research*, **19**, 671–676

74. Divry, P., David, M., Gregersen, N., Kolvraa, S. *et al.* (1983) Dicarboxylic aciduria due to medium-chain acylCoA dehydrogenase deficiency. *Acta Pediatrica Scandinavica*, **72**, 943–949

75. Roe, C.R., Millington, D.A.M. and Bohan, T.P. (1985) Diagnostic and therapeutic implications of medium-chain acylcarnitines in the medium-chain acylCoA dehydrogenase deficiency. *Pediatric Research*, **19**, 459–466

76. Rinaldo, P., O'Shea, J.J., Coates, P.M. *et al.* (1988) Medium-chain acylCoA dehydrogenase deficiency. *New England Journal of Medicine*, **319**, 1308–1313

77. Rhead, W.J., Amendt, B.A., Fritchman, K.S. *et al.* (1983). Dicarboxylic aciduria: deficient 1-^{14}C octanoate oxidation and medium-chain acylCoA dehydrogenase in cultured fibroblasts. *Science*, **221**, 73–75

78. Zierz, S., Engel, A.G. and Romshe, C.A. (1988) Assay for acylCoA dehydrogenase in muscle and liver and identification of four cases of medium-chain acylCoA dehydrogenase deficiency associated with systemic carnitine deficiency. *Advances in Neurology*, **48**, 231–237

79. Ikeda, Y., Hale, D.E., Keese, S.M. *et al.* (1986) Biosynthesis of variant medium chain acylCoA dehydrogenase in cultured fibroblasts from patients with medium chain acylCoA dehydrogenase deficiency. *Pediatric Research*, **20**, 843–847

80. Strauss, A.W., Duran, M., Zhang, Z. *et al.* (1990) Molecular analysis of medium-chain acylCoA dehydrogenase deficiency. In *Fatty Acid Oxidation: Clinical, Biochemical and Molecular Aspects* (eds P. Coates and K. Tanaka), Alan R. Liss Inc., New York, pp. 609–623

81. Matsubara, Y., Narisawa, K., Miyabayashi, S. *et al.* (1990) Molecular lesions in patients with medium chain acylCoA dehydrogenase deficiency. *Lancet*, **i**, 1589

82. Yokota, J., Tanaka, K., Coates, P.M. and Ugarte, M. (1990) Mutations in medium chain acylCoA dehydrogenase deficiency. *Lancet*, **ii**, 748

83. Yokota, I., Indo, Y., Coates, P.M. and Tanaka, K. (1990) Molecular basis of medium chain acylcoenzyme A dehydrogenase deficiency. *Journal of Clinical Investigation*, **86**, 1000-1003

84. Blakemore, A.I.F., Singleton, N., Pollit, R.J. *et al.* (1991) Frequency of the G985 MCAD mutation in the general population. *Lancet*, **i**, 298–299

85. Ding, J.H., Roe, C.R., Iafolla, A.K. and Chen, Y.T. (1991) Medium chain acylcoenzyme A dehydrogenase deficiency and sudden infant death. *New England Journal of Medicine*, **325**, 61–62

86. Turnbull, D.M., Bartlett, K. and Stevens, D.L. (1984) Short-chain acylCoA dehydrogenase deficiency associated with a lipid storage myopathy and secondary carnitine deficiency. *New England Journal of Medicine*, **311**, 1232–1236

87. Amendt, B.A., Green, C. and Sweetman, L. (1987) Short-chain acylcoenzyme A dehydrogenase deficiency. Clinical and biochemical studies in two patients. *Journal of Clinical Investigation*, **79**, 1303–1309

88. Coates, P.M., Hale, D.E. and Finocchiaro, G. (1988) Genetic deficiency of short-chain acylcoenzyme A dehydrogenase in cultured fibroblasts from a patient with muscle carnitine deficiency and severe muscle weakness. *Journal of Clinical Investigation*, **81**, 171–175

89. DiDonato, S., Cornelio, F., Gellera, C. *et al.* (1986) Short-chain acylCoA dehydrogenase deficient myopathy, with secondary carnitine deficiency. *Muscle & Nerve*, **9**, 178A

90. Bennet, M.J., Gray, R.G.F., Isherwood, D.M. *et al.* (1985) The diagnosis and biochemical investigation of a patients with a short-chain fatty acid oxidation defect. *Journal of Inherited Metabolic Disease*, **8**, 135–136

91. Farnsworth, L., Sheperd, I.M., Johnson, M.A. *et al.* (1990) Absence of immunoreactive enzyme protein in short-chain acylCoA dehydrogenase deficiency. *Annals of Neurology*, **28**, 717–720

92. DiDonato, S. and Gellera, C. (1990) Short-chain and medium-chain acylCoA dehydrogenases are lowered in riboflavin-responsive lipid myopathies with multiple acylCoA dehydrogenase deficiency.

In *Fatty Acid Oxidation: Clinical, Biochemical and Molecular Aspects* (eds K. Tanaka and P. Coates), Alan R. Liss Inc., New York, pp. 325–332

93. Wood, P.A., Amendt, B.A., Rhead, W.J. *et al.* (1989) Short-chain acyl-coenzyme A dehydrogenase deficiency in mice. *Pediatric Research*, **25**, 38–43

94. Naito, E., Indo, Y., Tanaka, K. *et al.* (1989) Short chain acylcoenzyme A dehydrogenase deficiency. Immunochemical demonstration of molecular heterogeneity due to variant SCAD with different stability. *Journal of Clinical Investigation*, **84**, 1671–1674

95. Osumi, T. and Hashimoto, T. (1980) Purification and properties of mitochondrial and peroxisomal 3-hydroxyacylCoA dehydrogenase from rat liver. *Archives of Biochemistry Biophysics*, **203**, 372–383

96. Wanders, R.J.A., Duran, M., Ijilst, L. *et al.* (1989) Sudden infant death and long chain 3-hydroxy-acyl-CoA dehydrogenase. *Lancet*, **ii**, 52–53

97. Hale, D.E., Thorpe, C., Braat, K. *et al.* (1990) The L-3-hydroxyacyl-CoA dehydrogenase deficiency. In *Fatty Acid Oxidation: Clinical, Biochemical and Molecular Aspects* (eds K. Tanaka and P.M. Coates), Alan R. Liss Inc., New York, pp. 503–510

98. Rocchiccioli, F., Wanders, R.J.A., Auburg, P. *et al.* (1990) Deficiency of long-chain 3-hydroxyacylCoA dehydrogenase: a cause of lethal myopathy and cardiomyopathy in early childhood. *Pediatric Research*, **28**, 657–662

99. Jackson, S., Bartlett, K., Land, J. *et al.* (1991) Long-chain 3-hydroxyacyl-CoA dehydrogenase deficiency. *Pediatric Research*, **29**, 406–411

100. Bertini, E., Dionisi-Vici, C., Garavaglia, B. *et al.* (1993) Peripheral sensory-motor neuropathy, pigmentary retinopathy, and fatal cardiomyopathy in long-chain 3-hydroxyacylCoA dehydrogenase deficiency. *European Journal of Pediatrics*, **151**, 121–126

101. Daum, R.S., Scriver, C.R., Mamer, O.A. *et al.* (1973) An inherited disorder of isoleucine catabolism causing accumulation of alpha-methylacetoacetate and alpha-methyl-β-hydroxybutyrate, and intermittent metabolic acidosis. *Pediatric Research*, **7**, 149–160

102. Saudubray, J.M., Specola, N., Middleton, B. *et al.* (1987) Hyperketotic states due to inherited defects of ketolysis. *Enzyme*, **38**, 80–90

103. Loehr, J.P., Goodman, S.I. and Frerman, F.E. (1990) Glutaric acidemia type II: heterogeneity of clinical and biochemical phenotypes. *Pediatric Research*, **27**, 311–315

104. Yamaguchi, S., Orii, T., Sakura, N., Miyazawa, S. and Hashimoto, T. (1988) Defect in biosynthesis of mitochondrial acetoacetyl coenzyme A thiolase in cultured fibroblasts from a boy with 3-ketothiolase deficiency. *Journal of Clinical Investigation*, **81**, 813–817

105. Jackson, S., Singh Kler, R., Bartlett, K. and Turnbull, D.M. (1991) Combined deficiency of long-chain 3-hydroxyacylCoA dehydrogenase and 3-oxoacylCoA thiolase presenting with severe muscle weakness. In *Proceedings of the 2nd International Symposium on Clinical, Biochemical and Molecular Aspects of Fatty Acid Oxidation* (eds P.M. Coates and K. Tanaka), abstract o-34

106. Przyrembel, H., Wendel, U., Becker, K. *et al.* (1976) Glutaric aciduria type II: report on a previously undescribed metabolic disorder. *Clinica Chimica Acta*, **66**, 227–239

107. Frerman, F.E. and Goodman, S.I. (1989) Glutaric aciduria type II and defects of mitochondrial respiratory chain. In *The Metabolic Basis of Inherited Disease* (eds C.R. Scriver, Beaudet, A.R., W.S. Sly and D. Valle), McGraw-Hill, New York, pp. 915–931

108. Gregersen, N. (1985) The acylCoA dehydrogenation deficiencies. *Scandinavian Journal of Clinical and Laboratory Investigation*, **45**, suppl. 174, 11–60

109. Rhead, W.J., Wolff, J.A., Lipson, M. *et al.* (1987) Clinical and biochemical variation and family studies in the multiple acylCoA dehydrogenation disorders. *Pediatric Research*, **21**, 371–376

110. Christensen, N., Kolvraa, S. and Gregersen, N. (1984) Glutaric aciduria type II: evidence for a defect related to the electron transfer flavoprotein or its dehydrogenase. *Pediatric Research*, **18**, 663–667

111. Frerman, F.E. and Goodman, S.I. (1985) Deficiency of electron transfer flavoprotein or electron transfer flavoprotein: ubiquinone oxidoreductase in glutaric aciduria type II fibroblasts. *Proceedings of the National Academy of Sciences, USA*, **82**, 4517–4520

112. Gregersen, N., Wintzensen, H., Christensen, S.K.E. *et al.* (1982) C_6-C_{10}-Dicarboxylic aciduria: investigations of a patient with riboflavin responsive multiple acyl-CoA dehydrogenation defects. *Pediatric Research*, **16**, 861–868

113. Ikeda, Y., Keese, S.M. and Tanaka, K. (1986) Biosynthesis of electron transfer flavoprotein in a cell-free system and in cultured fibroblasts. Defect in the alpha subunit synthesis is the primary lesion in glutaric aciduria type II. *Journal of Clinical Investigation*, **78**, 997–1002

114. Mantagos, S., Genel, M. and Tanaka, K. (1979) Ethylmalonic-adipic aciduria. In vivo and in vitro studies indicating deficiency of activities of multiple acylCoA dehydrogenases. *Journal of Clinical Investigation*, **64**, 1580–1589

115. DiDonato, S., Frerman, F.E., Rimoldi, M. *et al.* (1986) Systemic carnitine deficiency due to lack of electron transfer flavoprotein: ubiquinone oxidoreductase. *Neurology*, **36**, 957–963
116. Dusheiko, G., Kew, M.C., Joffe, B.I. *et al.* (1979) Recurrent hypoglycemia associated with glutaric aciduria type II in an adult. *New England Journal of Medicine*, **301**, 1405
117. Finocchairo, G., Ito, M., Ikeda, Y. and Tanaka, K. (1988) Molecular cloning and nucleotide sequence of cDNA encoding the α-subunit of human electron transfer flavoprotein. *Journal of Biological Chemistry*, **263**, 15773–15780
118. Finocchiaro, G., Ikeda, Y., Barton, D. *et al.* (1987) Molecular cloning and gene mapping of a α-subunit of human electron transfer flavoprotein. *American Journal of Human Genetics*, **41**, A214
119. Finocchiaro, G., Archidiacono, N., Gellera, C. *et al.* (1989) Molecular cloning and chromosomal localization of the β-subunit of human electron transfer flavoprotein. *American Journal of Human Genetics*, **45** (suppl. 1), A1386
120. Goodman, S., Bemelin, T. and Frerman, F. (1991) Human cDNAs encoding ETF dehydrogenase, and mutations in glutaric aciduria type II. In *Proceedings of the 2nd International Symposium on Clinical, Biochemical and Molecular Aspects of Fatty Acid Oxidation* (eds P.M. Coates and K. Tanaka), abstract o-20
121. Yamaguchi, S., Orh, T., Maeda, K., Oshima, M. and Hashimoto, T. (1990) A new variant of glutaric aciduria type II: deficiency of β-subunit of electron transfer flavoprotein. *Journal of Inherited Metabolic Disease*, **13**, 783–786
122. Carroll, J.E., Shumate, J.B., Brooke, M.H. and Hagberg, J.M. (1981) Riboflavin-responsive lipid myopathy and carnitine deficiency. *Neurology*, **31**, 1557
123. Harpey, J.-P., Charpentier, C., Goodman, S.I., Darbois, Y., Lefebvre, G. and Sebbah, J. (1983) Multiple acyl-CoA dehydrogenase deficiency occurring in pregnancy and caused by a defect in riboflavin metabolism in the mother. *Journal of Pediatrics*, **103**, 394
124. Green, A., Marshall, T.G., Bennet, M.J., Gray, R.G.F. and Pollit, R.J. (1985) Riboflavin-responsive ethylmalonic adipic aciduria. *Journal of Inherited Metabolic Disease*, **8**, 67–70
125. DeVisser, M., Scholte, H.R., Schutgens, R.B.H. *et al.* (1986) Riboflavin-responsive lipid storage myopathy and glutaric aciduria type II of early adult onset. *Neurology*, **36**, 367–372
126. DiDonato, S., Gellera, C., Peluchetti, D. *et al.* (1989) Normalization of short-chain acylcoenzyme A dehydrogenase after riboflavin treatment in a girl with multiple acylcoenzyme A dehydrogenase deficient myopathy. *Annals of Neurology*, **25**, 479–484
127. Hoppel, C., DiMarco, J.P. and Tandler, B. (1979) Riboflavin and rat hepatic cell structure and function: mitochondrial oxidative metabolism in deficient states. *Journal of Biological Chemistry*, **254**, 4164–4170
128. Veitch, K., Draye, J.P., Van Hoof, F. and Sherrat, H.S.A. (1988) Effects of riboflavin deficiency and clorfibrate treatment on the five acylCoA dehydrogenases in rat liver mitochondria. *Biochemical Journal*, **254**, 477–481

9
The use of tissue culture in the diagnosis of mitochondrial disease

Brian H. Robinson

INTRODUCTION

Despite the foundations laid for the study of mitochondrial myopathy and respiratory chain defects by the biochemical studies of mitochondria isolated from muscle biopsy tissue, there are some problems associated with mitochondrial isolation artifacts and a very confusing literature in which patients are categorized by biochemical phenotype. Although recent advances in molecular biology have gone some way to resolving this problem by describing the mitochondrial tRNA mutations responsible for MELAS and MERRF, by using tissue culture techniques we have been able both to describe a wide variety of defects in energy metabolism and perform studies that are not feasible with biopsy specimens. In this chapter the use of tissue culture techniques in the diagnosis and research of mitochondrial diseases will be reviewed and where possible compared with the results obtained from muscle biopsy and post-mortem material.

DETECTION STRATEGIES FOR DEFECTS OBSERVABLE IN CULTURED CELLS

Radioisotope flux studies

Our approach to the use of cultured cells, particularly skin fibroblasts in the detection of mitochondrial disease, stemmed initially from studies into the aetiology of lacticacidaemia in childhood rather than the study of mitochondrial myopathies. In 1980 we published a study of 40 skin fibroblast cultures from patients with lacticacidaemia and showed that we could demonstrate the presence of defects by a number of techniques [1]. The techniques used were a combination of direct enzyme assay – for the pyruvate dehydrogenase (PDH) complex and its components, pyruvate carboxylase, phosphoenolpyruvate carboxykinase and certain Krebs cycle enzymes. Also included were some other techniques for overall pathway assessment which included measurement of rates of $^{14}CO_2$ production by whole cells from $[1\text{-}^{14}C]$pyruvate, $[3\text{-}^{14}C]$pyruvate and $[U\text{-}^{14}C]$-

Table 9.1 Oxidative pathway assessment in cultured skin fibroblasts

Test	Defects	Problems
[1-^{14}C]Pyruvate oxidation	PDH complex (some) Respiratory chain defects (some)	Only detects about 50% in each category
[2-^{14}C]Pyruvate and [3-^{14}C]pyruvate oxidation	PDH complex (some) Respiratory chain defects (some) Krebs cycle defects (some)	Commercial [2-^{14}C]pyruvate and [3-^{14}C]-pyruvate contaminated [1-^{14}C]acetate and [2-^{14}C]acetate
[U-^{14}C]Glucose oxidation	Krebs cycle defects (some) Respiratory chain defects (some)	Pentose phosphate pathway interferes at variable rate with $^{14}CO_2$ production
[6-^{14}C]Glucose	Respiratory chain defects (some)	Low yield of $^{14}CO_2$ above background
Lactate and pyruvate production	PDH complex: pyruvate ↑ Respiratory chain defects: lactate ↑ pyruvate ↓	State of confluence of cells important
Lactate to pyruvate ratio (L/P) derived from above	PDH complex: L/P↓ Respiratory chain defects: L/P↑	State of confluence of cells important. Variable results with cells bearing mtDNA defects
ATP production from digitonin-treated cells or isolated mitochondria	PDH complex (some) Respiratory chain defects	Requires large numbers of cells
O$_2$ consumption in digitonin-treated cells	Respiratory chain defects	Requires large number of cells

glucose (Table 9.1). We found that the use of these latter techniques were not particularly useful for assessment of pathways because of interference from the pentose phosphate pathway in glucose oxidation and contamination of [2-^{14}C]- and [3-^{14}C]pyruvate with [2-^{14}C]- and [3-^{14}C]acetate in commercial preparations, up to 60% in some cases. However, in both our study and one reported by Miyabayashi *et al.* [2], [1-^{14}C]pyruvate oxidation to $^{14}CO_2$ in whole cells detected some but not all cases of PDH complex deficiency, some cases of NADH–CoQ reductase deficiency and some cases of cytochrome oxidase (COX) deficiency [3]. Both studies found cases in which [3-^{14}C]pyruvate oxidation was impaired, while [1-^{14}C]pyruvate oxidation was normal, which were deemed to be defects in the Krebs cycle. We have subsequently shown that some cases of NADH–CoQ reductase deficiency can give this combination of results. Defects in the PDH complex usually show low rates of [3-^{14}C]pyruvate oxidation.

Determination of lactate and pyruvate production rates and lactate to pyruvate ratios

As cultured skin fibroblasts, like most cultured cell types, are primarily glycolytic in their energy metabolism and therefore are not an ideal system for the study of mitochondrial defects, 84% of the added glucose is converted to pyruvic and lactic acids in the culture medium when cells are growing. Only when the glucose in the culture medium is exhausted do the cells become more oxidative and the lactate and pyruvate in the medium are consumed [4]. Skin fibroblasts also have a store of glycogen which can sustain them for about 3–5 h in the absence of added nutrients. Despite these drawbacks the simple incubation of a confluent cell culture with a standard glucose medium followed by measurement of pyruvate and lactate can provide a surprising amount of information. This is based on two premises: (a) that a cell having difficulty with oxidative metabolism will by default be forced to generate more ATP glycolytically and therefore will produce more lactic acid, and (b) that when such circumstances are caused by a defect in oxidative phosphorylation a cell will produce lactate and pyruvate at an abnormal molar ratio [4–6]. Table 9.2 shows the results of such experiments performed with fibroblast cultures bearing defects in the PDH complex and in Complex I (NADH–CoQ reductase) or Complex IV (cytochrome oxidase) of the respiratory chain. Control cells produce about 500 nmol of lactic acid/h, 25 nmol of pyruvate/h and a lactate to pyruvate ratio (L/P ratio) of between 20 and 30:1. In PDH complex deficiency, the amount of lactate produced is higher, but so is the amount of pyruvate, almost double, giving L/P ratios lower than the controls. In both Complex I and Complex IV deficiency, the lactate production is greatly increased but the pyruvate production is less than half of the controls. The resulting high L/P ratios range from twice to ten times the normal. Consideration of the mitochondrial and cytosolic NAD$^+$/NADH couples in the cell has shown that the cytosolic NAD$^+$/NADH ratio usually of the order of 300:1 is in equilibrium with the more reduced mitochondrial couple at NAD$^+$/NADH of 10:1 [7]. The L/P ratio by virtue of the equilibrium established at lactate dehydrogenase is a direct indication of the NAD$^+$/NADH ratio in the cytosolic compartment. Thus an L/P ratio of 25:1 is indicative of a ratio of 360:1 in the NAD$^+$/NADH ratio of the cytosol [3]. In normal circumstances we can write the equation such that:

Table 9.2 Lactate and pyruvate production in cultured skin fibroblasts

	Lactate produced (nmol/h per mg of protein)	*Pyruvate produced (nmol/h per mg of protein)*	*Lactate/pyruvate ratio*
Controls			
1	591±42(11)	28.5±5.6(11)	26.9±1.6(11)
2	537±66(8)	21.1±2.6(8)	24.1±2.4(8)
3	459±48(7)	24.5±4.3(7)	21.5±3.4(7)
4	465±24(7)	21.7±2.4(7)	23.4±2.5(8)
PDH-deficient			
5	710±58(8)	33.4±4.7(6)	21.8±3.8(6)
6	825±95(4)	49.4±15.3(4)	14.8±1.9(4)
7	565±73(3)	53.5±4.5(3)	10.9±2.1(3)
Cytochrome oxidase-deficient			
8	599±66(4)	5.4±1.9(4)	169±67(4)
9	1089±39(4)	5.1±0.6(3)	221±35(3)
10	910±224(4)	6.7±2.5(4)	189±92(4)
11	760±156(4)	13.9±1.6(4)	57±10(4)
NADH–CoQ reductase-deficient			
12	965±99(11)	9.7±1.7(11)	98±14(11)
13	921±74(4)	8.7±1.1(4)	104±19(4)
14	1188±103(7)	8.1±1.6(7)	182±45(7)
15	882±76(4)	9.1±1.8(4)	107±21(7)

Confluent skin fibroblast cultures were incubated with Krebs phosphate buffer for 1 h. At this time the buffer was siphoned off and replaced by Krebs phosphate containing 1 mM glucose. After 1 h perchloric acid was added to stop the reaction and the resulting extract assayed for lactate and pyruvate [6]. Results are means ± S.E.M. for the number of experiments given in parentheses.

$$\frac{\text{NAD}^+_c}{\text{NADH}_c} = \frac{1}{K_{\text{LDH}}} \times \frac{[\text{pyruvate}]}{[\text{lactate}]} = K_{en}\frac{\text{NAD}^+_m}{\text{NADH}_m}$$

where $\text{NAD}^+_c/\text{NADH}_c$ and $\text{NAD}^+_m/\text{NADH}_m$ are the nicotinamide nucleotide redox couples of the cytosol and mitochondria respectively, K_{LDH} is the equilibrium constant of the lactate dehydrogenase reaction (1.11×10^{-4}) and K_{en} is an energization constant that keeps the cytosolic compartment more oxidized than the mitochondrial compartment. It has a numerical value of about 30 in normal cells but this value may fall when mitochondria have a severe respiratory chain defect. The energization is provided to the equilibrium system by the electrogenic expulsion of aspartate from the mitochondria [8] (Figure 9.1). Thus a high L/P ratio observed in cultured cells is usually indicative of a problem in either electron transport or the shuttle system that equilibrates the intra- and extra-mitochondrial compartments. This system has been used to detect both COX deficiency and NADH–CoQ reductase deficiency, serving as a screening tool for the delineation of defects in energy metabolism [6,9].

Measurement of ATP synthesis in mitochondria

Another technique which is useful in the delineation of mitochondrial defects is the measurement of ATP synthesis in either isolated skin fibroblast mitochondria

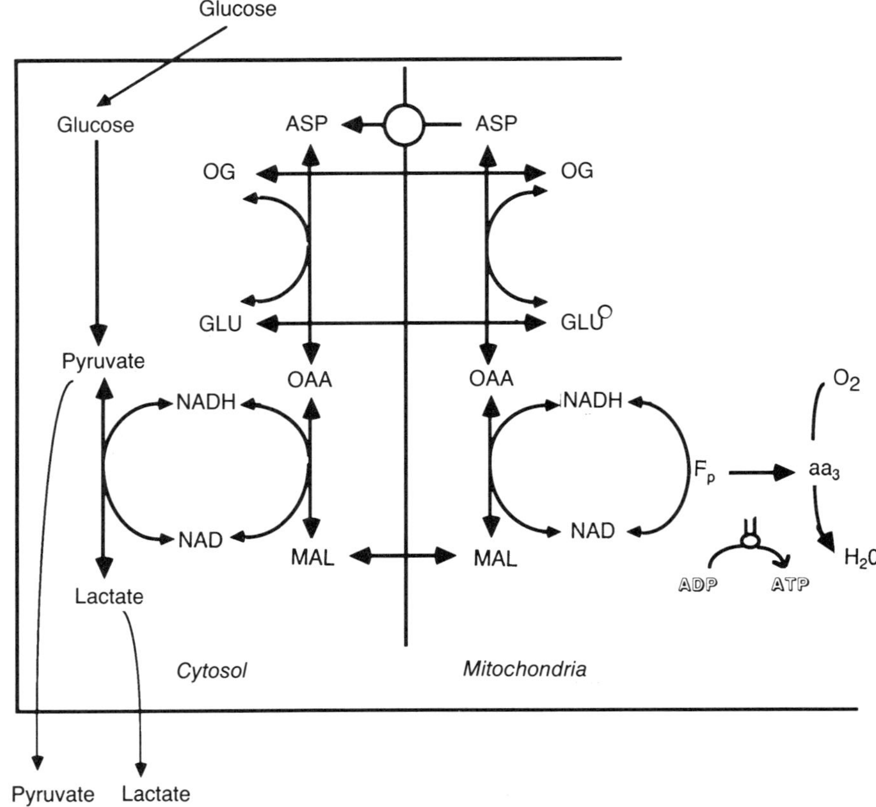

Figure 9.1 Equilibrium between intra- and extra-mitochondrial NAD⁺/NADH redox couples, the respiratory chain and lactate and pyruvate production from glucose. The equilibrium between cytosolic and mitochondrial NAD⁺/NADH couples is catalysed by the mitochondrial and cytosolic aspartate aminotransferases and malate dehydrogenases. Transfer of glutamate (GLU), aspartate (ASP), 2-oxoglutarate (OG) and malate (MAL) are thought to be equilibrium systems with the exception of the aspartate movement out of the mitochondria which is believed to be electrogenic

or digitonin-treated skin fibroblast preparations [5,6]. Figure 9.2 shows the results of a series of experiments on ATP synthesis performed by the digitonin method. In this procedure, confluent skin fibroblast cultures on Petri dishes are treated with a digitonin/sucrose solution to permeabilize the cell membranes, washed and then incubated with substrates for mitochondrial respiration, ADP and phosphate. If a defect is present, it can be shown by the rate at which mitochondria make ATP using different substrates. For instance, preparations of a cell line with a PDH complex defect will show poor ATP synthesis with pyruvate, but normal with all other substrates. A Complex I defect will show a poor rate of ATP synthesis with pyruvate or glutamate as substrates but a normal rate with succinate or ascorbate/TMPD (tetramethylphenylenediamine). A COX oxidase defect on the other hand will show reduced rates of ATP synthesis with all substrates [6]. This has been shown to work in the detection of both mild and severe defects (Figure 9.2)

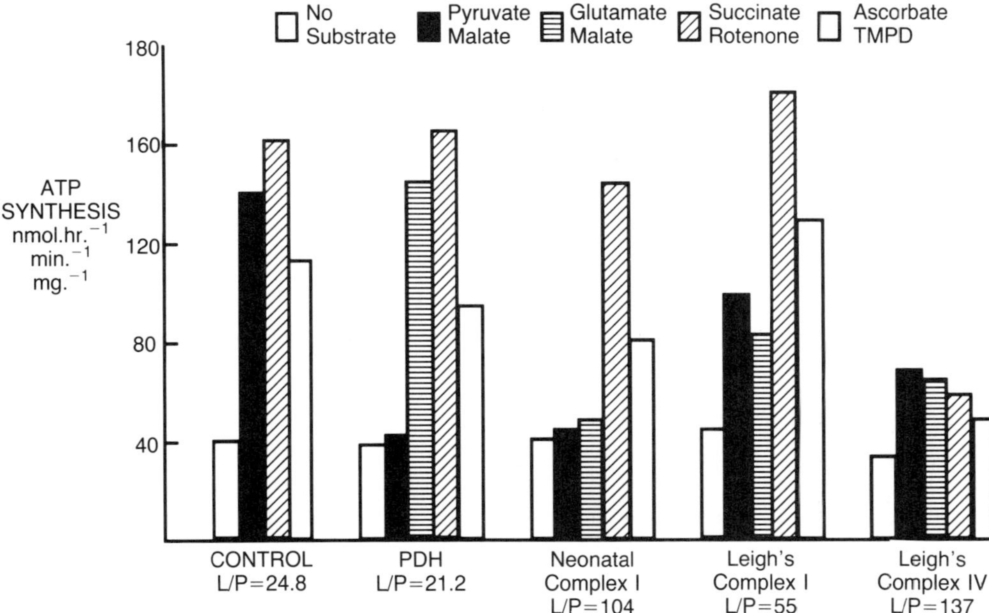

Figure 9.2 ATP synthesis in digitonin-treated fibroblasts from patients with mitochondrial defects. Rates of ATP synthesis are given as nmol/h per mg of cell protein and are shown with no substrate ($\square$); with 5 mM pyruvate plus 1mM L-malate ($\blacksquare$), with 5 mM glutamate plus 1 mM L-malate ($\boxminus$, with 5 mM succinate plus 1 μM rotenone ($\boxtimes$) and with 2 mM ascorbate plus 0.1 mM *NN'*-tetramethyl-phenylenediamine (TMPD) ($\square$). Fibroblast cell lines studies are control, PDH complex deficiency (E_1-type), Complex I deficiency with neonatal lactic acidosis, Complex I deficiency with Leigh's disease and Complex IV deficiency with Leigh's disease. The L/P ratios determined on these cells is given below each set of histograms

and also in the detection of defects in families where the inheritance is through mtDNA [10]. Booth *et al.* [10] showed that two brothers with the clinical presentation of MELAS, subsequently shown to have the mtDNA 3243 tRNA[Leu] mutation, had a defective rate of ATP synthesis in digitonin-treated mitochondria with substrates requiring Complex I activity. Digitonin-prepared fibroblasts can also be used to assess mitochondrial oxidative phosphorylation pathways by polarographic measurement of O_2 consumption [6].

Direct assay of skin fibroblast enzymes as a diagnostic procedure

Although procedures directed toward establishing that a patient skin fibroblast culture has a defect in metabolic turnover, lactate production or an abnormal L/P ratio, ultimately a defect should be established by direct enzyme assay. Table 9.3 delineates common methods used and some of the problems associated with the interpretation of results obtained by this methodology. The diagnosis of PDH complex deficiency is complicated because of the predominance of defects in the

Table 9.3 Enzyme assays commonly used for diagnosis of mitochondrial diseases in cultured skin fibroblasts

Disease	Reference for method	Problems
Pyruvate dehydrogenase complex, inactive and activated	[11]	PDH – $E_1\alpha$ subunit X-linked Lyonization in females gives misleading results. Partial defects may be present in some patients
Pyruvate carboxylase	[12]	K_m variant reported
Phosphoenolpyruvate carboxykinase	[12]	Two isoenzymic forms encoded by two genes, one mitochondrial the other cytosolic
NADH–cytochrome c reductase (rotenone-sensitive)	[13]	Use isolated mitochondria. Interference from diaphorases. Tissue variability
NADH–CoQ reductase (rotenone-sensitive)	[13]	Use isolated mitochondria. Interference from diaphorases. Tissue variability
Succinate–cytochrome c reductase	[14]	Partial defects may be present in some patients
Cytochrome c oxidase	[15]	Kinetic variants and partial defects common in the patient population

$E_1\alpha$ subunit which is encoded on the X chromosome [16,17]. Lyonization of the X chromosome in females leads to variable inactivation and in cases where there is one mutant and one normal allele this can lead to situations where PDH complex activity is variably deficient between tissues in the same patient and different between two females with the same mutation [18]. Such diagnostic problems do not occur in pyruvate carboxylase deficiency, where the activity in affected individuals is usually less than 5% of control activity [17]. This contrasts with PDH complex deficiency where individuals with 25–40% residual activity of the complex in all tissues may be symptomatic [19,20]. A K_m variant of pyruvate carboxylase was described by Brunette *et al.* [21].

The respiratory chain enzymes present a more difficult problem. Complex I can be assayed either as a single entity by rotenone-sensitive NADH–CoQ reductase activity or in combination with Complex III by measurement of rotenone-sensitive NADH–cytochrome c reductase activity. However, neither of these assays appears to work in whole cell homogenates, lysates or sonicates. This is due to the fact that these complexes are integral mitochondrial membrane components assembled in an ordered fashion. Any severe disruption of the complex disturbs its catalytic activity. Thus the best preparations for measurement of this complex are frozen–thawed isolated mitochondria. To complicate matters further, the initial part of the complex is a flavin-containing NADH dehydrogenase which then feeds electrons into a number of iron-sulphur proteins before coenzyme Q is reduced

Table 9.4 Cytochrome oxidase (COX) defects detectable in fibroblasts

Symptoms	L/P ratio in fibroblasts	COX activity			Fibroblast COX assembly
		Fibroblast	Muscle	Liver	
Classical Leigh's disease	2–4× elevated	10–25% of control	10–25% of control	10–25% of control	No
Partial Leigh's disease	2–4× elevated	35–50% of control	?	?	Yes
Liver form Leigh's disease, hepatomegaly	Normal or slightly elevated <×1.25	50% of control	50–75% of control	<10% of control	Yes
Cardiac form – cardiomyopathy, hypotonia, lactic acidosis	2×elevated	40–50% of control	14% of control	Normal	Yes

[22]. Once mitochondrial disruption occurs it is possible for other diaphorases to interact with the system, thus complicating matters further. This seems to be less of a problem with succinate–cytochrome c reductase measurement which is antimycin sensitive and can actually be carried out in whole cell sonicates.

COX DEFECTS

COX defects are fairly well characterized but their recognition using cultured cell systems is not simple. The most severe form of the COX deficiency causing neonatal death is one that is specific to muscle, kidney and heart, while brain, liver and skin fibroblast activities are normal [23]. The defect might be detected by examining cultured muscle cells. The mirror image of this defect, one in which liver and brain are severely COX deficient but muscle and kidney are normal, is detectable in fibroblasts which have 50% normal activity [24]. The commonest form of COX deficiency is expressed in all tissues, including skin fibroblasts [23,25], and has been definitively shown to be due to a nuclear gene by using mitochondrial transfer into SV40-transformed mitochondria-depleted fibroblasts [26] (Table 9.4). It has the neurodegenerative phenotype of Leigh's disease in affected patients. This phenotype also occurs in a partial deficiency [24,25]. In this form, the activity in fibroblasts may be only 50% of normal and the presence of the defect is betrayed by the Leigh's phenotype and the high L/P ratio present in fibroblasts [24]. In some of these cases, there are demonstrable abnormalities in the kinetics of COX c with respect to the substrate, reduced cytochrome c and also with respect to the product, oxidized cytochrome c [25,27]. A fourth variety of COX deficiency detectable in fibroblasts is the cardiomyopathic type. Patients with this defect have a cardiomyopathy associated with hypotonia and lactic acidosis. No brain pathology is evident and, while the enzyme defect can be seen in heart, muscle and skin fibroblasts, liver activity is normal. In general, the severe forms of both Complex

I and Complex IV deficiency show little or no assembly of the subunits of the complex [24,25] in cultured skin fibroblasts, while the milder forms of both of these defects show normal profiles of the subunits, as judged by immunoblotting experiments with cultured fibroblast mitochondrial proteins subjected to polyacrylamide gel electrophoresis.

COMPLEX I DEFECTS

While there are distinct varieties of COX deficiency that have been well characterized in their severity, symptoms and tissue distribution, the delineation of isolated complex deficiency has not been easy because of the confusion generated by the results from assay of mitochondrial enzymes in muscle biopsy specimens. In a recent study, Complex I deficiency in muscle was reported as an isolated defect in patients with Kearns–Sayre syndrome, MELAS syndrome, Alpers syndrome and Leigh's disease, and as a combined Complex I–IV defect in association with fatal neonatal encephalopathy and cardiomyopathy, MERRF syndrome and in myopathy with hepatopathy [28,29]. The wide variety of clinical manifestations are not present when isolated Complex I deficiency is diagnosed by assay of mitochondrial enzymes in muscle mitochondria in conjunction with measurement of the fibroblast L/P ratio and fibroblast enzymes [24,30]. This approach gives four basic clinical presentations but they all overlap with one another and may just be a spectrum of symptomatology based on the severity of the defect (Table 9.5). That this is so is suggested by the drop in L/P ratio going from the most to the least severe form.

There are several well-documented cases of Complex I deficiency with fatal neonatal lactic acidosis [5,6,13,30,31]. Usually in this form the brain pathology is that of a spongiform encephalopathy, but we have observed cases with the classical lesions of Leigh's disease in infants dying at 5–7 weeks of age. The other three presentations we have categorized as spastic quadriplegia, early- and late-onset Leigh's disease. In all of these forms of complex I deficiency there is increased blood lactic acid, psychomotor retardation and a tendency to exhibit either basal ganglia hypodensities on computed tomography (CT) scanning, hyperventricular cardiomyopathy or both [24,30]. The categorization emphasizes the rapidity of onset of the neurodegenerative disease so that the first group rarely show much motor development before becoming spastic. In the early group of Leigh's disease, neurodegeneration may start before 6 months of age and follow an unrelenting course until death occurs at 4–6 years of age. In the later form, onset is between 9 months and 2 years and the disease, although progressive, is not as severe, allowing continuation past puberty for the majority of cases [24]. In three of these categories there are individuals who develop a hypertrophic ventricular cardiomyopathy often with associated Wolff–Parkinson–White syndrome, a situation that is not seen in patients with Leigh's disease due to COX deficiency or PDH complex deficiency [24].

COMBINED DEFECT OF MITOCHONDRIAL ENZYMES

Although combined defects of respiratory chain enzymes in muscle mitochondria have frequently been reported in patients with MELAS, MERRF or Kearns–Sayre syndrome, they are not common in cultured skin fibroblasts.

Table 9.5 Complex I defects expressed in cultured skin fibroblasts

Fatal infantile	Moderate lactic acidosis, spastic quadriplegia	Leigh's (early)	Leigh's (late)
L/P124±13(8)	L/P119±34(4)	L/P84±10(4)	L/P75±14(8)
All die before 4 months of age	2 of 4 died before 4 years of age	1 of 4 died at 10 weeks	All alive
Severe acidosis	Moderate acidosis	Mild acidosis	Mild acidosis
Hypotonia, anorexia	Spastic quadriplegia	Nystagmus	2 Hypertonic, 6 hypertonic
	Psychomotor retardation	Psychomotor retardation	Psychomotor retardation
Cardiomyopathy in 3 of 8	Cardiomyopathy in 1 of 4	Cardiomyopathy in 3 of 4	Cardiomyopathy in 0 of 8
Basal ganglia disease in 2 of 8	Basal ganglia disease in 2 of 4	Basal ganglia disease in 2 of 4	Basal ganglia disease in 4 of 8

Table 9.6 Combined defects of mitochondrial enzymes in cultured skin fibroblasts

Type	Tissues affected	L/P ratio	Symptoms
1. Combined COX, Complex I, III and PDH complex [32]	Fibroblasts and liver (muscle normal)	162	Extreme hypotonia and weakness; 2 siblings died before 4 weeks
2. Combined Complex III and PDH complex (E$_1$), normal COX (Winter, Robinson *et al.*, unpublished	Muscle and liver	105	Neonatal lactic acidosis; 3 siblings died before 5 weeks

We have encountered two families with combined defects, one in which Complexes I, III and IV were deficient in association with PDH complex deficiency and a second family in which there was a combined deficiency of Complex III and the PDH complex (Table 9.6). Fibroblast cell lines from these families exhibited high L/P ratios, severely deficient enzyme activities and failure to assemble specific mitochondrial complexes [32]. In the first family the defect seemed to be present in liver as well as fibroblasts whereas in the second family, the liver and muscle were affected. The exact cause of the defects has not been elucidated in these families. The mtDNA is of normal size and does not carry the defects responsible for MELAS or MERRF.

mtDNA-ENCODED DEFECTS IN LYMPHOBLASTS AND CULTURED SKIN FIBROBLASTS

mtDNA-encoded defects resulting in respiratory chain defects are present in skin fibroblast mitochondria of the proteins concerned. We looked at a number of cell lines derived from patients with MELAS and found that, while some had elevated L/P ratios, many of them had normal L/P ratios indicating no respiratory chain problem [24]. As the tRNA^{Leu} defects that cause MELAS usually result in a defective respiratory chain activity in muscle, particularly in Complex I [28,29], this may at first be surprising . However, as these defects are almost always heteroplasmic in nature, it is not surprising that there is variability in expression. In a series of well-defined cases of MELAS, we recently showed that the tRNA^{Leu} defect (mtDNA 3243 A→G) was easily demonstrated in skin fibroblasts using amplification of the relevant mtDNA sequence. Similar results were obtained with skin fibroblasts from patients with Lebers hereditary optic neuropathy (LHON) who exhibited the mutation in ND4 of mtDNA at position 11877.

We also recently had the opportunity to investigate a family with a mutation at 8993 in the ATPase 6 gene of mtDNA [33,34] using tissue from affected individuals together with both cultured lymphoblasts and cultured skin fibroblasts. The family is depicted in Figure 9.3. The 8993 mutation is easily detected because it creates a new *Ava*I site. Amplification of ATPase 6 DNA followed by an *Ava*I digestion thus allows estimation of the percentage of abnormal mtDNA [33] (Figure 9.4). The index case who had almost undetectable wild-type ATPase 6 sequences in her mtDNA died of Leigh's disease at the age of seven months. A maternal aunt and uncle both died of similar causes before 1 year of age and

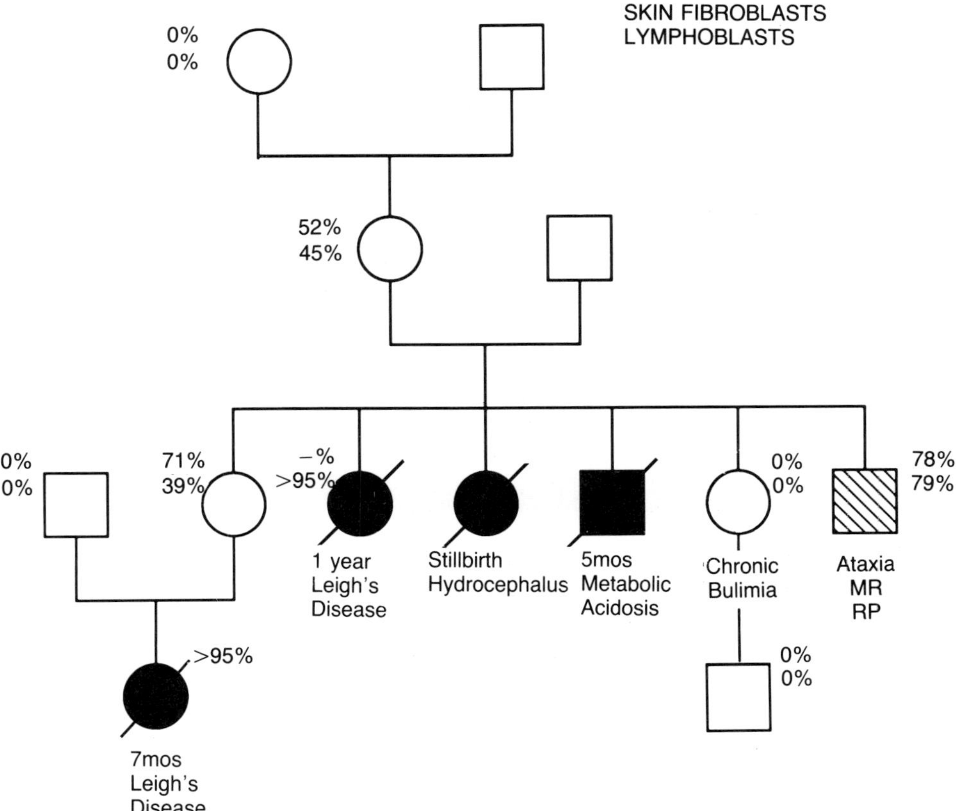

Figure 9.3 Pedigree of a family with a segregating heteroplasmic mutation at mtDNA 8993 in the ATPase 6 gene. The percentage of mutated mtDNA at 8993 is given for listed family members, the upper figure being for skin fibroblasts and the lower figure for cultured lymphoblasts. MR, mental retardation; RP, retinitis pigmentosa. Black symbols, affected; shaded symbol, partially affected

another maternal aunt was stillborn. A maternal uncle had retinitis pigmentosa, ataxia and mental retardation at the age of 34. The percentage of mutant mtDNA in cultured lymphoblasts obtained from five family members was the same as that determined from a white cell pellet taken directly from blood sample. Values obtained from skin fibroblasts were higher than the corresponding white cell values in two family members – the mother (71% vs. 39%) and the grandmother (52% vs. 45%) who were both asymptomatic. Values obtained from the fibroblasts were identical both at low and high passage numbers indicating that for this defect the mtDNA heteroplasmy did not change its percentage in continuous culture. Thus the difference that is observed between tissues and cultured cells in the percentage mtDNA heteroplasmic mutations is probably constitutive rather than an artifact brought about by the selective pressures of tissue culture techniques. We have shown this to be true of the 11778 mutation in LHON in culture but this remains to be rigorously tested for the MELAS and MERRF mutations.

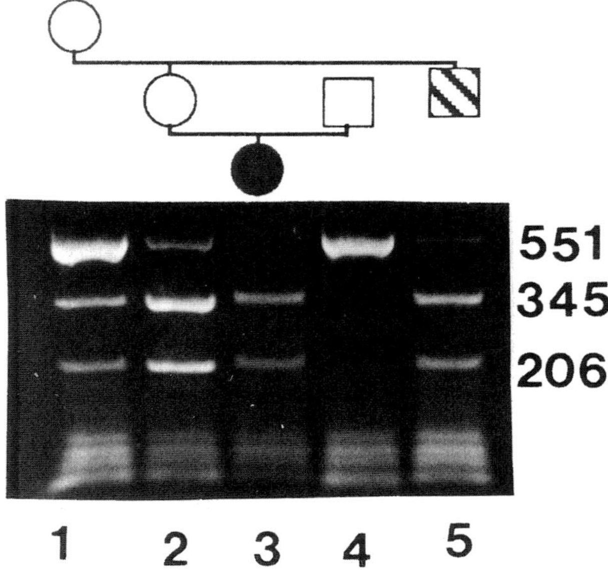

Figure 9.4 Demonstration of mtDNA mutation at 8993 (T→G) in polymerase chain reaction products from amplification of total skin fibroblast DNA. A 551 bp segment covering the ATPase 6 gene of human mitochondrial DNA was prepared from polymerase chain reaction amplification of total skin fibroblast DNA. This segment was then incubated with *Ava*I and run on a 2% agarose gel. Family members are indicated by the pedigree above. See text for further details

REFERENCES

1. Robinson, B.H., Taylor, J. and Sherwood, W.G. (1980) The genetic heterogeneity of lactic acidosis: occurrence of recognizable inborn errors of metabolism in a pediatric population with lacticacidosis. *Pediatric Research*, **14**, 956–962
2. Miyabayashi, S., Ito, T., Abukawa, D. *et al.* (1987) Immunochemical study in three patients with cytochrome *c* oxidase deficiency presenting Leigh's encephalomyelopathy. *Journal of Inherited Metabolic Disease*, **10**, 289–292
3. Robinson, B.H. (1988) Cell culture studies on patients with mitochondrial diseases. *Journal of Bioenergetics and Membranes*, **20**, 313–323
4. MacKay, N., Robinson, B.H., Brodie, R. and Rooke-Allen, N. (1983) Glucose transport and metabolism in cultured human skin fibroblasts. *Biochimica et Biophysica Acta*, **762**, 198–204
5. Robinson, B.H., MacKay, N., Goodyer, P. and Lancaster, G. (1985) Defective intramitochondrial NADH oxidation in skin fibroblasts from an infant with fatal neonatal lacticacidemia. *American Journal of Human Genetics*, **37**, 938–946
6. Robinson, B.H., Ward, J., Goodyer, P. and Baudet, A. (1986) Respiratory chain defects in the mitochondria of cultured skin fibroblasts from three patients with lacticacidemia. *Journal of Clinical Investigation*, **77**, 1422–1427
7. Krebs, H.A. (1973) Pyridine nucleotides and rate control. *Symposium of the Society of Experimental Biology*, **17**, 299–318
8. Williamson, J.R., Safer, B., LaNoue, K., Smith, C.M. and Walajtys, E. (1973) Mitochondrial cytosolic interaction in cardiac tissue; role of the malate–aspartate cycle in the removal of the glycolytic NADH from the cytosol. *Symposium of the Society of Experimental Biology*, **17**, 241–271
9. Wijburg, F.A., Feller, N., Ruitenbeek, W. *et al.* (1990) Detection of respiratory chain dysfunction by measuring lactate and pyruvate production in cultured fibroblasts. *Journal of Inherited Metabolic Disease*, **13**, 355–358

10. Booth, F.A., Haworth, J.C., Dilling, L.A. *et al.* (1989) Mitochondrial encephalomyopathy with associated aminoacidopathy in a male sibship. *Journal of Pediatrics*, **115**, 81–88

11. Sheu, K.F.-R., Hu, C.-W. C. and Utter, M.F. (1981) Pyruvate dehydrogenase activity in normal and deficient fibroblasts. *Journal of Clinical Investigation*, **67**, 1463–1471

12. Atkin, B.M., Utter, M.F. and Weinberg, M.D. (1979) Pyruvate carboxylase and phosphoenol-pyruvate carboxykinase activity in leukocytes and fibroblasts from a patient with pyruvate carboxylase deficiency. *Pediatric Research*, **13**, 38–43

13. Moreadith, R.W., Batshaw, M.L., Ohnishi, T. *et al.* (1984) Deficiency of the iron-sulphur clusters of mitochondrial reduced nicotinamide-adenine dinucleotide–ubiquinone oxidoreductase (Complex I) in an infant with congenital lactic acidosis. *Journal of Clinical Investigation*, **74**, 685–697

14. Fischer, J.C., Ruitenbeek, W., Stadhouders, A.M. *et al.* (1985) Investigation of mitochondrial metabolism in small human skeletal muscle biopsy specimens. Improvement of preparation procedure. *Clinica et Chimica Acta*, **145**, 89–94

15. DiMauro, S., Servidei, S., Zeviani, M. *et al.* (1987) Cytochrome *c* oxidase deficiency in Leigh syndrome. *Annals of Neurology*, **22**, 498–506

16. Brown, R.M., Dahl, H.H.-M. and Brown, G.K. (1989) X-Chromosome localization of the functional gene for the $E_1\alpha$ subunit of the human pyruvate dehydrogenase complex. *Genomics*, **4**, 174–181

17. Robinson, B.H. (1989) Lacticacidemia: biochemical, clinical and genetic considerations. In *Advances in Human Genetics* (eds H. Harris and K. Hirschhorn), Plenum Press, New York, pp. 151–179

18. Maragos, C., Hutchinson, W.M., Hayasaka, K., Brown, G.K. and Dalh, H.H.-M. (1989) Structural organization of the gene for the $E_1\alpha$ subunit of the human pyruvate dehydrogenase complex. *Journal of Biological Chemistry*, **264**, 12292–12298

19. Chun, K., MacKay, N., Petrova-Benedict, R. and Robinson, B.H. (1991) Pyruvate dehydrogenase deficiency due to a 20 bp deletion in Exon 11 of the pyruvate dehydrogenase (PDH) $E_1\alpha$ gene. *American Journal of Human Genetics*, **49**, 414–420

20. Robinson, B.H., MacMillan, H., Petrova-Benedict, R. and Sherwood, W.G. (1987) Variable clinical presentation in patients with defective E_1 component of the pyruvate dehydrogenase complex. *Journal of Pediatrics*, **111**, 525–533

21. Brunette, M.G., Delvin, E., Hazel, B. and Scriver, C.R. (1972) Thiamine-responsive lactic acidosis in a patient with deficient low-Km pyruvate carboxylase activity in liver. *Pediatrics*, **50**, 702–710

22. Hatefi, Y. (1988) The mitochondrial electron transport and oxidative phosphorylation system. *Annual Review of Biochemistry*, **54**, 1015–1069

23. DiMauro, S., Bonilla, E., Zeviani, M., Nakagawa, M. and DeVivo, D.C. (1985) Mitochondrial myopathies. *Annals of Neurology*, **17**, 521–538

24. Robinson, B.H., Glerum, D.M., Chow, W., Petrova-Benedict, R., Lightowlers, R. and Capaldi, R. (1990) The use of skin fibroblast cultures in the detection of respiratory chain defects in patients with lacticacidemia. *Pediatric Research*, **28**, 549–555

25. Glerum, D.M., Yanamura, W., Capaldi, R. and Robinson, B.H. (1988) Characterization of cytochrome *c* oxidase mutants in human fibroblasts. *FEBS Letters*, **236**, 100–104

26. Miranda, A.F., Ishii, S., DiMauro, S. and Shay, J.W. (1989) Cytochrome *c* oxidase deficiency in Leigh's syndrome: genetic evidence of a nuclear DNA-encoded mutation. *Neurology*, **39**, 697–702

27. Glerum, D.M., Robinson, B.H., Spratt, C., Wilson, J. and Patrick, D. (1987) Abnormal kinetic behaviour of cytochrome oxidase in a case of Leigh disease. *American Journal of Human Genetics*, **41**, 584–593

28. Tulinius, M.H., Holme, E., Kristiansson, B., Larsson, N.-G. and Oldfors, A. (1991) Mitochondrial encephalomyopathies in childhood. I. Biochemical and morphologic investigations. *Journal of Pediatrics*, **119**, 242–250

29. Tulinius, M.H., Holme, E., Kristiansson, B., Larsson, N.-G. and Oldfors, A. (1991) Mitochondrial encephalomyopathies in childhood. II. Clinical manifestations and syndromes. *Journal of Pediatrics*, **119**, 251–259

30. Robinson, B.H., DeMeirleir, L., Glerum, D.M., Sherwood, W.G. and Becker, L. (1987) Clinical presentation of mitochondrial respiratory chain defects in NADH coenzyme Q reductase and cytochrome oxidase. Clues to pathogenesis of Leigh disease. *Journal of Pediatrics*, **110**, 216–222

31. Hoppel, C.L., Kerr, D.S., Dahms, B. and Roessman, U. (1987) Deficiency of the reduced nicotinamide adenine dinucleotide dehydrogenase component of complex I. *Journal of Clinical Investigation*, **80**, 71–77

32. Robinson, B.H., Chow, W., Petrova-Benedict, R., Clarke, J., Van Allen, M., Becker, L. and Boulton, J. (1993) Fatal combined defects in mitochondrial multienzyme complexes in two siblings. *European Journal of Pediatrics*, **151**, 342–352

33. Holt, I.J., Harding, A.E., Petty, R.K.H. and Morgan-Hughes, J.A. (1990) A new mitochondrial disease associated with mitochondrial DNA heteroplasmy. *American Journal of Human Genetics*, **46**, 428–433
34. Tatuch, Y.M., Christodoulou, J., Feigenbaum, A. *et al.* (1993) Heteroplasmic mtDNA mutation (T→G) at 8993 can cause Leigh's disease when the percentage of abnormal mtDNA is high. *American Journal of Human Genetics*, **50**, 852–858

10
Leber's hereditary optic neuropathy

A.E. Harding and M.G. Sweeney

INTRODUCTION

Leber's hereditary optic neuropathy (LHON) has been recognized as a distinctive clinical entity for over 100 years [1,2]. Increased understanding of the genetic basis of this disease in recent years has to some extent changed the definition of its phenotype. What follows is a description of the classical clinical features of LHON, defined as an exclusively maternally transmitted disorder causing rapidly progressive blindness. The limits of the phenotype, and possible associations with other neurological deficits, will be described later in the context of defining LHON at a molecular genetic level.

CLINICAL, INVESTIGATIVE AND PATHOLOGICAL FEATURES

LHON characteristically gives rise to acute or subacute bilateral visual loss in males between the ages of 18 and 30 years, although earlier or later onset is well described [3–5]. It is one of the commonest causes of blindness in otherwise healthy young men. Central visual disturbance, usually described as fogging or blurring, is typically noted first in one eye and then the other, at an interval ranging from days to months. However, the condition occasionally remains monocular for more than 10 years [5]. Visual acuities deteriorate to 6/60 or less over several weeks and colour vision is lost early. The visual field loss consists initially of an enlarged blind spot; this increases to involve central vision, producing a large centrocaecal scotoma.

In most patients the nerve fibre layer around the optic disc is swollen, giving rise to a glistening white appearance, in the acute phase. Tortuous retinal arterioles and telangiectases are present in the peripapillary small vessels [6,7] (Figure 10.1). Superficial peripapillary retinal bleeding occurs occasionally. However, fluorescein angiography rarely shows leakage [8]. There is evidence that the appearances of capillary microangiopathy predate the onset of symptoms [7]. A few weeks after the onset of visual loss, the small vessels on the temporal side of the disc become attenuated and the papillomacular bundle exhibits pallor. Over several months more extensive evidence of axonal loss is seen, until the discs are

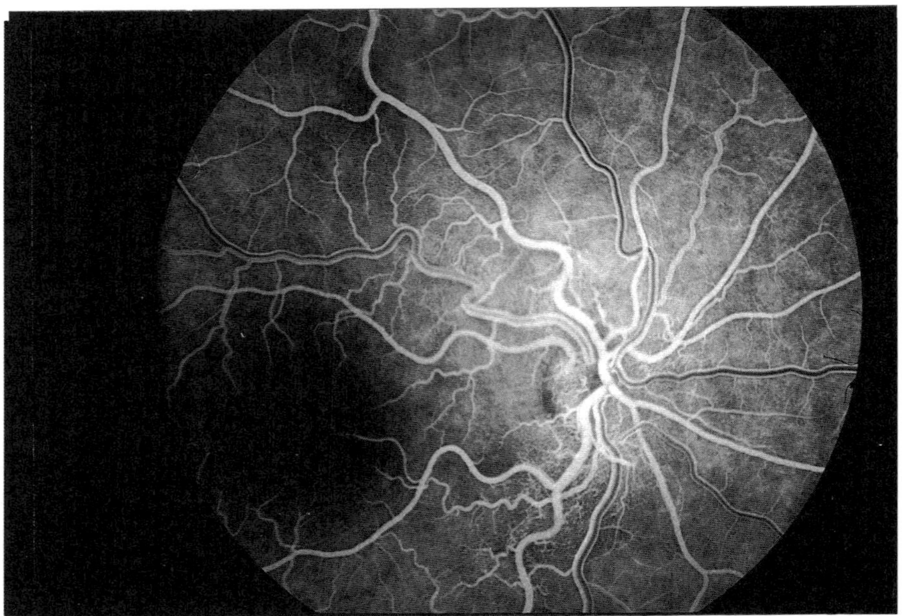

Figure 10.1 Arteriovenous phase fluorescein angiogram (right eye) of male with LHON at onset of visual loss in right eye. There is peripapillary tortuosity of arteries, veins and capillaries. No leakage of dye was seen in later photographs (Photograph courtesy of Mr M.D. Sanders)

atrophic and attenuation of the nasal retinal vessels has occurred [7]. The pupils usually exhibit a slow but symmetrical response to light, although a relative afferent defect is common at onset when one eye is predominantly affected. Loss of visual acuity generally remains severe, patients needing to be registered as blind, although a variable degree of improvement is sometimes observed after period of up to 5 or more years [3,9,10].

The age of onset of visual loss in affected women tends to be later than in men, with a range of 22–63 years reported by Nikoskelainen and colleagues [7] (compared with 16–33 years in males in the same kindreds). Other than this, there has been no marked difference in clinical manifestations between the sexes other than an association between LHON and a multiple sclerosis-like illness [11] which is discussed later.

Visual evoked potentials (VEPs) are usually absent in patients with severe visual impairment of long duration. In early or less severe cases, VEPs are reduced in amplitude, delayed and desynchronized; they tend to be less prolonged in latency and smaller than VEPs recorded in patients with known demyelinating optic neuropathy [5]. Also, patients with very early but definite visual impairment may have normal VEPs; these observations point against demyelination of the optic nerve, as opposed to axonal dysfunction, as the primary pathology in LHON. Electroretinography is normal [3].

Imaging studies using computerized tomography of the brain and optic nerves are usually normal in LHON, although distended optic nerve sheaths were reported in one case [12]. Magnetic resonance imaging (MRI) of the brain is also normal, but scans using short time inversion recovery sequences show unilateral or bilateral increased signal in the middle and posterior intraorbital parts of the optic nerves; this is thought to represent gliosis [13]. These findings are in contrast with those in acute optic neuritis, where high signal tends to affect the anterior and middle segments of the optic nerves. Autopsy studies in LHON are scarce, and there are no pathological data pertaining to early optic nerve disease. All the pathological reports concern individuals with long-standing blindness, showing marked axonal degeneration and associated loss of myelin concentrated mainly in the central part of the optic nerves [14].

FAMILY STUDIES

The features of telangiectatic microangiopathy were observed in the fundi of nine asymptomatic women from LHON families, including four of eight obligate carriers [15]. These changes were also seen in 14 of 21 males at risk of developing LHON, but with otherwise normal visual function, ranging in age between 9 and 40 years. Determining whether minor changes of microangiopathy are pathological can be difficult, as tortuosity of peripapillary retinal vessels occurs in healthy young people, but the fundus photographs in this study were mixed with those from unrelated normal and abnormal individuals and assessed blindly [15]. Carroll and Mastaglia [5] also observed one or more clinical abnormalities, including retinal vessel tortuosity, and/or VEP changes in 16 of 40 asymptomatic LHON family members. However, six of these (including one of the nine with possible microangiopathy) were not at risk of developing LHON as they were from the male lineage. Two of the four unaffected obligate female carriers had clinical and/or VEP abnormalities, as did five of 12 males at risk.

The pattern of inheritance of LHON has long been recognized as not following mendelian principles [16]. Although some reported families have shown paternal transmission [17], it is clear on analysing their clinical features that the diagnosis was not LHON but dominantly inherited optic atrophy. Classical LHON is only inherited through females. In Europe, about 85% of patients are male, and 18% of female carriers are affected [18,19]. The proportion of affected females is probably higher in Japanese families, but again many early reports included families with what appears to be dominantly inherited optic atrophy [20]. Paternal transmission of true LHON to children or grandchildren has never been described, making X-linked inheritance unlikely. Between 70 and 100% of daughters of female carriers are also carriers, and 50–100% of sons of carriers are affected [19]. Early theories proposed to explain these features of LHON families included transplacental or cytoplasmic transmission of an infective or genetic agent [16,21,22]. These were replaced more recently by the hypothesis of a pathogenic mitochondrial DNA (mtDNA) defect, given that mtDNA is exclusively maternally transmitted [18,23].

MOLECULAR GENETICS

In 1988, Wallace and colleagues [24] reported a point mutation at position 11778 of mtDNA in members of nine of 11 LHON pedigrees from the USA and Finland. This leads to an amino acid change in a mitochondrial respiratory chain Complex I subunit (ND4), and generates a restriction site loss for *Sfa*NI [24] and a site gain for *Mae*III [25]. The mutation was present in all maternally related individuals in these families, regardless of whether or not they were affected, but none of 45 control subjects. Wallace's group later provided evidence that this mutation is causative, rather than merely associated with LHON. MtDNA has a high mutation rate and exhibits considerable variation (polymorphism) in the normal population. Using restriction fragment analysis, Singh and colleagues [26] compiled mtDNA haplotypes for three patients with LHON (one American black and two European) and four control subjects. The simplest evolutionary tree which could be constructed for these mtDNAs indicated that the mutation at position 11778 must have occurred on two occasions independently. This mutation has subsequently been described in LHON families from all over Europe, the USA, Japan and Australia [9,27–32, D. Mackey, personal communication].

The initial report from Wallace's group [24] suggested that the 11778 bp mutation was homoplasmic, i.e. present in all mtDNA in LHON maternal lines, but studies in the UK showed that the 11778 bp mutation was heteroplasmic, i.e. there was a mixture of mutant and normal mtDNA, in leukocytes from the majority of subjects studied [9]. The degree of heteroplasmy appears to vary from family to family, some showing none [24,33] and others very rapid shifts in genotype from a high proportion of normal to nearly all mutant in two or three generations [28,34]. The latter observation may indicate that the mutation has occurred relatively recently in such families, again providing evidence that it is causative.

It is clear that the 11778 bp mutation is the most common one associated with LHON. However, the observations of Wallace and co-workers [24] showed that LHON was genetically heterogeneous, and this was confirmed in the UK and Finland, where only about 50–70% of LHON families have the 11778 bp mutation [9,35]. Subsequently five further mtDNA mutations (Table 10.1) have been described which are exclusively found in affected families [36–39B]. The mutation at position 3460 is in the ND1 subunit gene and has been described in Finnish,

Table 10.1 mtDNA mutations associated with LHON

mtDNA (bp)	RE site change	Amino acid change/ protein subunit	Origin	Incidence in control subjects (%)
1. Mutations found exclusively in LHON families				
11778	*Sfa*I/*Mae*III	His for Arg/ND4	Worldwide	0
3460	*Aha*II	Thr for AlaND1	Worldwide	0
4160	None*	Pro for Leu/ND1	Australia	0
5244	*Hap*II	Ser for Gly/ND2	USA	0
14484	none	Val for Met/ND6	UK/Australia	0
2. Mutations found in LHON families and controls				
4216	*Nla*III	His for Ala/ND1	USA	8
4917	*Mae*I	Asn for Asp/ND2	USA	4
7444	*Xba*I	Lys for stop/COI	USA	0.8
13708	*Eco*RII	Thr for Ala/ND5	Worldwide	4–8
15257	*Acc*I	Asn for Asp/cyt b	USA	0.3
15812	*Rsa*I	Met for Val/cytb	USA	0.1

Taken from references [24,25,30,36–39B].
RE, restriction endonuclease; ND, NADH dehydrogenase; CO, cytochrome oxidase; cyt, cytochrome.
*Can be detected using mismatched primer amplification and *Bst*XI [31].

British, American and Australian families [37,38]; it results in a restriction site loss of *Aha*II. The position 4160 mutation (also in the ND1 gene) has been reported exclusively in a large unusual Australian pedigree [36]. This family is of interest as some members had both LHON and other neurological abnormalities including a subacute infantile encephalopathy [21]. One branch of the kindred, containing patients with optic nerve disease alone, had an additional mutation in the ND1 gene (at position 4136). It was proposed that this second nucleotide substitution acted as an intragenic suppressor mutation which prevented the neurological, as opposed to ophthalmological, manifestations [36]. The fourth mutation has also only been reported in one family, but it is heteroplasmic (see below) and has not been detected in over 3000 control subjects [39]. It is at 5244 bp, in the ND2 gene. The 14484 mutation (ND6 gene) was originally described in two Australian pedigrees [39B], we have also detected it in 10 families with no other known LHON mutation (unpublished data).

Other mtDNA mutations have been reported in LHON pedigrees (Table 10.1) but their significance is unclear as they also occur in the normal population [30,39]. The one at bp 13708 (in the ND5 gene), detectable with *Eco*RII (site loss), appears to be common in familial LHON, either with or without the 11778 bp mutation, but it is found in 4–8% of control subjects [30,39–41]; whether it is significantly increased in LHON patients has yet to be established. The remaining five mutations (Table 10.1) are also seen in combination with others, and the one at bp 4216 was only observed as an 'additional' mutation in LHON [30].

One point against any of these six latter mutations being of primary pathogenic significance is that they are homoplasmic. Heteroplasmy has never been described in relation to harmless polymorphisms, despite the high mutation rate of mtDNA which implies a need for heteroplasmy during the transition from one genotype to another. Hauswirth and Laipis [42] demonstrated heteroplasmy in a single maternal line of Holstein cows, and suggested that the mtDNA could switch completely from one genotype to another in a single generation if the number of mtDNAs is greatly reduced at some point in oogenesis. On the other hand, heteroplasmy is a universal finding in at least some individuals with mtDNA mutations exclusively associated with disease, not only LHON but also in mitochondrial encephalomyopathies {43–45]. These observation suggest that heteroplasmy indicates the presence of a deleterious mutation. Johns and Berman [30] and Wallace and colleagues [39] have suggested that the above mutations, which occur at possibly a lower frequency in the healthy population, are only likely to cause optic nerve disease when they are associated with other mutations in the same individual. This is a difficult hypothesis to test, given the high degree of polymorphism in normal mtDNA and the mode of ascertainment of maternal lines in LHON families.

Although clearly associated with LHON, even heteroplasmic mtDNA mutations do not appear to be the sole determinant of the phenotype. In particular, they do not explain the excess of affected males. As has been mentioned, in many patients with the 11778 and 3460 bp mutations, and particularly their asymptomatic relatives, there is a degree of heteroplasmy in DNA from blood [9,31,38]. Although there is a tendency for individuals with a low proportion of mutant mtDNA to have a relatively low risk of developing or transmitting LHON [9], some males at risk with a high proportion of mutant mtDNA remain unaffected.

The existence of an X-linked visual loss susceptibility locus (VLSL) has recently been suggested, affected females needing to be homozygous for the susceptibility allele which appears to be common in the general population [46]. This is an attractive hypothesis as it accounts for the two observations drawn from pedigree data which are not explicable solely by mitochondrial inheritance, namely the excess of affected males with this disease and the fact that some males homoplasmic for any of the known mutations remain unaffected. An earlier study investigating the possibility of an X-linked locus in the pathogenesis of LHON did not support this [47], but the possibility of normal males transmitting the VSL allele to their daughters by virtue of marrying into LHON families was not taken into account. Linkage analysis in six Finnish families with LHON indicated the presence of a VLSL at or very close to DXS7 on the short arm of the X chromosome, with a maximum lod score (Z_{max} of 2.48 at this locus obtained by multipoint analysis. The Z_{max}-1 support interval was 9 cM distal and 7 cM proximal to DXS7 [46]. However, a study in one Italian and 12 British families with LHON using similar methodology excluded the presence of a VLSL in this interval, regardless of whether they had the 11778 or 3460 bp mutations or the polymorphism of unclear significance at 13708 bp [27,31].

The discrepancy between these two studies requires explanation. The Z_{max} obtained in the Finnish study was not particularly high, and it is possible that it is spurious, although detailed statistical analysis gave an empirical significance level of 0.0015 [46]. If there is a VLSL tightly linked to DXS7 in the Finnish population, one possible explanation for our findings is that of genetic heterogeneity, between either the LHON families in both series or the Finnish and British/Italian population. In relation to the former, both data sets contained similarly heterogeneous kindreds in relation to different mtDNA mutations, and it seems unlikely that this known mitochondrial genetic heterogeneity contributes to the different findings in linkage analysis. The Finnish population is relatively homogeneous compared with many European communities, and is genetically distinctive; there is a high incidence of several rare autosomal disorders in Finland whereas others common in the rest of Europe, such as cystic fibrosis and Huntington's disease, are very rare [48]. It is thus possible that there is an X-linked VLSL which is virtually confined to the Finnish population, and possibly another which maps elsewhere on the X chromosome in other parts of Europe. Although instinctively this hypothesis seems unlikely, it remains to be tested.

One feature of LHON pedigrees against a fully penetrant X-linked recessive visual loss susceptibility allele is that not all the sons, even homoplasmic males aged over 40, of affected females are affected. One explanation for this phenomenon is that some heterozygous females manifest the disease as a result of X chromosome inactivation. A recent statistical analysis of published LHON pedigrees suggested that this could account for 60% of affected females, with an estimated penetrance for heterozygous females of 11% [49]. Both our pedigrees and the Finnish ones are compatible with this model, but the significance of this is unclear.

mtDNA DEFECTS IN CLINICAL PRACTICE

The identification of presumed pathogenic mtDNA mutations in LHON families has provided diagnostic markers for the disease which have changed the definition of its phenotype. Previously it was difficult to make a diagnosis of LHON with certainty in the absence of affected relatives, particularly if the patient was not seen at onset when the typical fundal appearances may be observed. It has

Table 10.2 mtDNA mutations in LHON families studied in London

mtDNA (bp)	No. of index cases		
	FH +ve	FH −ve	Total
11778	19	21	40
3460	3	2	5
4160	0	0	0
13708*	4	?	4

Defined by the presence of a mutation or typical presentation and positive family history (FH).
*Also present in 8% of controls and patients with 11778 bp mutation.

since become clear that more than half of patients with LHON, as defined by subacute optic neuropathy in the presence of the 11778 or 3460 bp mtDNA mutations, have no known affected relatives (Table 10.2). Several of the patients investigated in London were thought to have bilateral optic neuritis or tobacco–alcohol amblyopia before molecular genetic diagnosis.

The clinical correlates of the 11778 bp mutation have been described in detail in a study of 49 pedigrees, 28 of which contained singleton cases [3]. The male to female ratio of patients was 59:13. Two-thirds developed visual loss between the ages of 16 and 37 years, with a range of 8–60 years. It is clear that onset of LHON can be much later than generally realized; one of our patients had onset at the age of 63 years and was thought initially to have an anterior ischaemic optic neuropathy [50]. Onset is simultaneous in both eyes in about half of patients, with an interval between eyes otherwise of up to 3 months in 25% and ranging from 3 to 9 months in the vast majority of the rest. Visual loss may be sudden and complete, or progress to stabilization over periods of up to 1 year, usually within 4 months. At onset, vision may fluctuate, and some patients experience Uhthoff's phenomenon (worsening of vision on exercise or after a hot bath) [51]. It has also become apparent that the 'classical' fundal appearances of LHON are not present acutely in all patients; the fundus was normal at presentation in nine of 49 cases [3]. Visual fields were uniformly described as showing central or centrocaecal scotomas in this study.

Holt and colleagues [9] showed that the 11778 bp mutation was associated with a poor prognosis; useful recovery of visual function was not observed in families with this mutation, but some improvement had occurred in members of four families (all of which have later been shown to have the 13708 bp mutation). This observation was confirmed by Newman and co-workers [3]. In 109 eyes, ultimate visual acuity was 20/200 or worse in all but six. Residual acuity was 20/50 in one eye in two patients. One female patient improved from 20/200 to 20/40 bilaterally over a few months, and in two males unilateral recovery to about 20/30 occurred after 12 and 30 months. Our current series of 40 families with the 11778 bp mutation shows a similarly poor prognosis, particularly in males, with significant recovery in only one female who had two episodes of severe bilateral visual loss with partial recovery on both occasions [11].

To date, there are few published data on the clinical characteristics of the 3460 bp mtDNA mutation, which has been reported in three Finnish, nine British and two Australian families [31,37,38]. The Finnish kindreds with this mutation were stated to be similar clinically to those with the 11778 bp mutation [37], and this is also our experience. No affected member of our five families with this genetic defect has shown significant recovery after up to 15 years; age of onset ranged from 10 to 40 years. The 14484 mutation appears to be associated with a better prognosis, particularly if onset is early [39B].

From the diagnostic point of view, the clinical observations in patients with the mtDNA mutations discussed above indicate that appropriate mtDNA analysis should be performed in patients of both sexes, at effectively any age, who present with simultaneous or sequential subacute bilateral optic neuropathy, even if there is no family history and the 'characteristic' fundal changes are absent. Using restriction endonuclease analysis of DNA amplified with the polymerase chain reaction (Figure 10.2), these analyses are quick, inexpensive and reliable. Apart from the diagnostic benefit of demonstrating any of the mtDNA mutations associated with LHON, there are important implications for genetic counselling which is discussed below.

BIOCHEMICAL OBSERVATIONS IN LHON

Early biochemical investigations of LHON focused on the possibility of an abnormality of cyanide metabolism, partly because it was thought that development of the disease could be related to cigarette smoking [52]. Tobacco smoke contains cyanide, which is mainly detoxified to thiocyanate by thiosulphate sulphurtransferase (rhodanese). Wilson [53] showed that thiocyanate concentrations in plasma and urine were lower in LHON patients who smoked than in control smokers. Plasma cyanocobalamin concentrations were found to be elevated in LHON patients, also compatible with the hypothesis of an inborn error of cyanide metabolism in this disease [54]. However, there is no evidence that cigarette smoking is more frequent in LHON patients than in the general population [3]. Although other sources of cyanide, such as urinary infection and a high dietary intake of thiocyanate-containing foods, have been proposed as relevant to the pathogenesis of LHON in non-smokers [14,53], the cyanide hypothesis has not been confirmed or refuted and most patients give no history suggesting excessive cyanide exposure. An early study of rhodanese activity in the liver of two patients yielded normal results [53]; subsequently there have been two reports of reduced rhodanese activity, in liver and rectal mucosa, in LHON [22,55], although others have reported normal activity in muscle [23]. Rhodanese is a ubiquitous enzyme which plays a role in the formation of iron-sulphur proteins, important molecules in electron transfer [56,57], which is of interest in relation to more recent data relating to respiratory chain function.

It is striking that nearly all the mtDNA mutations reported in LHON families produce amino acid substitutions in Complex I subunit genes (Table 10.1), although this may in part be due to sampling bias as investigators have concentrated on these genes in sequencing studies. The details of the functional effects of these mutations are not resolved, but several interesting observations have been made. Before the 4160 bp mutation (in the ND1 gene) was reported in the large Australian kindred referred to above [21,36], Parker and colleagues [58] found reduced rotenone-sensitive NADH–CoQ reductase (Complex I) activity in platelets from four affected male members of this family. Complex I deficiency, as measured by polarography, has also been described in patients' muscle in one Italian family [32] and both affected and unaffected members of a Swedish family [59] with the 11778 bp mutation. Spectrophotometric analyses of respiratory chain

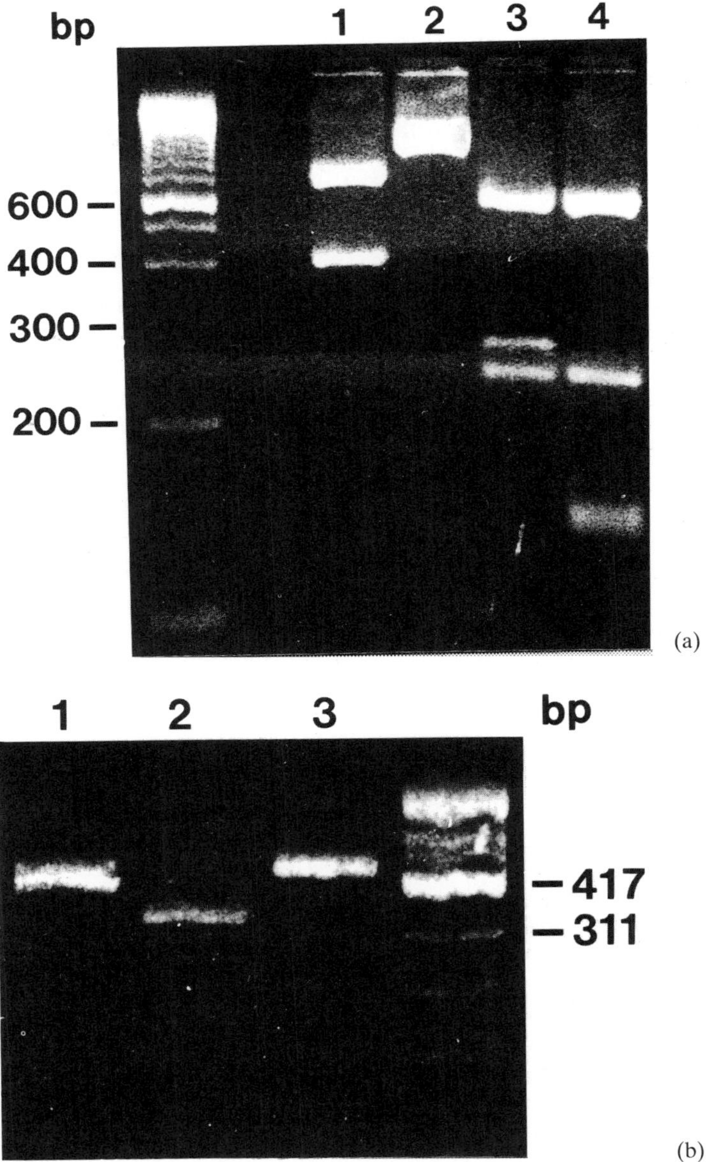

Figure 10.2 Fragments of polymerase chain reaction (PCR) product after amplification of the relevant region of mtDNA and (a) digestion with *Sfan*I and *Mae*III to show the 11778 bp mtDNA mutation in lanes 2 and 4, compared with control samples in lanes 1 and 3. The lane on the left is a bp ladder. This mutation causes loss of a restriction site for *Sfan*I, so the DNA fragment of 1.1 kb is cleaved into two of 687 and 413 bp in the normal subject (lane 1) but not the patient (lane 2). It creates a restriction site for *Mae*III; this enzyme yields visible fragments of 560, 255 and 228 bp in the control (lane 3), but in the patient the 255 bp fragment is cleaved into two of 128 and 127 bp (lane 4). In (b) with a bp ladder on the far right, the 3460 bp mutation is demonstrated in lane 2 after digestion with *Aha*II which cuts the mutant PCR product of 428 bp into two of 330 and 98 bp (latter not visible) by virtue of a gained restriction site; lane 1 contains uncut DNA from the patient and lane 3 a digested control sample

enzyme activities in isolated mitochondria were normal in the latter kindred. Both Nikoskelainen *et al.* [23] and Larsson and colleagues [59] have reported subsarcolemmal accumulations of enlarged mitochondria on ultrastructural studies of muscle. Lastly, Howell and colleagues [38] reported reduced Complex I activity in isolated platelet mitochondria from two affected and three unaffected members of a family with the 3460 bp mutation. This deficiency was not more severe in the patients. Given these observations in platelet and muscle mitochondria, it is difficult to explain the tissue (and patient) specificity of LHON.

ASSOCIATION BETWEEN LHON AND OTHER DISORDERS

Apart from other neurological disease (see below), the only feature of LHON pedigrees suggesting an association with non-ophthalmological disease is the observation of electrocardiographic (ECG) changes suggesting a pre-excitation syndrome in about 30% of both affected and unaffected members of Finnish pedigrees [60]. Nikoskelainen and colleagues [19] found these in half of the descendants of female carriers, compared with only one of 21 subjects in paternal lines in the same families. Individuals in maternal lines also had other less specific ECG changes such as deep Q and tall R waves in the inferior and lateral chest leads; similar findings were described in two males with LHON [61]. Very few of the Finnish patients had ever had significant cardiac symptoms; one patient and a relative of another in the series of Newman and colleagues [3] had a history of supraventricular tachycardia. The significance of these observations, particularly that of a short P–R interval without other abnormalities, is still unclear and this area deserves further study.

An association between LHON and numerous neurological disorders was suggested in the early literature; many of these reports can be discarded as the diagnosis of LHON was clearly, in retrospect, incorrect, and the patients had optic atrophy as a feature of a wide variety of neurodegenerative disease [17]. It is striking that only five of the 72 patients with LHON and the 11778 bp mutation had other neurological abnormalities such as seizures or mental retardation [3]; these were probably coincidental. Patients reported with the typical clinical features of LHON (subacute severe optic neuropathy with exclusively maternal transmission) include the following. A family with Charcot–Marie–Tooth disease and LHON has been described, but it appeared that the two disorders were segregating independently in this pedigree [62]. The cases described by Wilson [52] had prominent dementia, extrapyramidal and anterior horn cell disease.

In the large Queensland family reported by Wallace [21], which has the 4160 bp mutation, nearly half of the patients had neurological features. Nine had a severe encephalitic illness in childhood which resulted in death in three. This syndrome consisted of vomiting, headaches, neck stiffness, seizures, coma, irregular respiration, papilloedema, hemiplegia and ataxia. Two patients had polydipsia. Cerebrospinal fluid (CSF) showed a slight increase in protein and variable mild pleocytosis. Autopsy data are not detailed but no specific changes were described. The aetiology of this syndrome is obscure but it is possible that these patients had Leigh's syndrome, particularly given the prominence of respiratory abnormalities; this syndrome has been reported in families with other mtDNA mutations [43,63]. Patients who recovered had residual optic atrophy but minimal other neurological deficits. Nine other patients had dysarthria, occasional distal wasting, mild

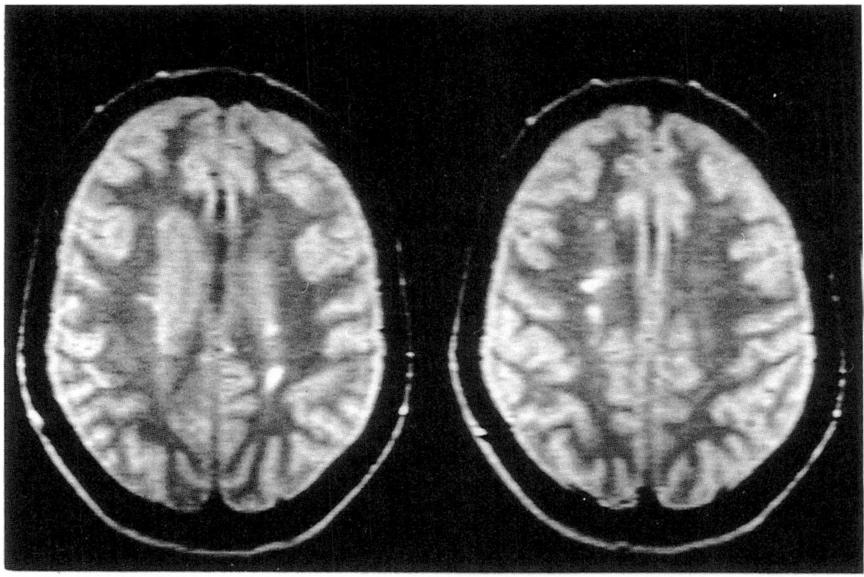

Figure 10.3 T_2 weighted MRI in a female with LHON shows multiple cerebral white matter lesions, characteristic of multiple sclerosis; she had other neurological features consistent with this diagnosis (Photograph courtesy of Dr D.H. Miller)

spasticity and movements of the face and distal limbs described as athetosis developing in childhood but these were not progressive.

Novotny and colleagues [64] reported an interesting family in which a variable neurological disorder was maternally transmitted and thought to be mitochondrial in origin. Some patients had clinical features consistent with a diagnosis of LHON. Others had childhood onset dystonia, of variable severity, often with intellectual impairment. Both conditions were present in one patient. Muscle biopsies in two cases did not show any evidence of mitochondrial disease, but sophisticated biochemistry was not performed. CT scan showed low-density lesions in the basal ganglia, chiefly involving the striatum. This family does not have the 11778 bp mutation [51]. Two other patients with the features of LHON and dystonia have been described [65], and a possible X-linked dystonic syndrome was reported in another LHON kindred [66].

An association between LHON and multiple sclerosis (MS) was suggested long ago [67], but this was difficult to substantiate before the availability of modern investigative techniques. A recent report described six females, with a family history suggesting LHON, who presented with bilateral optic neuropathy but later developed other neurological features compatible with a diagnosis of MS [11]. This diagnosis was supported by investigations including MRI (Figure 10.3). A further two females with LHON in the absence of other neurological symptoms had the characteristic features of MS on MRI. All had the 11778 bp mtDNA mutation. One particularly striking feature of this series is that all the patients were female. We have not observed other neurological abnormalities in over 50 males with LHON, and brain MRI performed in 13 of them was normal [13]. Of interest is the finding that four of six patients investigated had the HLA DR2 and/or DQw6 alleles, which are strongly associated with MS in the caucasian population [68].

Other reported cases fulfilling current diagnostic criteria for LHON and clinically definite MS include one reported by Lees and colleagues [67] and another by Ferguson and Critchley [69]; these were both female. One male patient with LHON and the 11778 bp mutation had oligoclonal bands in his CSF and white matter lesions on MRI; two other patients had a family history of MS, but the sex of these relatives was not stated [3]. Thus this association appears to be rare. However, the diagnosis of each disorder was difficult to confirm without modern methods of investigation such as MRI, CSF immunoglobulin electrophoresis and, in the case of LHON, mtDNA analysis.

These observations prompt speculation about the role of immunological factors in the pathogenesis of LHON, which are discussed below, and also that of mitochondrial genes in MS. Susceptibility to MS is multifactorial, involving an interplay between genes encoded at different loci (including one or more genes encoded in the class II region of the histocompatibility complex locus on chromosome 6) and environmental triggers [70]. These genes do not account entirely for the degree of heritability implicated in several population-based studies of MS, and it is presumed that other, as yet unidentified, genes contribute to susceptibility; these could include one or more mitochondrial genes. In favour of this hypothesis is the observation that females with MS are more likely to have affected children than males, even taking the female preponderance of MS into account [71].

POSSIBLE ROLE OF AUTOIMMUNITY IN LHON

There are several clinical observations in LHON that are surprising for a genetic disease. The onset of blindness is subacute and the optic disc appearances at the time of presentation are reminiscent of an inflammatory microangiopathy, although it is notable that the vessels do not usually leak fluorescein [8]. It is also not clear how fairly conservative amino acid changes in a respiratory chain subunit cause subacute blindness in young adult life. Many patients have increased immunoglobulins in their CSF [72]. These observations, together with the association between LHON and MS referred to above, suggest that immunological factors could be relevant in the pathogenesis of LHON.

There is evidence that mtDNA products play a part in the immune response in rodents, in the form of a maternally transmitted minor murine histocompatibility antigen (maternally transmitted factor (MTF) [73,74]). The inheritance of this factor is stable over 20 generations and is not affected by foster nursing, embryo transfer to surrogate mothers or prolonged cell contact in bone marrow chimeras [73], excluding viral transmission. Comparison of the mitochondrial genomes encoding the four allelic forms of MTF showed the least number of amino acid substitutions in the ND1 gene, suggesting that this gene determines MTF antigenicity. The α- and β-forms of ND1 differ only at position 6 of the first 17 amino acids, indicating that this residue confers antigenicity and cytotoxicity.

It is not clear how mitochondrially encoded peptides are able to play a role in the immune response. T cell helper and cytotoxic epitopes generally contain a charged residue or glycine followed by two hydrophobic residues [75]. The amino acid sequence in ND1-MTF does not carry this pattern in any of its allelic forms. In fact, the N-terminal sequence, which confers cytotoxicity, is strongly hydrophobic and contains no charged residues. However, the conservation of this sequence suggests an important role, probably in anchoring ND1 in the lipid membrane. An amino acid change may alter the orientation of the extramembranous residues to induce cytotoxic activation.

The 33 kDa polypeptide of ND1 is part of NADH–CoQ reductase (Complex I), the first enzyme protein of the mitochondrial respiratory chain. ND1 forms the major part of the binding site of the Complex I inhibitor rotenone and thus clearly plays an important role in the function of Complex I. ND1 is translated on mitochondrial ribosomes and integrated into the Complex I holoenzyme on the inner mitochondrial membrane. Its presence on the plasma membrane as a histocompatibility antigen suggests either export of mRNA or mature product from the mitochondrion, or direction to the plasma membrane following recycling of autophaged mitochondria [74].

Although it seems unlikely that the part of ND4 affected in LHON by the 11778 bp mutation is homologous to MTF, it is possible that the resulting amino acid change similarly elicits an abnormal tissue-specific immune response which could effect the development of blindness in LHON. The mutation converts a conserved arginine at position 340 to a histidine; both these amino acids are basic and hydrophilic, are likely to be part of an extramembranous sequence and as such may form an antigenic determinant. It is of interest that the two other, less common, mtDNA mutations associated exclusively with LHON each give rise to an amino acid substitution in a hydrophilic loop of the ND1 protein and could thus generate autoantigens [38].

Pathogenic mtDNA mutations are sometimes associated with circulating antibodies directed against mitochondrial proteins [76]. A specific antibody to a 41 kDa mitochondrial matrix protein was observed in a patient with the MELAS (mitochondrial encephalopathy, lactic acidosis and stroke-like episodes) phenotype of mitochondrial myopathy. This patients had the MELAS mtDNA mutation in the tRNA^{Leu} gene [43,77]. Suborganelle fractionation and immunoblotting studies suggested that the 41 kDa antigen was a matrix protein, not part of the respiratory chain, and thus would be nuclearly encoded. The relevance of the autoantibody to this patient's disease is difficult to establish. The same applies to the antibodies to mitochondrial proteins which occur in primary biliary cirrhosis (PBC). The mitochondrial M2 subclass of antibodies in PBC is directed against a heterogeneous groups of antigens, the most important of which are subunits of pyruvate dehydrogenase [78]. It is of interest that steroid responsiveness has been described in MELAS patients, suggesting that immunity may play some part in the development and expression of this disease which is associated with a point mutation in the mtDNA tRNA^{Leu} gene in most cases [43,77].

Thus there are clinical and theoretical observations which indicate that the possible role of immunological factors in the pathogenesis of LHON are worth investigating. In some individuals with the 11778 bp mutation, particularly females, an immune-mediated process appears to involve other myelinated axons in the central nervous system, producing a disorder indistinguishable from MS. The identification of these mechanisms could have important implications in terms of management; it is possible that the aetiology of blindness in LHON families is even more complex that that of the proposed interaction between mitochondrial and nuclear genes.

MANAGEMENT

There have been no large studies evaluating the effects of any therapies in LHON. Hydroxocobalamin has a high affinity for cyanide and was suggested as a treatment for the disease on this basis [14]. Although suggested as conferring benefit initially [79], treatment with hydroxocobalamin, even in the very early stages of visual loss, was subsequently reported to be ineffective [7]. There is also no evidence that high-dose corticosteroids alter the course of the disease [7]. Given that respiratory chain cofactors have been suggested to be helpful in patients with mitochondrial myopathies and respiratory chain defects [80], the use of these, such as ubiquinone and riboflavin, is worth considering although it is not clear that respiratory chain dysfunction is the primary cause of visual loss in LHON. Patients and individuals at risk should be encouraged not to smoke or drink alcohol excessively, despite the rather tenuous connection between the disease and abuse of tobacco and alcohol.

Management of blind individuals in LHON families should include provision of low vision aids and social service support, much of which is made accessible to people registering as blind in the UK. Not surprisingly, as visual loss in LHON often occurs in adults with developing careers and young families, unemployment, financial difficulties and reactive depression are common. Retraining is important, and full productive employment can be regained by individuals with this disease.

Genetic counselling is also essential. Males, either affected or unaffected, in LHON families can be reassured that there is a negligible risk of transmitting the disease to their children of either sex. This is only complicated occasionally by LHON patients having blind partners who (perhaps unknowingly) also have LHON; this has been described in one family, giving an erroneous impression of paternal transmission [31,33].

The identification of mtDNA mutations has made it possible to identify families who need genetic counselling for LHON which previously may have gone unrecognized because there was only one affected member [3]. Unfortunately, molecular genetic analysis has not made a major impact on refining genetic risks in LHON families, because of the observation that some males with a high proportion of mutant mtDNA never develop the disease and genetically similar females do not always transmit it. Based on pedigree data, about 50% of the sons of a normal female who is an obligate carrier (with an affected son and brother or uncle) are affected, and about 50% of her daughters transmit the disease themselves. These risks are rather higher (probably about 75%) for the offspring of affected females [4,19,81,82]. These are approximate figures, as no genetic counselling data have been provided from a large series of unselected families, including those containing singleton cases, taking age-related penetrance into account. Also, it is possible that penetrance varies between families. It is even more difficult to discern what the risk of developing the disease is in women, but in the sisters of male patients and the daughters of obligate carriers it is about 12% in European populations, and twice this if the individual's mother is affected [4,19].

The significance of microangiography is uncertain in terms of predicting which males will develop the disease and which females will transmit it. The same applies to the detection of any of the associated mtDNA mutations if they are homoplasmic or nearly so, as is most often the case in maternal relatives of patients. However, relatively low amount of mutant mtDNA (less than 75%) may confer a lower risk of developing or transmitting the disease [9], although the amount of mutant mtDNA in a woman's ova may not be reflected faithfully by that in her leucocytes and platelets. Some women with less than 75% mutant mtDNA in blood have had affected, almost homoplasmic, sons [28,83], and the proportions of normal and mutant mtDNA may vary between blood and hair roots [83]. In view of these problems, analysis of mtDNA in at-risk males is unhelpful for clinical purposes; it usually only serves to confirm maternity and often enhances anxiety. This particularly applies to such investigations in children which are arguably dubious from the ethical point of view and thus best avoided. The same difficulties arise in molecular genetic prenatal diagnosis. As it is not known what proportion of mutant mtDNA in chorionic villus material reduces the genetic risk, or whether this proportion may change during pre- and post-natal development, the only option for female carriers who wish to avoid a high-risk pregnancy is abortion of male fetuses. However, the daughters of these women have a tangible risk of being affected and some probable carriers may choose alternatives such as *in vitro* fertilization of donated ova. A high risk of transmitting mitochondrial genetic disease is probably one of the most compelling genetic indications for this approach [84].

REFERENCES

1. Leber, T. (1871) Ueber hereditäre und congenital-angelegte Sehnervenleiden. *Archiv für Ophthalmologie*, **17**, 249–291
2. von Graefe, A. (1858) Ein ungewohnlicher Fall von hereditärer Amaurose. *Archiv für Ophthalmologie*, **4**, 266–268
3. Newman, N.J., Lott, M.T. and Wallace, D.C. (1991) The clinical characteristics of pedigrees of Leber's hereditary optic neuropathy with the 11778 mutation. *American Journal of Ophthalmology*, **111**, 750–762
4. van Senus, A.H.C. (1963) Leber's disease in the Netherlands. *Documenta Ophthalmologica*, **17**, 1–162
5. Carroll, W.M. and Mastaglia, F.L. (1979) Leber's optic neuropathy. A clinical and visual evoked potential study of affected and asymptomatic members of a six generation family. *Brain*, **102**, 559–580
6. Smith, J.L., Hoyt, W.F. and Susav, J.O. (1973) Ocular fundus in acute Leber optic neuropathy. *Archives of Ophthalmology*, **90**, 349–354
7. Nikoskelainen, E., Hoyt, W.F. and Nummelin, K. (1983) Ophthalmoscopic findings in Leber's hereditary optic neuropathy. The fundus findings in the affected family members. *Archives of Ophthalmology*, **101**, 1059–1068
8. Nikoskelainen, E., Hoyt, W.F., Nummelin, K. and Schatz, H. (1984) Fundus findings in Leber's hereditary optic neuroretinopathy III. Fluorescein angiographic studies. *Archives of Ophthalmology*, **102**, 981–989
9. Holt, I.J., Miller, D.H. and Harding, A.E. (1989) Genetic heterogeneity and mitochondrial DNA heteroplasmy in Leber's hereditary optic neuropathy. *Journal of Medical Genetics*, **26**, 739–743
10. Lessell, S., Gise, R.L. and Krohel, G.B. (1983) Bilateral optic neuropathy with remission in young men. Variation on a theme by Leber? *Archives of Neurology*, **40**, 2–6
11. Harding, A.E., Sweeney, M.G., Miller, D.H., Mumford, C.J., Kellar-Wood, H., Menard, D., McDonald, W.I. and Compston, D.A.S. (1992) Occurrence of a multiple sclerosis-like illness in women who have a Leber's hereditary optic neuropathy mitochondrial DNA mutation. *Brain*, **115**, 979–989
12. Smith, J.L., Tse, D.T., Byrne, S.F., Johns, D.R. and Stone, E.M. (1990) Optic nerve sheath distention in Leber's optic neuropathy and the significance of the 'Wallace mutation'. *Journal of Clinical Neuro-ophthalmology*, **10**, 231–238
13. Kermode, A.G., Moseley, I.F., Kendall, B.E., Miller, D.H., Macmanus, D.G. and McDonald, W.I. (1989) Magnetic resonance imaging in Leber's optic neuropathy. *Journal of Neurology, Neurosurgery and Psychiatry*, **52**, 671–674
14. Adams, J.H., Blackwood, W. and Wilson, J. (1966) Further clinical and pathological observations on Leber's optic atrophy. *Brain*, **89**, 15–26

15. Nikoskelainen, E., Hoyt, W.F. and Nummelin, K. (1982) Ophthalmoscopic findings in Leber's hereditary optic neuropathy. Fundus findings in asymptomatic family members. *Archives of Ophthalmology*, **100**, 1597–1602
16. Imai, Y. and Moriwaki, D. (1936) A probable case of cytoplasmatic inheritance in man: a critique of Leber's disease. *Journal of Genetics*, **33**, 163–187
17. Bell, J. (1931) Hereditary optic atrophy (Leber's disease). In *The Treasury of Human Inheritance, Volume II, Anomalies and Diseases of the Eye* (ed. K. Pearson), Cambridge University Press, London, pp. 325–423
18. Nikoskelainen, E.K. (1984) New aspects of the genetic, etiologic and clinical puzzle of Leber's disease. *Neurology*, **34**, 1482–1484
19. Nikoskelainen, E.K., Savontaus, M-L., Wanne, O.P., Katila, M.J. and Nummelin, K.U. (1987) Leber's hereditary optic neuroretinopathy, a maternally inherited disease. A genealogic study of four pedigrees. *Archives of Ophthalmology*, **105**, 665–671
20. Kawakami, R. (1926) Beiträge zur Vererbung der familiären Sehnervenatrophie. *Archiv für Ophthalmologie*, **66**, 568–595
21. Wallace, D.C. (1970) A new manifestation of Leber's disease and a new explanation for the agency responsible for its unusual pattern of inheritance. *Brain*, **93**, 121–132
22. Erickson, P. (1972) Leber's optic atrophy: a possible example of mitochondrial inheritance. *American Journal of Human Genetics*, **24**, 348–349
23. Nikoskelainen, E.K., Hassinen, I.E., Paljarvi, L., Lang, H. and Kalimo, H. (1984) Leber's hereditary optic neuroretinopathy, a mitochondrial disease? *Lancet*, **ii**, 1474
24. Wallace, D.C., Singh, G., Lott, M.T., Hodge, J.A., Schurr, T.G., Lezza, A.M., Elsas, L.J. and Nikoskelainen, E.K. (1988) Mitochondrial DNA mutation associated with Leber's hereditary optic neuropathy. *Science*, **242**, 1427–1430
25. Johns, D.R. (1990) Improved molecular-genetic diagnosis of Leber's hereditary optic neuropathy. *New England Journal of Medicine*, **323**, 1488–1489
26. Singh, G., Lott, M.T. and Wallace, D.C. (1989) A mitochondrial DNA mutation as a cause of Leber's hereditary optic neuropathy. *New England Journal of Medicine*, **320**, 1300–1305
27. Sweeney, M.G., Davis, M.B., Brockington, M., Toscano, A. and Harding, A.E. (1991) Is visual loss in Leber's disease partly determined by an X-linked locus? *American Journal of Human Genetics*, **49** (Suppl.), 361
28. Bolhuis, P.A., Bleeker Wagemakers, E.M., Ponne, N.J., Van Schooneveld, M.J., Westerveld, A., Van den Bogert, C. and Tabak, H.F. (1990) Rapid shift in genotype of human mitochondrial DNA in a family with Leber's hereditary optic neuropathy. *Biochemical and Biophysical Research Communications*, **170**, 994–997
29. Yoneda, M., Tsuji, S., Yamauchi, T., Inuzuka, T., Miyatake, T., Horai, S. and Ozawa, T. (1989) Mitochondrial DNA mutation in family with Leber's hereditary optic neuropathy. *Lancet*, **i**, 1076–1077
30. Johns, D.R. and Berman, J. (1991) Alternative, simultaneous complex I mitochondrial DNA mutations in Leber's hereditary optic neuropathy. *Biochemical and Biophysical Research Communications*, **174**, 1324–1330
31. Sweeney, M.G., Davis, M.B., Lashwood, A.M., Brockington, M., Toscano, A. and Harding, A.E. (1992) Evidence against a locus close to DXS7 determining visual loss in Italian and British families with Leber's hereditary optic neuropathy. *American Journal of Human Genetics*, **51**, 741–748
32. Toscano, A., Harding, A.E., Castagna, I. *et al.* (1990) Familial Leber's hereditary optic neuropathy: morphology, biochemistry and genetics in blood and skeletal muscle. *Journal of the Neurological Sciences*, **98** (Suppl.), 366
33. Poulton, J., Deadman, M.E., Bronte-Stewart, J., Foulds, W.S. and Gardiner, R.M. (1991) Analysis of mitochondrial DNA in Leber's hereditary optic neuropathy. *Journal of Medical Genetics*, **28**, 765–770
34. Vilkki, J., Savontaus, M.L. and Nikoskelainen, E.K. (1990) Segregation of mitochondrial genomes in a heteroplasmic lineage with Leber hereditary optic neuroretinopathy. *American Journal of Human Genetics*, **47**, 95–100
35. Vilkki, J., Savontaus, M.L. and Nikoskelainen, E.K. (1989) Genetic heterogeneity in Leber hereditary optic neuroretinopathy revealed by mitochondrial DNA polymorphism. *American Journal of Human Genetics*, **45**, 206–211
36. Howell, N., Kubacka, I., Xu, M. and McCullough, D.A. (1991) Leber hereditary optic neuropathy: involvement of the mitochondrial ND1 gene and evidence for an intragenic suppressor mutation. *American Journal of Human Genetics*, **48**, 935–942
37. Huoponen, K., Vilkki, J., Aula, P. and Nikoskelainen, E.K. (1991) A new mtDNA mutation associated with Leber hereditary optic neuroretinopathy. *American Journal of Human Genetics*, **48**, 1147–1153

38. Howell, N., Bindoff, L.A., McCullough, C.A., Kubacka, I., Poulton, J., Mackey, D., Taylor, L. and Turnbull, D.M. (1991) Leber hereditary optic neuropathy: identification of the same mitochondrial ND1 mutation in six pedigrees. *American Journal of Human Genetics*, **49**, 939–950

39A. Wallace, D.C., Brown, M.D., Lott, M.T., Voljavec, A.S., Torroni, A. and Yang, C.-C. (1992) Multiple mitochondrial DNA mutations associated with Leber's hereditary optic neuropathy. *Cytogenetics and Cell Genetics*, **58**, 2121

39B. Mackey, D. and Howell, N. (1992) A variant of Leber hereditary optic neuropathy characterized by recovery of vision and by an unusual mitochondrial genetic etiology. *American Journal of Human Genetics*, **51**, 1218–1228

40. Holt, I.J., Harding, A.E. and Morgan-Hughes, J.A. (1988) Mitochondrial DNA polymorphism in mitochondrial myopathy. *Human Genetics*, **79**, 53–57

41. Holt, I.J., Miller, D.H. and Harding A.E. (1988) Restriction endonuclease analysis of leukocyte mitochondrial DNA in Leber's optic atrophy. *Journal of Neurology, Neurosurgery and Psychiatry*, **51**, 1075–1077

42. Hauswirth, W.W. and Laipis, P.J. (1985) Transmission genetics of mammalian mitochondria: a molecular model and experimental evidence. In *Achievements and Perspectives of Mitochondrial Research* (eds E. Quagliariello, E.C. Slater, F. Palmieri, C. Saccone and A.M. Kroon), Elsevier, Amsterdam, pp. 49–60

43. Hammans, S.R., Sweeney, M.G., Brockington, M., Morgan-Hughes, J.A. and Harding A.E. (1991) Mitochondrial encephalopathies: molecular genetic diagnosis from blood samples. *Lancet*, **337**, 1311–1313

44. Holt, I.J., Harding, A.E., Petty, R.K.H. and Morgan-Hughes, J.A. (1990) A new mitochondrial disease associated with mitochondrial DNA heteroplasmy. *American Journal of Human Genetics*, **46**, 428–433

45. Shoffner, J.M., Lott, M.T., Lezza, A.M., Seibel, P., Ballinger, S.W. and Wallace, D.C. (1990) Myoclonic epilepsy and ragged-red fiber disease (MERRF) is associated with a mitochondrial DNA tRNA(Lys) mutation. *Cell*, **61**, 931–937

46. Vilkki, J., Ott, J., Savontaus, M.L., Aula, P. and Nikoskelainen, E.K. (1991) Optic atrophy in Leber hereditary optic neuroretinopathy is probably determined by an X-chromosomal gene closely linked to DXS7. *American Journal of Human Genetics*, **48**, 486–491

47. Chen, J.D., Cox, I. and Denton, M.J. (1989) Preliminary exclusion of an X-linked gene in Leber optic atrophy by linkage analysis. *Human Genetics*, **82**, 203–207

48. Ikonen, E., Palo, J., Ott, J., Gusella, J.F., Somer, H., Karila, L., Palotie, A. and Peltonin, L. (1990) Huntington disease in Finland: linkage disequilibrium of chromosome 4 RFLP haplotypes and exclusion of a tight linkage between the disease and D4S43 locus. *American Journal of Human Genetics*, **46**, 5–11

49. Bu, X. and Rotter, J.I. (1991) X chromosome-linked and mitochondrial gene control of Leber hereditary optic neuropathy: evidence from segregation analysis for dependence on X chromosome inactivation. *Proceedings of the National Academy of Science USA*, **88**, 8198–8202

50. Borruat, F.-X., Green, W.T., Graham, E.M., Sweeney, M.G., Morgan-Hughes, J. and Sanders, M.D. (1992) Late onset Leber's optic neuropathy: a case confused with ischaemic optic neuropathy. *British Journal of Ophthalmology*, **76**, 571–573

51. Newman, N.J. and Wallace, D.C. (1990) Mitochondria and Leber's hereditary optic neuropathy. *American Journal of Ophthalmology*, **109**, 726–730

52. Wilson, J. (1963) Leber's hereditary optic atrophy. Some clinical and aetiological considerations. *Brain*, **86**, 347–362

53. Wilson, J. (1965) Leber's hereditary optic atrophy: a possible defect of cyanide metabolism. *Clinical Science*, **29**, 505–515

54. Wilson, J., Linnell, J.C. and Matthews, D.M. (1971) Plasma-cobalamins in neuro-ophthalmological diseases. *Lancet*, **i**, 259–261

55. Poole, C.J.M. and Kind, P.R.N. (1986) Deficiency of thiosulphate sulphurtransferase (rhodanese) in Leber's hereditary optic neuropathy. *British Medical Journal*, **292**, 1229–1230

56. Cerletti, P. (1986) Seeking a better job for an underemployed enzyme: rhodanese. *Trends in Biochemical Sciences*, **11**, 369–372

57. Hatefi, Y. (1985) The mitochondrial electron transfer and oxidative phosphorylation system. *Annual Review of Biochemistry*, **54**, 1015–1069

58. Parker, W.D., Jr., Oley, C.A. and Parks, J.K. (1989) A defect in mitochondrial electron-transport activity (NADH–coenzyme Q oxidoreductase) in Leber's hereditary optic neuropathy. *New England Journal of Medicine*, **320**, 1331–1333

59. Larsson, N.-G., Anderson, O., Holme, E., Oldfors, A. and Wahlstrom, J. (1991) Leber's hereditary optic neuropathy and complex I deficiency in muscle. *Annals of Neurology*, **30**, 701–708

60. Nikoskelainen, E., Wanne, O. and Dahl, M. (1985) Pre-excitation syndrome and Leber's hereditary optic neuroretinopathy. *Lancet*, **i**, 696
61. Rose, F.C., Bowden, A.N. and Bowden, P.M.A. (1970) The heart in Leber's optic atrophy. *British Journal of Ophthalmology*, **54**, 388–393
62. McLeod, J.G., Low, P.A. and Morgan-Hughes, J.A. (1978) Charcot–Marie–Tooth disease with Leber optic atrophy. *Neurology*, **28**, 179–184
63. Tatuch, Y., Christodoulou, J., Feigenbaum, A., Clarke, J.T.R., Wherret, J., Smith, C., Rudd, N., Petrova-Benedict, R. and Robinson, B.H. (1992) Heteroplasmic mtDNA mutation (T→G) at 8993 can cause Leigh's disease when the percentage of abnormal mtDNA is high. *American Journal of Human Genetics*, **50**, 852–858
64. Novotny, E.J., Jr., Singh, G., Wallace, D.C., Dorfman, L.J., Louis, A., Sogg, R.L. and Steinman, L. (1986) Leber's disease and dystonia: a mitochondrial disease. *Neurology*, **36**, 1053–1060
65. Marsden, C.D.,Lang, A.E., Quinn, N.P., McDonald, W.I., Abdallat, A. and Nimri, S. (1986) Familial dystonia and visual failure with striatal CT lucencies. *Journal of Neurology, Neurosurgery and Psychiatry*, **49**, 500–509
66. Bruyn, G.W. and Went, L.N. (1964) A sex-linked heredo-degenerative neurological disorder associated with Leber's optic atrophy Part 1. Clinical studies. *Journal of Neurological Sciences*, **1**, 59–80
67. Lees, F., Macdonald, A.-M.E. and Aldren Turner, J.W. (1964) Leber's disease with symptoms resembling disseminated sclerosis. *Journal of Neurology, Neurosurgery and Psychiatry*, **27**, 415–421
68. Olerup, O. and Hillert, J. (1991) HLA class II-associated genetic susceptibility in multiple sclerosis: a critical evaluation. *Tissue Antigens*, **38**, 1–15
69. Ferguson, F.R. and Critchley, M. (1928) Leber's optic atrophy and its relationship with the heredo-familial ataxias. *Journal of Neurology and Psychopathology*, **9**, 120–132
70. Compston, D.A.S. (1991) Genetic susceptibility to multiple sclerosis. In *McAlpine's Multiple Sclerosis* (ed. W.B. Matthews), Churchill-Livingstone, Edinburgh, pp. 301–319
71. Sadovnick, A.D., Bulman, D. and Ebers, G.C. (1991) Parent–child concordance in multiple sclerosis. *Annals of Neurology*, **29**, 252–255
72. Pallini, A.F., Manneschi, L., Annuziata, P. *et al.* (1988) Is Leber's hereditary optic atrophy (LHOA) a mitochondrial disease? *Journal of Neurology*, **235**, S7
73. Fischer-Lindahl, K. and Burki, K. (1982) MTA, a maternally inherited surface antigen of mouse, is transmitted in the egg. *Proceedings of the National Academy of Sciences USA*, **79**, 5362–5366
74. Loveland, B., Wang, C.-R., Yonekawa, H., Hermel, E. and Fischer Lindahl, K. (1990) Maternally transmitted histocompatibility antigen of mice: a hydrophobic peptide of a mitochondrially encoded protein. *Cell*, **60**, 971–980
75. Rothbard, J. and Taylor, W.R. (1988) A sequence pattern common to T cell epitopes. *EMBO Journal*, **7**, 93–100
76. Schapira, A.H.V., Cooper, J.M., Manneschi, L., Vital, C., Morgan-Hughes, J.A. and Clark, J.B. (1990) A mitochondrial encephalomyopathy with specific deficiencies of two respiratory chain polypeptides and a circulating antibody to a mitochondrial matrix protein. *Brain*, **113**, 419–432
77. Goto, Y., Nonaka, I. and Horai, S. (1990) A mutation in the tRNA$^{Leu(UUR)}$ gene associated with the MELAS subgroup of mitochondrial encephalomyopathies. *Nature*, **348**, 651–653
78. Yeaman, S., Fussey, S.P.M., Danner, D.J., James, O.F.W., Mutimer, D.J. and Bassendine, M.F. (1988) Primary biliary cirrhosis: identification of two major M2 mitochondrial autoantigens. *Lancet*, **i**, 1067–1070
79. Foulds, W.S. (1969) Visual disturbances in systemic disorders: optic neuropathy and systemic disease. *Transactions of the Ophthalmological Society of the United Kingdom*, **89**, 125–146
80. Bresolin, N., Doriguzzi, C., Ponzetto, C., Angelini, C., Moroni, I., Castelli, E., Cossutta, E., Binda, A., Gallenti, A., Gabellini, S., Piccolo, G., Martinuzzi, A., Ciafaloni, E., Arnaudo, E., Liciardello, L., Carenzi, A. and Scarlato, G. (1990) Ubidecarenone in the treatment of mitochondrial myopathies: A multi-center double-blind trial. *Journal of the Neurological Sciences*, **100**, 70–78
81. Seedorff, T. (1968) Leber's disease. *Acta Ophthalmologica*, **46**, 4–25
82. Seedorff, T. (1985) The inheritance of Leber's disease. A genealogical follow-up study. *Acta Ophthalmologica*, **63**, 135–145
83. Lott, M.T., Voljavec, A.S. and Wallace, D.C. (1990) Variable genotype of Leber's hereditary optic neuropathy patients. *American Journal of Ophthalmology*, **109**, 625–631
84. Harding, A.E., Holt, I.J., Sweeney, M.G., Brockington, M. and Davis, M.B. (1992) Prenatal diagnosis of mitochondrial DNA$^{8993T→G}$ disease. *American Journal of Human Genetics*, **50**, 629–633

11
Recent advances in mitochondrial genetics

John M. Shoffner and Douglas C. Wallace

INTRODUCTION

Normal ATP generation by oxidative phosphorylation (OXPHOS) is a complex process requiring the coordinated expression of two genomes: the nuclear DNA (nDNA) and the mitochondrial DNA (mtDNA). Much of our knowledge as well as many recent questions concerning how the nDNA and mtDNA interact comes from the detailed clinical, biochemical and genetic analysis of OXPHOS diseases.

OXPHOS is carried out by five enzyme complexes assembled from subunits encoded by the mtDNA. The first four complexes (I–IV) create the electron-transport chain which oxidizes electrons from NADH or $FADH_2$ and uses the energy to pump protons out of the mitochondrial inner membrane. This electrochemical gradient is utilized by Complex V to synthesize ATP from $ADP+P_i$. The ATP is then exchanged across the mitochondrial inner membrane for cytosolic ADP by the adenine nucleotide translocator (ANT) [1]. Complex I (NADH dehydrogenase) has over 30 polypeptides, seven from the mtDNA; Complex II (succinate dehydrogenase) four nuclear polypeptides; Complex III (ubiquinol–cytochrome c oxidoreductase) 10 polypeptides, one from the mtDNA; Complex IV (cytochrome c oxidase) 13 polypeptides, three from the mtDNA; and Complex V (ATP synthase) 12 polypeptides, two from the mtDNA. ANT is a homodimer that is encoded by three different nuclear isoform genes. ANT1 gene expression is confined to heart and muscle, ANT3 is preferentially expressed in brain and kidney but also transcribed in heart and skeletal muscle, and ANT2 is cell cycle controlled [2–7]. ANT1 and the catalytic ATP synthase β subunit share a muscle-specific *cis* element, the OXBOX, which results in their coordinated elevated expression in heart and skeletal muscle [7]. In addition to the 13 OXPHOS polypeptides, the mtDNA also encodes the rRNAs and tRNAs of mitochondrial protein synthesis.

The first pathogenic mutations in the mtDNA [8,9] were discovered almost 7 years after the complete human mtDNA sequence was published [10]. This relatively long lag time reflected, in part, some of the controversies involved in recognizing OXPHOS disease phenotypes and the complexities of mtDNA analysis. Since that time, the number of pathogenic mtDNA mutations has increased dramatically, resulting in a deeper understanding of how OXPHOS genetics apply to human disease. This chapter examines some of the recent advances in mitochondrial genetics and how they have altered our understanding of rare and common age-related diseases.

MERRF: A MODEL DISEASE FOR THE PRINCIPLES OF MITOCHONDRIAL GENETICS

Since the identification of pathogenic mtDNA mutations, mitochondrial geneticists have invoked a number of mechanisms to explain the unusual degrees of phenotypic heterogeneity associated with mitochondrial diseases. The principles of mitochondrial genetics account for a considerable proportion of this variability. These principles include: (A) maternal inheritance of the mtDNA, (B) replicative segregation of mutant and normal mtDNAs, (C) threshold expression of disease manifestations and (D) a high mutation rate for the mtDNA relative to the nuclear DNA. In depth study of a large myoclonic epilepsy and ragged red fibre (MERRF) pedigree has given insight into how these principles explain the clinical manifestations, variable expression and age-related progression of the disease.

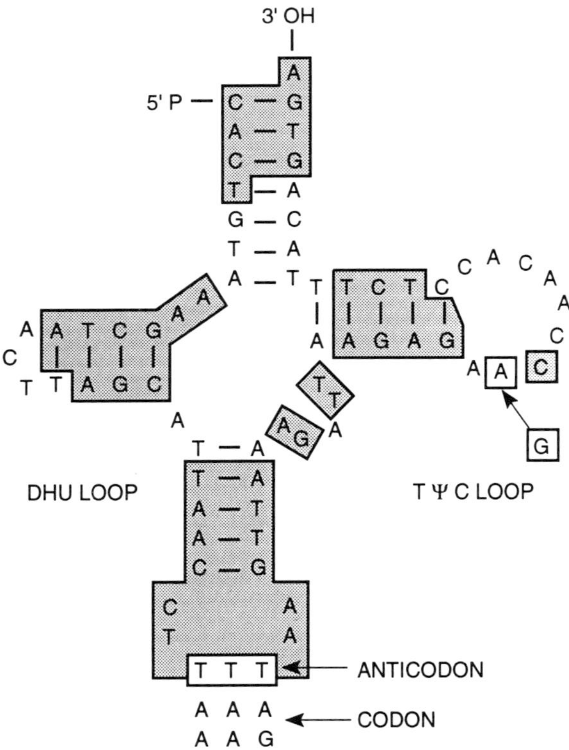

Figure 11.1 Secondary structure for the tRNA^{Lys} showing the bp 8344 A to G mutation in the TψC loop of MERRF patients. Shaded regions show 100% evolutionary conservation between human, cow and mouse

MERRF is associated with an A to G transition mutation at nucleotide 8344 in the mtDNA [11]. This mutation occurs in the TψC loop of the mitochondrial tRNA[Lys] (Figure 11.1), is heteroplasmic (both mutant and wild-type mtDNAs are present within an individual) and produces defective mitochondrial protein synthesis. It was originally identified by sequencing 94% of the protein coding ribosomal RNA and tRNA sequence from a large Georgia MERRF pedigree. The tRNA[Lys] mutation was present in two other MERRF pedigrees and absent in 75 ethnic controls [11]. Enzymological analysis of OXPHOS revealed that Complexes I and IV were severely affected [12]. This finding correlated with the decrease in mitochondrial protein synthesis of the Complex I and IV subunits [13].

Maternal inheritance

Maternal inheritance is the transmission of the mtDNA along the maternal lineage by the mother to her children and by her daughters to their children. The egg contains several hundred thousand mtDNAs which overwhelm the small number of mtDNAs contributed by the sperm [14,15]. Although mtDNA has been shown to be maternally inherited without exception, the phenotypes associated with mtDNA mutations are variable. Therefore, to demonstrate maternal inheritance of a disease, it is necessary to exclude paternal transmission. This necessitates the analysis of large pedigrees. Such pedigrees have been observed for MERRF, Leber's hereditary optic neuropathy (LHON), and LHON + infantile bilateral striatal necrosis [9,12,16,17].

Replicative segregation

When at least two types of mtDNA exist within a cell (heteroplasmy), cytokinesis produces the random partitioning of these mtDNAs to the daughter cells, resulting in variation of genotypes as well as phenotypes [18]. Over many cell generations, heteroplasmic mtDNA genotypes drift toward either pure mutant or normal mtDNA populations (homoplasmy) [18–20].

In the MERRF pedigree, replicative segregation produced variable proportions of mutant and wild-type mtDNA along the maternal lineage. The mutant mtDNA was 73–97% of the total muscle mtDNA while the normal mtDNA varied between 3 and 27% [11]. The proband (III-1) who had a severe mitochondrial myopathy was nearly homoplasmic (94%) in muscle and blood for the mutant mtDNA with only 6% wild-type mtDNA (Figure 11.2). The asymptomatic cousin (III-4) who also had a normal muscle biopsy had 15% wild-type mtDNA. Hence, the tRNA[Lys] function in MERRF is only compromised and not abolished. The large protective effect of a small percentage of wild-type mtDNAs emphasizes the importance of replicative segregation in determining phenotype.

Threshold effect

The phenotypic consequences of a mtDNA mutation are a function of the severity of the OXPHOS defect and the differing energetic requirements of human organs and tissues. Each tissue requires a different minimum level (threshold) of mitochondrial ATP production to sustain normal cellular functions {1,18]. In

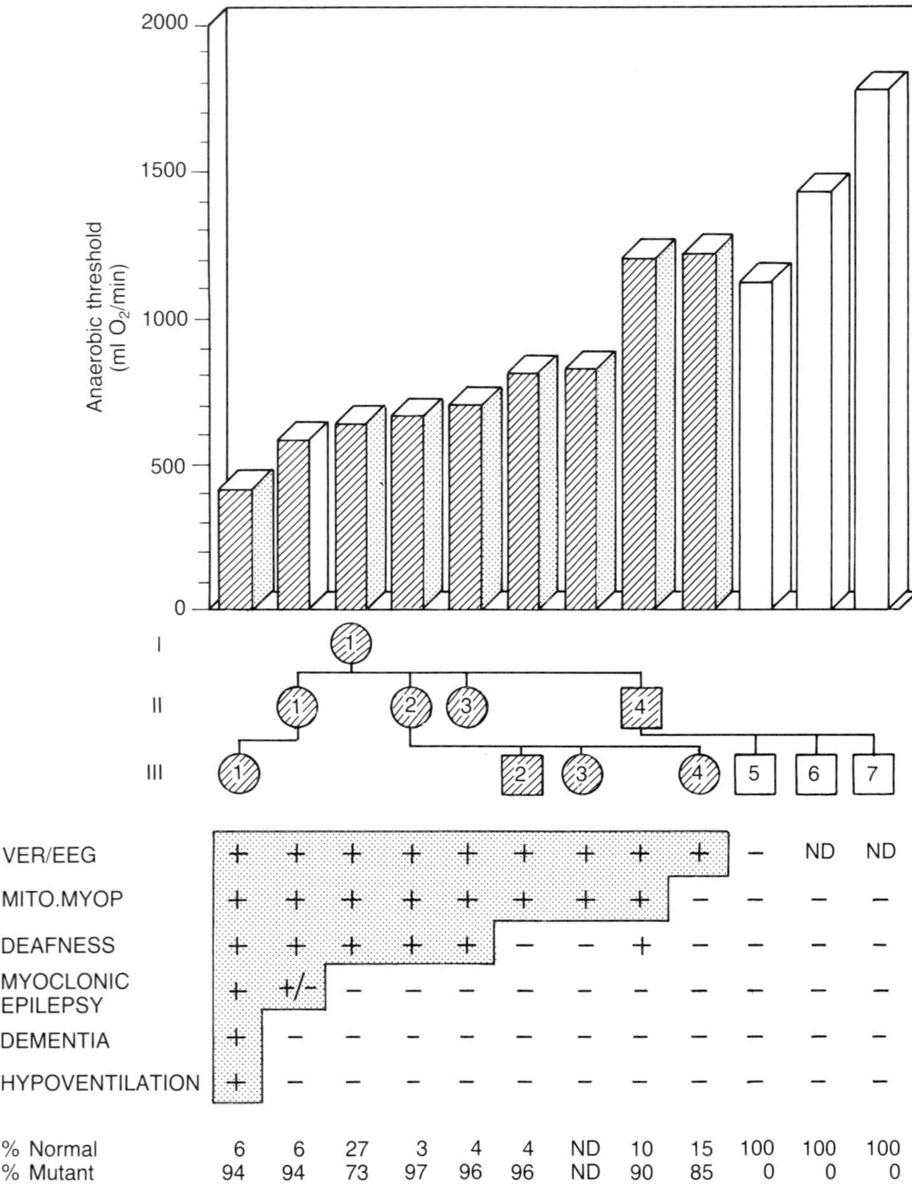

Figure 11.2 Phenotypic variability across the maternal lineage of a large MERRF pedigree harbouring the bp 8344 tRNA^{Lys} mutation. Shaded regions represent individuals with the MERRF mutation. VER, visual evoked response; EEG, electroencephalogram; MITO. MYOP., mitochondrial myopathy with ragged red fibres. ND, not investigated. The genotype for each individual (%Normal and %Mutant mtDNA) is given at the bottom of the figure

families with heteroplasmic mtDNA mutations, the severity of the OXPHOS defect varies with the percentage of normal mtDNAs. As different family members can inherit different percentages of mutant mtDNAs, they will present with different clinical manifestations. In general, the central nervous system is affected first, followed by skeletal muscle, heart, kidney and finally liver [1,11,12].

The clinical, physiological and biochemical characteristics within MERRF pedigrees illustrate this effect [12]. The maternal lineage showed significant heterogeneity for all parameters tested while the paternal lineage was unaffected (Figure 11.2). For example, the anaerobic threshold which quantifies the oxidative (type I) muscle fibre work capacity varied between 409 ml O_2/min for the severely affected proband (III-1) and 1212 ml O_2/min for the mildly affected cousin (III-4) (Figure 11.2). The enzyme activities for Complexes I and IV as determined by polarographic analysis, correlated highly with these anaerobic threshold changes ($r = 0.84$ and 0.94 respectively).

Although phenotypic parameters within the pedigree correlated highly with the severity of the OXPHOS defect, they did not correlate well with the percentages of mutant mtDNAs found in the skeletal muscle. This is because OXPHOS as defined by maximum oxygen consumption during exercise, basal metabolic rate, and muscle enzyme activity declines with age [21–23]. Therefore, an individual's OXPHOS capacity and phenotype are a product of the percentage of mutant mtDNAs they inherit and their age. When we placed the MERRF pedigree members into age groups, the improved association between the percentage of mutant mtDNAs and the severity of the clinical manifestations became clear. Thus, MERRF family members are phenotypically normal during childhood. Later in life, symptoms appear and progressively worsen. The age that manifestations appear within an organ depends on the combined effects of the genotype and the age-specific decline in OXPHOS.

The interplay between tissue-specific energetic thresholds, mtDNA mutations and age-related OXPHOS decline in creating disease phenotypes is demonstrated in Figure 11.3. Curve A represents the decline of OXPHOS function with aging. Curves B to F represent the added influence of various mtDNA mutations on the age-related decline in OXPHOS function. These curves are conceptually useful for considering the influence of heteroplasmy and mutation pathogenicity on the natural history of a disease.

The more severe mutations would push the curves down and to the left (Figure 11.3, curves B, C and D) while less severe mutations would expand toward the normal on the right. This delayed expression of inherited OXPHOS defects could account for the late onset of a variety of disorders. The slope of each disease curve could be further modified by environmental factors and other genetic elements.

High mtDNA mutation rate

The last unique principle of mtDNA genetics is its high mutation rate. The mtDNA sequence changes about 10–20 times faster than comparable nDNA genes [2,24]. This high mtDNA mutation rate has created extensive sequence variation between populations [25]. Consequently, when two independent mtDNA sequences are compared, they will differ on average by 3 nucleotides in 1000 which translates into about 50 nucleotide substitutions per mtDNA genome [25–31]. The

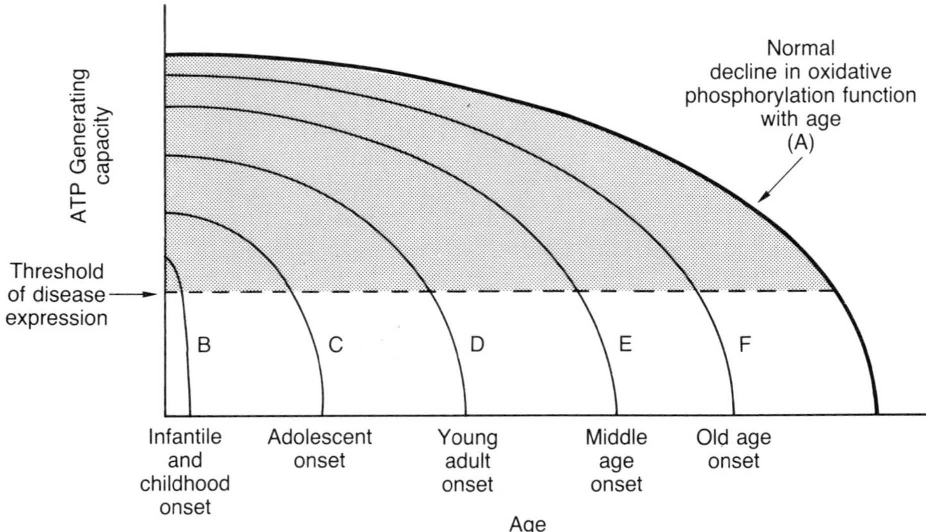

Figure 11.3 Paradigm for phenotype expression in disorders caused by mutations in the mtDNA. Curve A shows the natural decline in ATP-generating capacity with aging. Curves B to F show the decreased ATP-generating capacity caused by inherited mutations in the mtDNA. Subsequent decline is hypothesized to result from the accumulation of additional somatic mutations. The shaded area represents levels of ATP generation that can sustain normal cellular function and thus account for the delay in the appearance of clinical manifestations in late-onset diseases

combined effect of a high mutation rate and a highly constrained genome results in a high frequency of deleterious mtDNA mutations.

LEBER'S HEREDITARY OPTIC NEUROPATHY (LHON)

High-risk mtDNA mutations

LHON was the first human disease to be associated with a mtDNA point mutation [9]. Patients with LHON present with acute or subacute, painless loss of central visual acuity which usually occurs between 12 and 30 years of age [32]. The typical ophthalmoscopic features of acute LHON include circumpapillary telangiectatic microangiopathy and swelling of the nerve fibre layer around the optic disc [33,34]. Once visual loss has occurred, spontaneous recovery is uncommon but has been reported in a few patients [35,36]. Up to 80–90% of males within Caucasian pedigrees develop visual loss, whereas only 8–32% of females are affected [32]. This male bias in expression of the disease has been attributed to the influence of an X-linked locus, as yet undefined [37–39].

To determine if the primary cause of LHON involved a mtDNA mutation, the mtDNA of an affected proband from a large LHON family from Georgia was sequenced [9]. A missense mutation (G→A) at nucleotide 11 778 (LHON [11 778]) of the mtDNA was identified as the cause for the disease in this family (Table 11.1).

Table 11.1 High-risk mtDNA mutations for LHON

Locus/allele	Nucleotide	NT	NT homology H/B/M/X	AA	AA homology H/B/M/X	Controls	Homoplasmy	Heteroplasmy
MTND1/LHON[3460]	3460	G→A	G/G/G/G	Ala→Thr	Ala/Ala/Ala/Ala	0/107 (0%)	+	−
MTND1/LHON[4160]	4160	T→C	T/T/T/T	Leu→Pro	Leu/Leu/Leu/Leu	0/18 (0%)	+	−
MTND4/LHON[11 778]	11 778	G→A	G/G/G/G	Arg→His	Arg/Arg/Arg/Arg	0/507 (0%)	+	+
MTCYB/LHON[15 257]	15 257	G→A	G/G/G/G	Asp→Asn	Asp/Asp/Asp/Asp	1/362 (0.3%)	+	−

NT, nucleotide transition; AA, amino acid substitution or homology. H, human; B, bovine; M, mouse, X, *Xenopus*.

This point mutation changed a highly conserved arginine to a histidine at the 340th amino acid of Complex I, subunit 4 (ND4). This high-risk mutation seems sufficient in itself to produce the clinical manifestations of the disease, and approximately 40–60% of clinically recognized cases have the LHON[11 778] mutation. In large LHON pedigrees, the LHON[11 778] mutation is essentially homoplasmic [9,40], but approximately 58% of LHON pedigrees are singleton cases with 14% of these being heteroplasmic [20,41]. In heteroplasmic pedigrees, replicative segregation can result in shifts in the mutant mtDNA percentages between pedigree members as well as the differences in the proportion of mutant mtDNA between tissues of a single individual [20,42]. In fact, mtDNA segregation is so rapid that the proportion of LHON[11 778] in blood can shift from 50% to essentially pure mutant in a single generation [20]. Although most individuals with the LHON[11 778] mutation generally have an uncomplicated clinical presentation, a Swedish LHON family that was homoplasmic for the LHON[11 778] mutation had complex neurological manifestations in addition to the classic ophthalmological findings [43]. These included bilateral lesions in the putamen on MRI, tremor, ataxia, posterior column dysfunction, dystonia, corticospinal tract dysfunction and extrapyramidal rigidity. Muscle biopsy showed a subsarcolemmal increase in mitochondria as well as a few fibres with paracrystalline inclusions. This family represents a unique and surprising phenotype associated with the LHON[11 778] mutation. To determine if this is a part of the normal spectrum of LHON[11 778] expression, it will be necessary to sequence the proband's mtDNA to rule out possible exacerbating mutations.

Although the presence of LHON[11 778] places an individual at risk for blindness, quantification of heteroplasmy versus homoplasmy has not been very useful in predicting clinical outcome for an individual patient [20,40,42]. This clinical variability which includes age of onset and male predilection has been hypothesized to result from sex-related physiological differences, environmental stressors such as smoking which could reduce ATP generation by carbon monoxide inhibition of Complex IV (cytochrome *c* oxidase) [44], cyanocobalamin accumulation due to abnormal cyanide detoxification [45], X-linked alleles that modulate disease expression [37], and other mtDNA mutations [46–48].

Investigation of LHON pedigrees that lack the LHON[11 778] mutation have led to the identification of three other high-risk mutations that cause blindness: bp 3460 (LHON[3460]) [49,50]; bp 4160 (LHON[4160]) [51] and bp 15 257 (LHON[15 257]) [47,48] (Table 11.1). The LHON[3460] and LHON[4160] mutations are both located in the ND1 subunit gene of Complex I, but have been associated with strikingly different phenotypes. LHON[3460] is a G→A transition mutation that converts an alanine to threonine at amino acid 52 of the ND1 polypeptide [49,50]. This mutation was recently reported in three large Finnish pedigrees [50] and six pedigrees from Australia and the United Kingdom [49]. LHON[3460] and LHON[11 778] have several features in common. Both typically present as classical LHON with no or only minor neurological manifestations, they can be heteroplasmic or homoplasmic, and both are found in pedigrees with phenotypic expression along the maternal lineage as well as in singleton cases [9,12,20,49,50,52].

LHON[4160] is a T→C mutation in the ND1 gene which replaces a highly conserved leucine with a proline at amino acid 285 [51]. Pedigree members can display LHON in isolation or in association with severe neurological degeneration which includes dysarthria, deafness, ataxia, tremor, posterior column dysfunction, corticospinal tract dysfunction and skeletal deformities [51B]. A severe childhood encephalopathy was observed in 9 of 56 pedigree members and was characterized by headache, vomiting, focal or generalized seizures with a hemiparesis that gradually resolved and cerebral oedema [51B]. Specific neuropathological abnormalities were not identified in the three individuals that died. One female who developed the infantile encephalopathy recovered and is clinically normal as an adult [51]. As predicted by maternal inheritance, she was positive for the LHON[4160] mutation and had four affected children. The description of these transient neurological deficits has similarities to cases with mitochondrial encephalopathy, lactic acidosis and stroke-like episodes (MELAS) [1]. The extreme clinical variability of this disease cannot be attributed to replicative segregation, as the LHON[4160] mutation was found to be homoplasmic [51]. As an alternative explanation, it was hypothesized that the variable phenotype could be the result of a second 'suppressor mutation' at bp 4136 (LHON[4136]). This mutation was an A→G transition which resulted in the substitution of a cysteine for tyrosine at amino acid 277 in the ND1 polypeptide. The LHON[4136] mutation was homoplasmic, evolutionarily conserved from human to *Xenopus*, and present in a subset of individuals within the pedigree who were less severely affected. As one branch of the pedigree had the LHON[4136] mutation, it must be a new mutation which rapidly segregated to homoplasmy. More in depth clinical studies are needed to clarify the role of this mutation and other factors which may create the clinical variability seen in pedigrees harbouring the LHON[4160] and LHON[11 778] mutations [43,51].

Complete mtDNA sequencing of a familial LHON case without the LHON[11 778] mutation revealed a G→A transition at bp 15 257 (LHON[15 257]) which changed a highly conserved aspartate to asparagine at amino acid position 171 in the cytochrome *b* polypeptide (Table 11.1) [48]. The LHON[15 257] mutation was homoplasmic and present in 4 of 23 patients (17.4%) with LHON and in 1 of 362 (0.3%) population controls. Three pedigrees had familial expression of LHON and one pedigree represented a singleton case. Complex neurological manifestations were not present.

The mechanism(s) by which the ND4 (LHON[11 778]), the ND1 (LHON[3460], LHON[4160]) and the cytochrome *b* (LHON[15 257]) mutations cause LHON is unclear. The ND4 subunit is located in the large hydrophobic protein domain of Complex I, but its function is unknown [53]. Patients with the LHON[11 778] mutation have abnormal NADH-linked substrate utilization by polarographic measurements [43,54] while measurement of Complex I specific activity is normal (unpublished result; [43,54]). NADH is bound to various NAD$^+$-linked dehydrogenases within the mitochondrial matrix and appears to be delivered to Complex I via direct enzyme–enzyme interactions (NAD$^+$-linked dehydrogenase to Complex I) rather than by diffusion of NADH from an aqueous pool [55–57]. Therefore, it has been hypothesized that the polarographic measurements may be reflecting altered interactions between Complex I and other NAD$^+$-linked dehydrogenases which would be detected by polarography but not by direct enzyme assays.

The ND1 subunit LHON[3460] and LHON[4160] mutations both produce defects in Complex I specific activity [54,58]. ND1 is located within the large hydrophobic domain of Complex I, contains the rotenone-binding site and may be involved in electron transfer to ubiquinone [53,59]. As assays that measure Complex I specific activity use an analogue of ubiquinone, mutations in the ND1 subunit could yield low enzyme activities.

Synergistic mtDNA point mutations

Although the LHON[3460], LHON[4160], LHON[11 778] and LHON[15 257] mutations account for many cases of LHON, still other patients lack these mutations. Efforts to identify mtDNA mutations that explain these cases have led to the discovery that several partially deleterious mtDNAs can act synergistically to cause blindness. These mtDNA point mutations occur at bps 4216 (LHON[4216]), 4917 (LHON[4917]), 5244 (LHON[5244]), 13 708 (LHON[13 708]) and 15 812 (LHON[15 812]) (Table 11.2) [46–48]. Several combinations of these mutations have been observed in association with the LHON phenotype (Figure 11.4, mutation groups A to E).

The LHON[13 708] mutation has been found in conjunction with almost all of the other synergistic point mutations and appears to be an important premutation for the development of the LHON phenotype [46–48]. The LHON[13 708] mutation is a G→A transition which changes an alanine in ND5 at amino acid 458 to a polar threonine (Table 11.2). By itself, this mutation is usually not associated with vision loss. Screening of the general Caucasian population for this mutation has revealed a prevalence rate of 5% (16/320) for the LHON[13 708] mutation [46,48]. By contrast, approximately 26% (29/112) of patients with LHON (LHON[11 778]-positive = 11/59 or 19% and LHON[11 778]-negative = 18/53 or 34%) harbour this mutation, which is significantly higher than its frequency in the general population [46,48].

The best example of the synergistic interaction of LHON mutations has come from the complete mtDNA sequencing of a patient and the assessment of candidate mutations for evolutionary conservation, for prevalence of the mutation in diseased individuals and controls from multiple ethnic groups, and phylogenetic analysis of patients and controls [48]. LHON[15 257] was found to be linked to three additional mutations, LHON[13 708], LHON[15 812] and LHON[5244] in some mtDNAs [48] (Figure 11.4, mutation groups A to C). Haplotype analysis showed that these four mutations belonged to the same mtDNA lineage and population studies suggested that the accumulation of these mutations within an individual's mtDNA increased the risk of developing blindness. The LHON[15 812] mutation is a homoplasmic G→A transition in the cytochrome *b* gene which converts amino acid 356 from a valine to a methionine (Table 11.2). The LHON[5244] mutation is a heteroplasmic G→A transition which substitutes a serine for a glycine at an evolutionarily conserved position of the ND2 subunit of Complex I. Population studies and phylogenetic analysis revealed that as a mtDNA lineage acquired each of these mutations (mutation groups A to C), the probability of finding the haplotype group in the normal population decreased (Figure 11.5). Five percent of the control population harboured only the LHON[13 708] mutation while 0.3% had LHON[13 708] + LHON[15 257], 0.1% had LHON[13 708] + LHON[15 257] + LHON[15 812] + LHON[5244] [30,31,48,60,61]. This decline in mutation frequency in a large number of normals along with their concentration within a LHON cohort supports the paradigm that these mutations act together to reduce mitochondrial ATP production and, therefore, enhance the probability of optic nerve death.

mtDNA MUTATIONS AND LEIGH'S DISEASE

Leigh's disease or subacute necrotizing encephalomyelopathy (SNE) begins in infancy or childhood as psychomotor regression, brainstem abnormalities and lactic acidosis. Symmetrical abnormalities are seen on MRI in the brainstem and

Table 11.2 Synergistic mtDNA mutations for LHON

Locus/allele	Nucleotide	NT	NT homology H/B/M/X	AA	AA homology H/B/M/X	Controls	Homoplasmy	Heteroplasmy
MTND1/LHON[4216]	4216	T→C	T/C/C/C	Tyr→His	Tyr/His/His/His	4/49 (8%)[1]	+	−
MTND2/LHON[4917]	4917	A→G	A/A/A/A	Asn→Asp	Asn/Asn/Asn/Asn	2/49 (4%)[1]	+	−
MTND2/LHON[5244]	5244	G→A	G/G/G/G	Gly→Ser	Gly/Gly/Gly/Gly	0/2103 (0%)[2]	−	+
MTND5/LHON[13 708]	13 708	G→A	G/C/G/G	Ala→Thr	Ala/Leu/Ala/Ala	16/320 (5%)[1,2]	+	−
MTCYB/LHON[15 812]	15 812	G→A	G/G/A/G	Val→Met	Val/Val/Ile/Val	1/759 (0.1%)[2]	+	−

Categories are defined in Table 11.1.
H, human; B, bovine; M, mouse; X, *Xenopus*. [1][46]; [2][48].

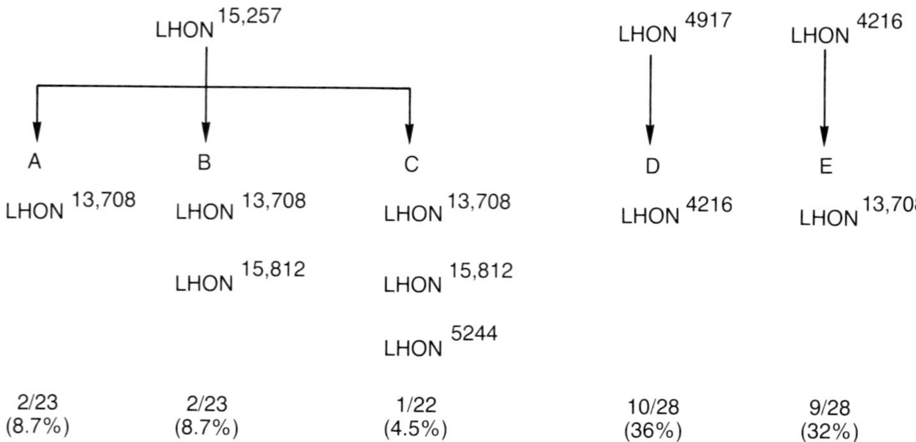

Figure 11.4 Associations between LHON mutations on different mtDNA haplotypes. Because of the synergistic interaction of LHON mutations, patients that harbour more than one LHON mutation have an increased probability of visual loss. Percentages of patients with these genotypes are from refs. [48] (groups A–C), [47] (groups A and B), and [46] (groups D and E)

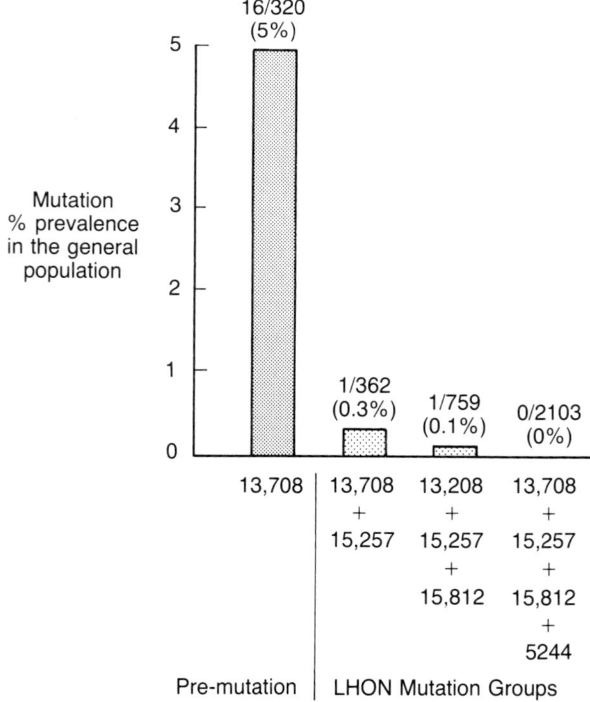

Figure 11.5 Evidence of increased pathology as LHON mutations accumulate along a mtDNA lineage. The prevalence of each LHON haplotype in the unaffected population declines as the number of mutations increases [48]. The LHON[13 708] mutation appears to be a premutation that in most individuals does not cause blindness. The other three mutation groups show extremely low prevalence in the general population, but increased prevalence in LHON patients

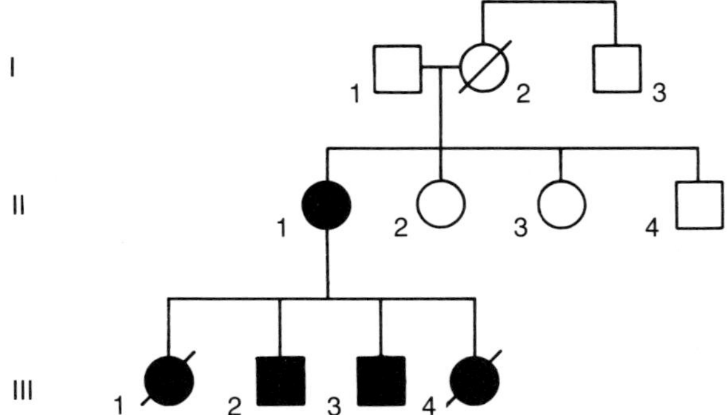

Figure 11.6 Pedigree harbouring the ATPase 6, position 8993 mutation. Filled symbols have the mutation as well as clinical manifestations. Only individuals II-1 to II-3 and III-1 to III-4 were available for mtDNA analysis

basal ganglia [62] correlate pathologically with areas of neuropil rarefaction, capillary proliferation and astrocytic gliosis [63,64]. Biochemically, SNE is associated with defects in three enzyme complexes of energy metabolism: cytochrome *c* oxidase (Complex IV) [65–72], NADH-ubiquinone oxidoreductase (Complex I) [73] and pyruvate dehydrogenase [74–76].

We recently encountered a pedigree in which two female children died of SNE which was caused by a mtDNA point mutation at bp 8993 in the ATP synthase, subunit 6 gene [77] (Figure 11.6). This mutation (NARP[8993]) was originally reported in a pedigree with varying combinations of neurogenic muscle weakness, ataxia, retinitis pigmentosa with classic bone spicule formation, sensory neuropathy, seizures and mental retardation or dementia [19]. In our family, the clinical presentations for both infants were dominated by brainstem abnormalities. A unique aspect of their case was that they both had psychomotor retardation rather than the usual developmental regression and also had a pigmentary retinopathy. However, on autopsy, the brain lesions were classical for SNE. The mother (II-1) of the pedigree had two sons with less severe disease manifestations. These two individuals who are now 11 (III-2) and 14 (III-3) years old have psychomotor retardation, an attention-deficit disorder, a mild pigmentary retinopathy, episodic ataxia and mild lactic acidaemia. Muscle biopsy of both of these sons showed normal histochemistry and electron microscopy. OXPHOS biochemistry demonstrated a Complex I defect in both brothers (III-2 and III-3) and an additional defect in Complex III in patient III-2. The mother (II-1) of these four children has very subtle manifestations which included a mild pigmentary retinopathy and migraine headaches. The retinal abnormalities defined in the two sons (III-2 and III-3) and the mother (II-1) required detailed ophthalmological examinations for diagnosis and were much milder than the retinal changes reported in the original pedigree [19].

Southern blot analysis of the mtDNA in buffy coat from the mother (II-1) demonstrated 86% mutant mtDNAs (Figure 11.7). As predicted from the clinical presentation, higher levels of the NARP[8993] mutation were found in the muscle (92% and 95%) and in the buffy coats (>95%) for the two sons (III-2 and III-3). Polymerase chain reaction amplification of DNA isolated from paraffin-embedded

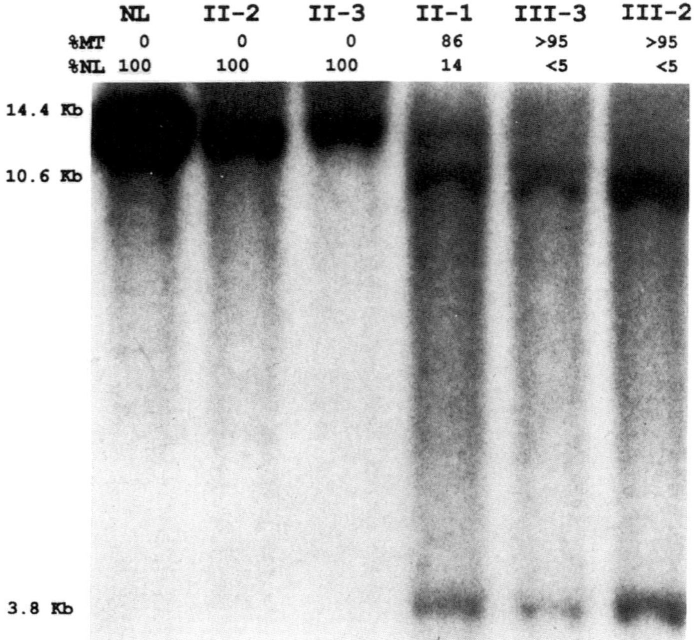

Figure 11.7 Southern blot of buffy coat mtDNA digested with *Ava*I. Normal mtDNA is cut into 14.4 and 2.1 kb fragments. The mutant mtDNA is cut into 10.6, 3.8 and 2.1 kb fragments. The 2.1 kb fragment is not shown in the figure. The pedigree numbers correspond to individuals in the figure 11.6 pedigree. %MT, %mutant mtDNA; %NL, %normal mtDNA

brain from the two daughters with SNE also demonstrated high levels of the NARP[8993] mutation. This analysis demonstrated that the clinical manifestations in the mother's children were the result of an increased concentration of the NARP[8993] mutation. As predicted by the principles of mitochondrial genetics, the central nervous system and retina were the most severely affected tissues. A unique aspect of the pedigree was the extremely rapid segregation of the NARP[8993] mutation. The mother of the affected children (II-1) had two unaffected sisters (II-2 and II-3), each with three clinically normal children. The NARP[8993] mutation was undetectable in the buffy coats of these two individuals (Figure 11.7).

This pedigree provides an excellent example of the wide variety of clinical phenotypes that can be associated with a mtDNA mutation. Like the large Georgia MERRF pedigree with a tRNA[Lys] mutation [11], this pedigree harbours a continuum of phenotypes, ranging from normal to Leigh's disease. In addition, this pedigree emphasizes the difficulties encountered in assigning specific phenotypes to a particular mtDNA mutation. Rigid correlations between genotype and phenotype are in many aspects contrary to the principles of mitochondrial genetics which predict clinical heterogeneity. This association of the NARP[8993] mutation with SNE has also been noted in another study [78]. The presence of SNE and MERRF pedigrees with the tRNA[Lys] mutation at bp 8334 [71] and in MELAS pedigrees with the tRNA[Leu(UUR)] mutation at bp 3243 (unpublished results) indicates that this clinical and pathological phenotype is a final common pathway for severe defects in ATP production.

MATERNALLY TRANSMITTED mtDNA DELETIONS AND DIABETES MELLITUS

Collectively, diabetes mellitus (DM) is one of the most common chronic disorders. Between 5 and 10% of individuals in the Western world are afflicted with glucose intolerance. Classification of DM is complex and based on phenotypic characteristics such as age of onset of symptoms, pancreatic islet cell function and inheritance. This has resulted in dividing DM into idiopathic (type I and type II DM) and hereditary forms of the disease [79]. Type I DM is a chronic autoimmune disorder with childhood or adolescent onset, low or absent blood insulin levels, pancreatic islet cell antibodies, HLA DR3 and/or DR4 linkage, and episodes of ketoacidosis. Type II DM usually appears after age 40, has normal to low blood insulin levels, is ketoacidosis resistant and is frequently associated with obesity. Although this system offers clinical guidelines for patient classification, genetic predictions are difficult. For example, epidemiological studies of DM patients older than 25 years indicate a 2–3-fold increase in maternal transmission of the disease suggesting that mitochondrial DNA mutations could be an important cause for DM in this age group [80–83].

mtDNA mutations have not typically been associated with the expression of DM as the primary phenotypic trait. Spontaneously occurring mtDNA deletions and duplications are associated with the development of DM in patients with severe, systemic manifestations of their OXPHOS disease [84–87]. Associated manifestations include ophthalmoparesis, ptosis, mitochondrial myopathy and lactic acidaemia which permit relatively straightforward clinical recognition [1,25,88–92]. We recently ascertained a family whose primary clinical manifestations were DM and deafness (Figure 11.8) [88]. The proband had DM, hearing loss and several strokes in the distribution of both large and small cerebral vessels. A left parietal infarction and multiple lacunar infarctions involving the basal ganglia, thalamus and internal capsules were present. Metabolic analysis and muscle pathology were normal. Quadriceps muscle biopsy had no evidence of mitochondrial myopathy by histochemistry or electron microscopy. OXPHOS biochemistry of muscle mitochondria demonstrated defects in Complexes I, III and IV specific activities.

Pedigree analysis revealed that the phenotypes of diabetes and deafness were maternally inherited. The pedigree matriarch (I-2) who was married twice had DM, hearing loss and strokes. During her first marriage, she had one daughter (II-1) with only DM and hearing loss. Her second marriage resulted in six other children (II-2 to II-7), including the proband, who had DM and hearing loss. The hearing loss generally began first, with onset between early childhood and approximately 20 years of age. DM occurred later in these individuals, between 20 and 43 years old. Three siblings experienced frequent episodes of diabetic ketoacidosis (II-3, II-4 and II-5) while their other brothers and sisters (II-1, II-2, II-6 and II-7) did not have this complication.

Southern blot analysis of blood from individuals (II-1 to II-5, III-1 and III-2) and muscle from individual II-1 revealed a heteroplasmic rearrangement which consisted of a duplicated segment of mtDNA as well as a segment with a 10.4 kb deletion in the mtDNA (Figure 11.8). This complex deletion/duplication

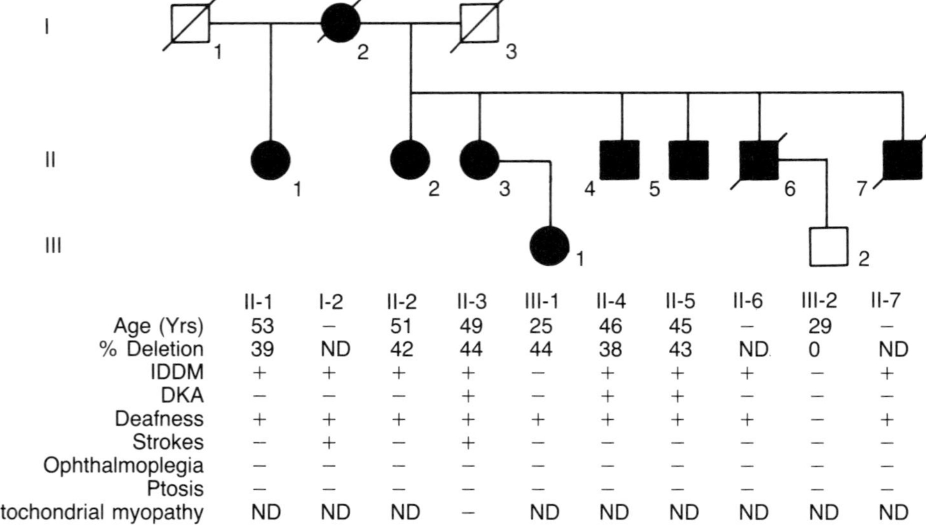

	II-1	I-2	II-2	II-3	III-1	II-4	II-5	II-6	III-2	II-7
Age (Yrs)	53	–	51	49	25	46	45	–	29	–
% Deletion	39	ND	42	44	44	38	43	ND	0	ND
IDDM	+	+	+	+	–	+	+	+	–	+
DKA	–	–	–	+	–	+	+	–	–	–
Deafness	+	+	+	+	+	+	+	+	–	+
Strokes	–	+	–	+	–	–	–	–	–	–
Ophthalmoplegia	–	–	–	–	–	–	–	–	–	–
Ptosis	–	–	–	–	–	–	–	–	–	–
Mitochondrial myopathy	ND	ND	ND	–	ND	ND	ND	ND	ND	ND

Figure 11.8 Clinical manifestations and genotype for the diabetes and deafness pedigree with a maternally inherited 10.4 kb deletion/duplication in the mtDNA. Filled symbols, affected individuals along the maternal lineage. +, present; –, absent; deceased. IDDM, insulin-dependent diabetes mellitus; DKA, diabetic ketoacidosis; ND, not investigated

mutation was found exclusively along the maternal lineage with no evidence of paternal transmission. Direct sequencing across the deletion breakpoint showed that a 10 423 bp segment had been removed between bp 4398 in the tRNA[Gln] gene and bp 14 822 in the cytochrome *b* gene. As is common for mtDNA deletions [1,25], the deletion was flanked by two 10 bp direct repeats (5'-CACCCCATCC-3') at bp 14 812–14 821 and 4389–4398. The deletion removed 2 bp at the 5' end of the tRNA[Gln] gene, 11 of 13 OXPHOS subunit genes, 15 of 22 tRNAs, the origin of light strand replication (O_L), and 75 bp at the 5' end of cytochrome *b*. Mitochondrial protein synthesis analysis further demonstrated the functional relevance of this mutation. In Epstein–Barr virus-transformed lymphoblasts and primary myoblast cell lines from the proband (II-3), the amount of protein synthesis inhibition correlated with the mutant mtDNA deletion percentage. The lymphoblasts with 53% mutant mtDNAs had a severe impairment of mitochondrial protein synthesis, incorporating only 52% of the control [35S]-methionine into mitochondrial translation products. The proband's myoblast cell lines demonstrated replicative segregation of the mutant mtDNA from 69% in the muscle biopsy to 17% in a myoblast clone. This decrease in the concentration of mutant mtDNAs resulted in normalization of mitochondrial protein synthesis. This novel association with diabetes and hearing loss implies that a wide variety of late-onset degenerative diseases may be caused by mtDNA mutations.

mtDNA DELETIONS: A RELATIONSHIP TO AGING AND DEGENERATIVE DISEASES

Aging and heart disease

Aging in humans is associated with an increase in the number of cytochrome *c* oxidase-deficient fibres in skeletal muscle and heart [93,94] and the decline in both the respiration rate and OXPHOS enzyme activities of Complex I and IV in human skeletal muscle [23] and liver [95]. This age-related decline is also correlated with increased mtDNA damage in senescent rats [96] and the accumulation of 5 kb [97,98] and 7.4 kb [99] mtDNA deletions in aging humans.

A possible cause for this progressive OXPHOS dysfunction and accumulation of mtDNA deletions is damage to the OXPHOS system by oxygen free radicals. Oxygen radicals result from the transfer of electrons from reduced mitochondrial flavins, coenzyme Q and cytochrome *b* to oxygen. Between 1 and 4% of oxygen utilized by normal mitochondria is converted to superoxide anion and H_2O_2 [100–103]. Any inhibition of the electron-transport chain increases the electronegativity of the early stages of OXPHOS, greatly stimulating oxygen radical generation [103]. Increased oxygen radicals further damage mitochondria through lipid peroxidation, protein cross-linking, sulphydryl bond oxidation and DNA base oxidation to form 8-hydroxydeoxyguanosine, thymine glycols and thymidine glycols. The mtDNA is preferentially mutated by reactive oxygens. Nuclear DNA in normal tissues has about 1/130 000 8-hydroxydeoxyguanosine adducts per base whereas mtDNA has 1/8000 adducts or about two per genome. The extreme sensitivity of the mtDNA to oxygen damage may stem from its close proximity to OXPHOS (the source of reactive oxygens), a lack of protective histones and limited DNA-repair systems [104–107].

In ischaemic heart disease, atherosclerotic plaques in the coronary arteries deprive heart mitochondria of substrates and oxygen. Therefore, coronary artery heart disease (CAHD) is an ideal system for assessing mtDNA damage from free radical production because of the recurrent episodes of ischaemia and reperfusion that characterize the disease. Early in cardiac ischaemia, a reversible impairment of Complex I function occurs creating a defective coupling of Complex I to ADP phosphorylation [108–111]. As hypoxia continues, Complexes II and III are affected. Cellular ATP availability is further compromised by depletion of mitochondrial adenine nucleotides [111–115], impairment of ANT activity by increased levels of long-chain acyl-CoA esters [116–118], and energy utilization by futile cycling of Ca^{2+} across the inner mitochondrial membrane [119–121]. Once the electrochemical gradient is dissipated by severe hypoxia, the mitochondrial ATP synthase (Complex V) acts as an ATP hydrolase and may break down up to 80% of ATP during the first 20 min of ischaemia [122,123]. In ischaemic hearts, oxygen radical formation is greatly increased because of abnormalities in the respiratory chain induced by ischaemia and subsequent reperfusion which increases the electronegativity of the flavins and coenzyme Q [124–126]. Ischaemic hearts also have decreased levels of oxidation protection systems such as glutathione, superoxide dismutase, catalase and glutathione peroxidase [127–129]. The oxygen free radicals generated by this process could block mtDNA replication and cause deletions.

Chronic inhibition of OXPHOS would also be expected to reduce mitochondrial ATP production and elicit a compensatory induction of OXPHOS gene expression. In patients with chronic progressive external ophthalmoplegia, muscle fibres having high proportions of deleted mtDNAs also show high levels of mtDNA transcript levels [130]. In cancer cell lines which are generally glycolytic, both nDNA and mtDNA OXPHOS gene transcript levels are elevated [131]. The proposed interrelationship between OXPHOS inhibition, mtDNA damage and compensatory OXPHOS gene induction is summarized in Figure 11.9.

In order to test this hypothesis, we looked for mtDNA damage and OXPHOS gene induction in ten control hearts from patients of various ages, ten hearts from patient with non-CAHD pathology and seven hearts with CAHD [98]. mtDNA damage was estimated by quantifying the proportion of the mtDNA molecules which contained the 5 kb deletion (mtDNA^{4977}), the most frequently occurring mtDNA deletion (Figure 11.10). This deletion is flanked by 13 bp direct repeats at bp 8470 and 13 447 and represents between 30 and 50% of the deletions found in patients with chronic progressive external ophthalmoplegia [25,89,90,132]. In somatic tissue, the mtDNA^{4977} deletion probably represents only a fraction of the total mtDNA mutations. Therefore, it was used in this study as an index for total mtDNA damage.

The mtDNA^{4977} deletion was present in three of ten control hearts (Figure 11.11). Hearts from patients under age 40 had essentially no detectable deletions whereas hearts from patients over age 40 had deletions that increased to a

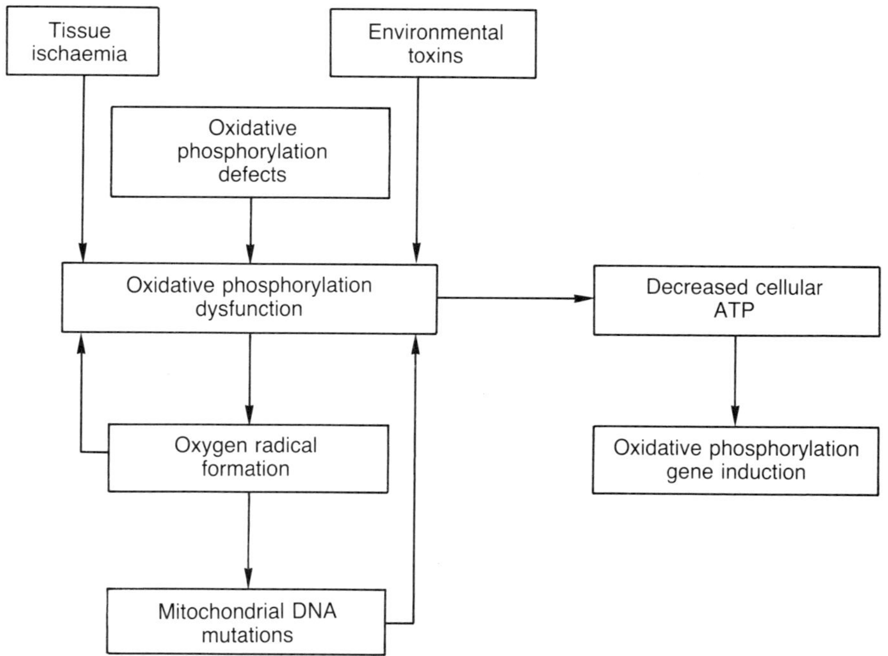

Figure 11.9 Paradigm for the progressive age-related decline of OXPHOS resulting from inherited mtDNA defects, tissue ischaemia or environmental insults. The initial OXPHOS defect increases oxygen radical production which damages mtDNA thus further reducing OXPHOS. Concomitant ATP deficiency is partially compensated by induction of OXPHOS gene expression

maximum of 0.0035%. These data support the hypothesis that mtDNA damage accumulates in tissues after the age of 40 and is consistent with other reports [97,99,133,134]. For the seven CAHD hearts, the mtDNA^{4977} deletion was present as 0.02–0.85% of the total mtDNA, representing an 8 to 2200 times increase over controls. The presence of other mtDNA deletions has also been confirmed in these CAHD hearts. The non-CAHD hearts were more heterogeneous in their accumulation of mtDNA damage. A group of six hearts from patients ranging from infants to 71 years of age did not differ from the control hearts in their degree of mtDNA damage. This group had a wide range of pathologies with two hearts from patients with familial cardiomyopathy and hearts of patients with myocarditis, endocardial fibroelastosis, cardiac failure due to sequellae from a large ventriculoseptal defect and hypertrophic cardiomyopathy. Four hearts (ages 16–84 years) with various degrees of ischaemia had increased mtDNA damage relative to controls. These hearts included two with idiopathic dilated cardiomyopathy and mild CAHD, one with hypertrophic cardiomyopathy associated with the MELAS mutation at bp 3243, and one with severe brown atrophy and moderate CAHD. Both CAHD and non-CAHD diseased hearts showed significant increases in the induction of nuclear and mitochondrial transcripts, independent of the deletion level, indicating that it is a general response to cardiac overload.

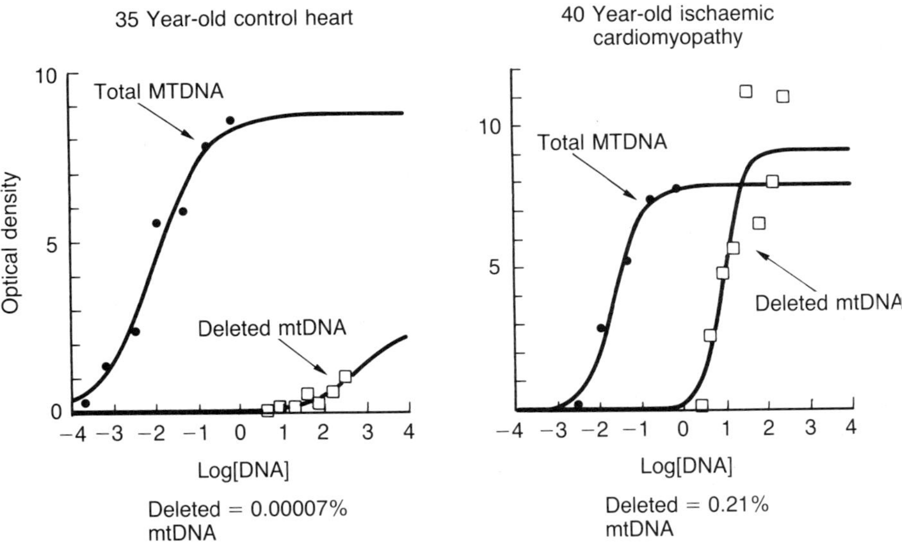

Figure 11.10 Comparison of mtDNA deletion accumulation in a normal 35-year-old heart and in a 40-year-old heart with ischaemic cardiomyopathy. DNA was extracted from these hearts, serial dilutions performed, and PCR fragments representing total mtDNA and the mtDNA^{4977} deletion were amplified. Densitometric analysis is performed on the dilution curves and plotted against the log of the dilution. As one type of mtDNA is diluted out, the PCR amplification signal is lost. The ratio of dilutions at which the PCR signal for the mtDNA^{4977} deletion is lost relative to the dilutions where the total mtDNA signal is lost gives the percentage of the mtDNA^{4977} deletion. This procedure allows construction of curves for the high-concentration normal mtDNA for comparison with the low-concentration mtDNA^{4977} deletion [98]

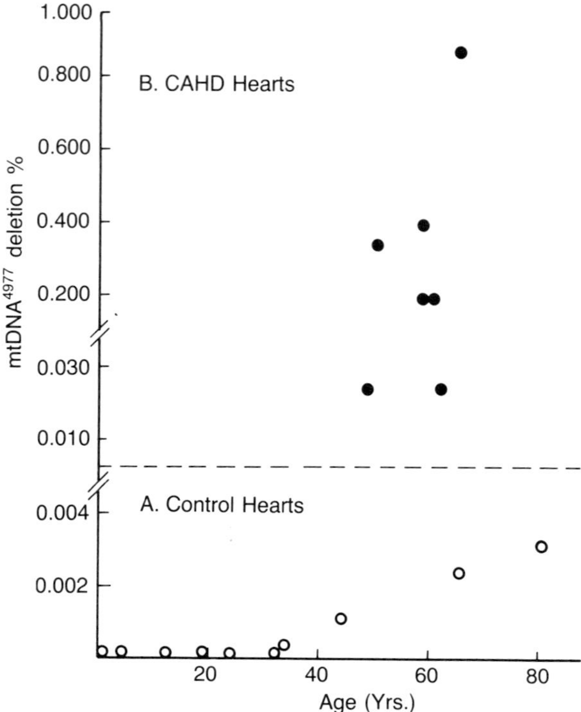

Figure 11.11 The age-related accumulation of the mtDNA^{4977} deletion in control and chronically ischaemic hearts. O, Non-ischaemic hearts; (●), CAHD hearts

These results support the paradigm that OXPHOS impairment is associated with the accumulation of mtDNA damage, possibly caused by oxygen free radicals. This may explain some portion of the decline in OXPHOS function with aging (Figure 11.3, curve A). Factors such as CAHD would augment the decline in ATP-generating capacity and lead to a rapid progression of cardiac dysfunction in many cases. The total contribution of mtDNA deletions to the aging process or a degenerative disease like CAHD is unknown. However, if all the possible mtDNA mutations (deletions, duplications, point mutations) from abnormal mtDNA replication could be detected, it seems likely that the extent of mtDNA damage could be highly significant. Moreover, experiments investigating the effects of low levels of damaged mtDNAs on the cell implied that a relatively few damaged mtDNAs could be deleterious [135]. A rapid loss of cell viability was observed within 4 days of injecting approximately 10–15 mitochondria from senescent rats into human diploid WI-38 cells. No effect on cell viability was seen when mitochondria from young rats were injected.

Parkinson's disease and Alzheimer's disease

OXPHOS defects have been identified in tissues of patients with age-related disorders such as Parkinson's disease (PD) and Alzheimer's disease (AD). PD is an adult-onset disease of unknown aetiology associated with degeneration of the nigrostriatal dopaminergic pathway. Toxins that inhibit OXPHOS, such as the Complex I inhibitor MPTP (1-methyl-4-phenyl-1,2,3,6-tetrahydropyridine) [136–138] and the Complex IV inhibitors carbon monoxide [139] and cyanide [140], produce disorders that are in many respects similar to PD. Tissues that have been analysed from PD patients include substantia nigra [141], platelets [142] and muscle [143,144]. A detailed analysis of muscle OXPHOS enzymes of PD patients revealed various defects in Complex I, Complex I and III and Complex IV [144]. These results imply significant heterogeneity in PD cases. Brains of Parkinson's patients have been shown to harbour low levels of mtDNA deletions [133]. Hence, PD could be a systemic disorder of OXPHOS function that has its most severe manifestations in the basal ganglia.

A similar process has also been suggested for AD patients. AD is a progressive neurodegenerative disorder characterized clinically by progressive cognitive decline which usually begins after the age of 60 years. It is characterized neuropathologically by senile plaques, many with abnormal neurites or 'neuritic plaques', neurofibrillary tangles within neuronal perikarya and amyloid angiopathy. Cholinergic deficiency frequently occurs in AD and has been associated with cell loss in the nucleus basalis of Meynert [145]. The substantial clinical, pathological, biochemical and genetic heterogeneity has complicated current classifications schemes as well as genetic approaches to AD. OXPHOS defects have been identified in various tissues of AD patients. Complex I defects have been found in platelets [146] and abnormalities in mitochondrial respiration in neocortex [147] and fibroblasts [148]. Treatment of normal human fibroblasts with the OXPHOS uncoupler CCCP (carbonyl cyanide m-chlorophenylhydrazone) results in a 10-fold increase in the proportion of cells reacting with an antibody to paired helical filaments and a 157-fold increase in cells reacting to the Alzheimer's monoclonal antibody-50 [149], suggesting a linkage between OXPHOS defects and AD pathology. Abnormal mitochondria with paracrystalline inclusions, like those frequently encountered in the muscle from patients with mtDNA deletions and point mutations [1], have been described in the brain of a patient with both AD and PD pathology [150] and in patients with AD [151]. Because of the unique genetics of OXPHOS, mtDNA mutations could explain a number of the genetic, pathological and biochemical features of AD and PD cases. The decline of OXPHOS function [23,95] and accumulation of mtDNA damage with age [97–99] could augment the effects of a mtDNA mutation, contributing to the age-dependent expression of these diseases (Figure 11.3, curve F). Like the various mutations associated with LHON [12,46,48–51], a variety of mtDNA mutations could result in AD or PD phenotypes.

REFERENCES

1. Shoffner, J.M. and Wallace, D.C. (1990) Oxidative phosphorylation diseases. Disorders of two genomes. *Advances in Human Genetics*, **19**, 267–330

2. Neckelmann, N., Li, K., Wade, R.P., Shuster, R. and Wallace, D.C. (1987) cDNA sequence of a human skeletal muscle ADP/ATP translocator: lack of a leader peptide, divergence from a fibroblast translocator cDNA, and coevolution with mitochondrial DNA genes. *Proceedings of the National Academy of Sciences of the USA*, **84**, 7580–7584
3. Battini, R., Ferrari, S., Kaczmarek, L. *et al.* (1987) Molecular cloning of a cDNA for a human ADP/ATP carrier which is growth-related. *The Journal of Biological Chemistry*, **262**, 4355–4359
4. Houldsworth, J. and Attardi, G. (1988) Two distinct genes for ADP/ATP translocase are expressed at the mRNA level in adult human liver. *Proceedings of the National Academy of Sciences of the USA*, **85**, 377–381
5. Cozens, A.L., Runswick, M.J. and Walker, J.E. (1989) DNA sequences of two expressed nuclear genes for human mitochondrial ADP/ATP translocase. *The Journal of Molecular Biology*, **206**, 261–280
6. Li, K., Warner, C.K., Hodge, J. *et al.* (1989) A human muscle adenine nucleotide translocator gene has four exons, is located on chromosome 4, and is differentially expressed. *The Journal of Biological Chemistry*, **264**, 13998–14004
7. Li, K., Hodge, J.A. and Wallace, D.C. (1990) OXBOX, a positive transcriptional element of the heart-skeletal muscle ADP/ATP translocator gene. *Journal of Biological Chemistry*, **265**, 20585–20588
8. Holt, I.J., Harding, A.E. and Morgan, H.J.A. (1988) Deletions of muscle mitochondrial DNA in patients with mitochondrial myopathies. *Nature*, **331**, 717–719
9. Wallace, D.C., Singh, G., Lott, M.T. *et al.* (1988) Mitochondrial DNA mutation associated with Leber's hereditary optic neuropathy. *Science*, **242**, 1427–1430
10. Anderson, S., Bankier, A.T., Barrell, B.G. *et al.* (1981) Sequence and organization of the human mitochondrial genome. *Nature*, **290**, 457–465
11. Shoffner, J.M., Lott, M.T., Lezza, A.M. *et al.* (1990) Myoclonic epilepsy and ragged-red fiber disease (MERRF) is associated with a mitochondrial DNA tRNA(Lys) mutation. *Cell*, **61**, 931–937
12. Wallace, D.C., Zheng, X., Lott, M.T. *et al.* (1988) Familial Mitochondrial Encephalomyopathy (MERRF): genetic, pathophysiological and biochemical characterization of a mitochondrial DNA disease. *Cell*, **61**, 601–610
13. Wallace, D.C., Yang, J., Ye, J. *et al.* (1986) Computer prediction of peptide maps: assignment of polypeptides to human and mouse mitochondrial DNA genes by analysis of two-dimensional-proteolytic digest gels. *The American Journal of Human Genetics*, **38**, 461–481
14. Michaels, G.S., Hauswirth, W.W. and Laipis, P.J. (1982) Mitochondrial DNA copy number in bovine oocytes and somatic cells. *Developmental Biology*, **94**, 246–251
15. Gyllensten, U., Whaton, D., Josefsson, A. and Wilson, A.C. (1991) Paternal inheritance of mitochondrial DNA in mice. *Nature*, **352**, 255–257
16. Rosing, H.S., Hopkins, L.C., Wallace, D.C., Epstein, C.M. and Weidenheim, K. (1985) Maternally inherited mitochondrial myopathy and myoclonic epilepsy. *Annals of Neurology*, **17**, 228–237
17. Novotny, E.J.J., Singh, G., Wallace, D.C. *et al.* (1986) Leber's disease and dystonia: a mitochondrial disease. *Neurology*, **36**, 1053–1060
18. Wallace, D.C. (1986) Mitotic segregation of mitochondrial DNAs in human cell hybrids and expression of chloramphenicol resistance. *Somatic Cell Molecular Genetics*, **12**, 41–49
19. Holt, I.J., Harding, A.E., Petty, R.K.H. and Morgan-Hughes, J.A. (1990) A new mitochondrial disease associated with mitochondrial DNA heteroplasmy. *The American Journal of Human Genetics*, **46**, 428–433
20. Lott, M.T., Voljavec, A.S. and Wallace, D.C. (1990) Varible genotype of Leber's Hereditary Optic Neuropathy patients. *American Journal of Ophthalmology*, **109**, 625–631
21. Astrand, I., Astrand, P.O., Hallback, I. and Kilborn, A. (1973) Reduction in maximal oxygen uptake with age. *Journal of Applied Physiology*, **35**, 649–654
22. Lentner, C., Lentner, C. and Wink, A. (1981) *Geigy Scientific Tables*, Vol. 1. Ciba-Geigy Corporation, New Jersey, pp. 228–231
23. Trounce, I., Byrne, E. and Marzuki, S. (1989) Decline in skeletal muscle mitochondrial respiratory function: possible factor in ageing. *The Lancet*, **i**, 637–639
24. Wallace, D.C., Ye, J., Neckelmann, S.N. *et al.* (1987) Sequence analysis of cDNAs for the human and bovine ATP synthase β subunit: miotchondrial DNA genes sustain seventeen times more mutations. *Current Genetics*, **12**, 81–90
25. Wallace, D.C., Lott, M., Torroni, A. and Shoffner, J.M. (1992) The human mitochondrial DNA. *Cell Genetics and Cytogenetics*, **58**, 1103–1123
26. Johnson, M.J., Wallace, D.C., Ferris, S.D., Rattazzi, M.C. and Cavalli-Sforza, L.L. (1983) Radiation of human mitochondrial DNA types analyzed by restriction endonuclease cleavage patterns. *The Journal of Molecular Evolution*, **19**, 255–271

27. Cann, R.L., Stoneking, M. and Wilson, A.C. (1987) Mitochondrial DNA and human evolution. *Nature*, **325**, 31–36
28. Singh, G., Lott, M.T. and Wallace, D.C. (1989) A mitochondrial DNA mutation as a cause of Leber's hereditary optic neuropathy. *New England Journal of Medicine*, **320**, 1300–1305
29. Schurr, T.G., Ballinger, S.W., Gan, Y. *et al.* (1990) Amerindian mitochondrial DNAs have rare Asian mutations at high frequencies, suggesting they derived from four primary maternal lineages. *The American Journal of Human Genetics*, **46**, 613–623
30. Ballinger, S.W., Schurr, T.G., Torroni, A. *et al.* (1992) Southeast Asian mitochondrial DNA analysis reveals genetic continuity of ancient Mongoloid migrations. *Genetics*, **130**, 139–152
31. Torroni, A., Schurr, T.G., Yang, C. *et al.* (1992) Native American mitochondrial DNA analysis indicates that the Amerind and the Nadene populations were founded by two independent populations. *Genetics*, **130**, 153–162
32. Newman, N.J. (1991) Leber's hereditary optic neuropathy. *Ophthalmology Clinics of North America*, **4**, 431–447
33. Smith, J.L., Hoyt, W.F. and Susac, J.O. (1973) Ocular fundus in acute Leber's optic neuropathy. *Archives of Ophthalmology*, **90**, 349–354
34. Smith, J.L., Tse, D.T., Byrne, S.F. *et al.* (1990) Optic nerve sheath distention in Leber's optic neuropathy and the significance of the 'Wallace mutation'. *The Journal of Clinical Neuro-Ophthalmology*, **10**, 231–238
35. Lessell, S., Gise, R.L. and Krohel, G.B. (1983) Bilateral optic neuropathy with remission in young men. Variation on a theme by Leber? *Archives of Neurology*, **40**, 2–6
36. Stone, E.M., Newman, N.J., Miller, N.R. *et al.* (1992) Visual recovery in patients with Leber's hereditary optic neuropathy and the 11778 mutation. *Journal of Clinical Neuro-Ophthalmology*, **12**, 10–14
37. Vilkki, J., Ott, J., Savontaus, M.-L. *et al.* (1991) Optic atrophy in Leber hereditary optic neuroretinopathy is probably determined by an X-chromosome gene closely linked to DXS7. *The American Journal of Human Genetics*, **48**, 486–491
38. Sweeney, M.G., Davis, M.B., Brockington, M., Toscano, A. and Harding, A.E. (1991) Is visual loss in Leber's disease partly determined by an X-linked locus? *American Journal of Human Genetics*, **49**, 361
39. Sweeney, M.G., Davis, M.B., Lashwood, A.M., Brockington, M., Toscano, A. and Harding, A.E. (1992) Evidence against a locus close to DXS7 determining visual loss in Italian and British families with Leber's hereditary optic neuropathy. *American Journal of Human Genetics*, **51**, 741–748
40. Vilkki, J., Savontaus, M.-L. and Nikoskelainen, E.K. (1989) Genetic heterogeneity in Leber hereditary optic neuroretinopathy revealed by mitochondrial DNA polymorphism. *The American Journal of Human Genetics*, **45**, 206–211
41. Holt, I.J., Miller, D.H. and Harding, A.E. (1989) Genetic heterogeneity and mitochondrial DNA heteroplasmy in Leber's hereditary optic neuropathy. *Journal of Medical Genetics*, **26**, 739–743
42. Bolhuis, P.A., Bleeker-Wagemakers, E.M., Ponne, N.J. *et al.* (1990) Rapid shift in genotype of human mitochondrial DNA in a family with Leber's hereditary optic neuropathy. *Biochemical and Biophysical Research Communications*, **170**, 994–997
43. Larsson, N., Andersen, O., Holme, E., Oldfors, A. and Wahlstrom, J. (1991) Leber's hereditary optic neuropathy and complex I deficiency in muscle. *Annals of Neurology*, **30**, 701–708
44. Wilson, J. (1963) Leber's hereditary optic atrophy: some clinical and aetiological considerations. *Brain*, **83**, 347–362
45. Wilson, J., Linnell, J.C. and Matthews, D.M. (1971) Plasma cobalamins in neuro-ophthalmological diseases. *The Lancet*, **i**, 259–261
46. Johns, D.R. and Berman, J. (1991) Alternative, simultaneous complex I mitochondrial DNA mutations in Leber's hereditary optic neuropathy. *Biochemical and Biophysical Research Communications*, **174**, 1324–1330
47. Johns, D.R. and Neufeld, M.J. (1991) Cytochrome b mutations in Leber hereditary optic neuropathy. *Biochemical and Biophysical Research Communications*, **181**, 1358–1364
48. Brown, M.D., Voljavec, A.S., Lott, M.T. *et al.* (1992) Mitochondrial DNA complex I and III mutations associated with Leber's hereditary optic neuropathy. *Genetics*, **130**, 163–173
49. Howell, N., Bindoff, L.A., McCullough, D.A. *et al.* (1991) Leber hereditary optic neuropathy: identification of the same mitochondrial ND1 mutation in six pedigrees. *The American Journal of Human Genetics*, **49**, 939–950
50. Huoponen, K., Vilkki, J., Aula, P., Nikoskelainen, E.K. and Savontaus, M.L. (1991) A new mutation associated with Leber hereditary optic neuroretinopathy. *The American Journal of Human Genetics*, **48**, 1147–1153

51A. Howell, N., Kubacka, I., Xu, M. and McCullough, D.A. (1991) Leber hereditary optic neuropathy: involvement of the mitochondrial ND1 gene and evidence for an intragenic suppressor mutation. *The American Journal of Human Genetics*, **48**, 935–942

51B. Wallace, D.C. (1970) A new manifestation of Leber's disease and a new explanation for the agency responsible for its unusual pattern of inheritance. *Brain*, **93**, 121–132

52. Newman, N.J., Lott, M.T. and Wallace, D.C. (1991) The clinical characteristics of pedigrees of Leber's hereditary optic neuropathy with the 11,778 mutation. *The American Journal of Ophthalmology*, **111**, 750–762

53. Ragan, C.I. (1987) Structure of NADH–ubiquinone reductase (Complex I). *Current Topics in Bioenergetics*, **15**, 1–36

54. Majander, A., Huoponen, K., Sanontaus, M.-L., Nikoskelainen, E. and Wikstrom, M. (1991) Electron transfer properties of NADH: ubiquinone reductase in the ND1/3460 and ND4/11,778 mutations of the Leber hereditary optic neuroretinopathy (LHON). *FEBS Letters*, **292**, 289–292

55. Sumegi, B. and Srere, P.A. (1984) Complex I binds several mitochondrial NAD-coupled dehydrogenases. *The Journal of Biological Chemistry*, **259**, 15040–15045

56. Srivastava, D.K. and Bernhard, S.A. (1986) Metabolite transfer via enzyme-enzyme complexes. *Science*, **234**, 1081–1086

57. Fukushima, T., Decker, R.V., Anderson, W.M. and Spivey, H.O. (1989) Substrate channeling of NADH and binding of dehydrogenases to Complex I. *The Journal of Biological Chemistry*, **264**, 16483–16488

58. Parker, W.D., Oley, C.A. and Parks, J.K. (1989) A defect in mitochondrial electron transport activity (NADH–coenzyme Q oxidoreductase) in Leber's hereditary optic neuropathy. *The Lancet*, **320**, 1331–1333

59. Earley, F.G.P., Patel, S.D., Ragan, C.I. and Attardi, G. (1987) Photolabeling of a mitochondrially encoded subunit of NADH degydrogenase with [³H] dihydrorotenone. *FEBS Letters*, **219**, 108–112

60. Horai, S. and Matsunaga, E. (1986) Mitochondrial DNA polymorphism in Japanese. II. Analysis with restriction enzymes of four or five base pair recognition. *Human Genetics*, **68**, 324–332

61. Stoneking, M., Jorde, L.B., Bhatia, K. and Wilson, A.C. (1990) Geographic variation in human mitochondrial DNA from Papua New Guinea. *Genetics*, **124**, 717–723

62. Davis, P.C., Hoffman, J.C., Braun, I.F., Ahmann, P. and Krawiecki, N. (1987) MR of Leigh's disease (subacute necrotizing encephalomyelopathy). *American Journal of Neuroradiology*, **8**, 71–75

63. Montpetit, V.J.A., Andermann, F., Carpenter, S. *et al.* (1971) Subacute necrotizing encephalomyelopathy: a review and a study of two families. *Brain*, **94**, 1–30

64. Pincus, J.H. (1972) Subacute necrotizing encephalomyelopathy (Leigh's disease): a consideration of clinical features. *Developmental Medicine and Child Neurology*, **14**, 87–101

65. Willems, J.L., Monnens, L.A., Trijbels, J.M. *et al.* (1977) Leigh's encephalomyelopathy in a patient with cytochrome *c* oxidase deficiency in muscle tissue. *Pediatrics*, **60**, 850–857

66. Miyabayashi, S., Ito, T., Narisawa, K., Iinuma, K. and Tada, K. (1985) Biochemical study in 28 children with lactic acidosis, in relation to Leigh's encephalomyelopathy. *The European Journal of Pediatrics*, **143**, 278–283

67. Robinson, B.H., De, M.L., Glerum, M., Sherwood, G. and Becker, L. (1987) Clinical presentation of mitochondrial respiratory chain defects in NADH–coenzyme Q reductase and cytochrome oxidase: clues to pathogenesis of Leigh disease. *The Journal of Pediatrics*, **110**, 216–222

68. Arts, W.F., Scholte, H.R., Loonen, M.C. *et al.* (1987) Cytochrome *c* oxidase deficiency in subacute necrotizing encephalomyelopathy. *The Journal of Neurological Science*, **77**, 103–115

69. DiMauro, S., Servidei, S., Zeviani, M. *et al.* (1987) Cytochrome *c* oxidase deficiency in Leigh syndrome. *Annals of Neurology*, **22**, 498–506

70. Glerum, M., Robinson, B.H., Spratt, C., Wilson, J. and Patrick, D. (1987) Abnormal kinetic behavior of cytochrome oxidase in a case of Leigh disease. *The American Journal of Human Genetics*, **41**, 584–593

71. Berkovic, S.F., Carpenter, S., Evans, A. *et al.* (1989) Myoclonus epilepsy and ragged-red fibres (MERRF) 1. A clinical, pathological, biochemical, magnetic resonance and positron emission tomographic study. *Brain*, **112**, 1231–1260

72. Miranda, D.F., Ishii, S., DiMauro, S. and Shay, J.W. (1989) Cytochrome *c* oxidase deficiency in Leigh's syndrome: genetic evidence for a nuclear DNA-encoded mutation. *Neurology*, **39**, 697–702

73. Van Erven, P.M.M., Gabreels, F.J.M., Weevers, R.A. *et al.* (1987) Intravenous pyruvate loading test in Leigh syndrome. *The Journal of Neurological Science*, **77**, 217–227

74. Stansbie, D., Wallace, S.J. and Marsac, C. (1986) Disorders of the pyruvate dehydrogenase complex. *The Journal of Inherited Metabolic Disease*, **9**, 105–109

75. Kretzschmar, H.A., DeArmond, S.J., Koch, T.K. *et al.* (1987) Pyruvate dehydrogenase complex deficiency as a cause of subacute necrotizing encephalopathy (Leigh disease). *Pediatrics*, **79**, 370–373

76. Robinson, B.H. (1988) Cell culture studies on patients with mitochondrial diseases: molecular defects in pyruvate dehydrogenase. *The Journal of Bioenergetics and Biomembranes*, **20**, 313–323

77. Shoffner, J.M., Fernhoff, P.M., Krawiecki, N.S. *et al.* (1992) Subacute necrotizing encephalopathy: oxidative phosphorylation defects and the ATPase 6 point mutation. *Neurology*, **42**, 2168–2174

78. Tatuch, Y., Christodoulou, J., Feigenbaum, A. *et al.* (1992) Heteroplasmic mitochondrial DNA mutation [T to G] at 8993 can cause Leigh disease when the percentage of abnormal mitochondrial DNA is high. *The American Journal of Human Genetics*, **50**, 852–858

79. Vadheim, C., Rimoin, D. and Rotter, J. (1990) Diabetes mellitus. In *Principles and Practice of Medical Genetics*, 2nd edn (eds A.E.H. Emery and D.L. Rimoin), Churchill-Livingstone, New York, Vol. 2, pp. 1521–1558

80. Dorner, G. and Mohnike, A. (1976) Further evidence for a predominantly maternal transmission of maturity-onset type diabetes. *Endokrinologie*, **68**, 121–124

81. Dorner, G., Mohnike, A. and Steindel, E. (1975) On possible genetic and epigenetic modes of diabetes transmission. *Endokrinologie*, **66**, 225–227

82. Dorner, G., Plagemann, A. and Reinagel, H. (1987) Familial diabetes aggregation in type I diabetics: gestational diabetes an apparent risk factor for increased diabetes susceptibility in the offspring. *Experimental and Clinical Endocrinology*, **89**, 84–90

83. Freinkel, N., Metzger, B.E., Phelps, R.L. *et al.* (1986) Gestational diabetes mellitus: a syndrome with phenotypic and genotypic heterogeneity. *Hormones and Metabolic Research*, **18**, 427–430

84. Piccolo, G., Aschei, M., Ricordi, A. *et al.* (1989) Normal insulin receptors in mitochondrial myopathies with ophthalmoplegia. *The Journal of Neurological Science*, **94**, 163–172

85. Poulton, J., Deadman, M.E. and Gardiner, R.M. (1989) Duplications of mitochondrial DNA in mitochondrial myopathy [see comments]. *The Lancet*, **i**, 236–240

86. Eviatar, L., Shanske, S., Gauthier, B. *et al.* (1990) Kearns–Sayre syndrome presenting as renal tubular acidosis. *Neurology*, **40**, 1761–1763

87. Rotig, A., Bessi, J.-L., Romero, N. *et al.* (1992) Maternally inherited duplication of the mitochondrial genome in a syndrome of proximal tubulopathy, diabetes mellitus, and cerebellar ataxia. *The American Journal of Human Genetics*, **50**, 364–370

88. Ballinger, S.W., Shoffner, J.M., Hedaya, E.V. *et al.* (1992) Maternally transmitted diabetes and deafness associated with a 10.4 kb mitochondrial DNA deletion. *Nature Genetics*, **1**, 11–15

89. Moraes, C.T., DiMauro, S., Zeviani, M. *et al.* (1989) Mitochondrial DNA deletions in progressive external ophthalmoplegia and Kearns–Sayre syndrome. *New England Journal of Medicine*, **320**, 1293–1299

90. Shoffner, J.M., Lott, M.T., Voljavec, A.S. *et al.* (1989) Spontaneous Kearns–Sayre/chronic external ophthalmoplegia plus syndrome associated with a mitochondrial DNA deletion: a slip-replication model and metabolic therapy. *Proceedings of the National Academy of Sciences of the USA*, **86**, 7952–7956

91. Zeviani, M., Gellera, C., Pannacci, M. *et al.* (1990) Tissue distribution and transmission of mitochondrial DNA deletions in mitochondrial myopathies. *Annals of Neurology*, **28**, 94–97

92. Larsson, N.-G., Eiken, H.G., Boman, H. *et al.* (1992) Lack of transmission of deleted mtDNA from a woman with Kearns–Sayre syndrome to her child. *The American Journal of Human Genetics*, **50**, 360–363

93. Muller-Hocker, J. (1989) Cytochrome *c* odixase deficient cardiomyocytes in the human heart. An age related phenomenon. *American Journal of Pathology*, **134**, 1167–1173

94. Muller-Hocker, J. (1990) Cytochrome *c* oxidase deficient fibres in the limb muscle and diaphragm of man without muscular disease: an age-related alteration. *The Journal of Neurological Science*, **100**, 14–21

95. Yen, T.C., Su, J.H., King, K.L. and Wei, Y.H. (1991) Ageing-associated 5 kb deletion in human liver mitochondrial DNA. *Biochemical and Biophysical Research Communications*, **178**, 124–131

96. Piko, L., Hougham, A.J. and Bulpitt, K.J. (1988) Studies of sequence heterogeneity of mitochondrial DNA from rat and mouse tissues: evidence for an increased frequency of deletions/additions with aging. *Mechanisms of Ageing and Development*, **43**, 279–293

97. Cortopassi, G.A. and Arnheim, N. (1990) Detection of a specific mitochondrial DNA deletion in tissues of older humans. *Nucleic Acids Research*, **18**, 6927–6933

98. Corral-Debrinski, M., Stepien, G., Shoffner, J.M. *et al.* (1991) Hypoxemia is associated with mitochondrial DNA damage and gene induction. *The Journal of the American Medical Association*, **266**, 1812–1816

99. Hattori, K., Tanaka, M., Sugiyama, S. *et al.* (1991) Age-dependent increase in deleted mitochondrial DNA in the human heart: possible contributory factor to presbycardia. *American Heart Journal*, **121**, 1735–1742

100. Cadenas, E., Boveris, A., Ragan, C.I. and Stoppani, A.O.M. (1977) Production of O_2 radicals and H_2O_2 by NADH–ubiquinone reductase and ubiquinol–cytochrome *c* reductase from beef heart mitochondria. *Archives of Biochemistry and Biophysics*, **180**, 248–257

101. Berry, E.A. and Trumpower, B.L. (1985) Pathways of electrons and protons through the cytochrome bc_1 complex of the mitochondrial respiratory chain. In *Coenzyme Q* (ed. G. Lenaz), John Wiley, New York, pp. 365–389

102. Nohl, H. (1986) Oxygen radical release in mitochondria: influence of age. In *Free Radicals, Aging, and Degenerative Diseases* (eds J.E. Johnson, Jr., R. Walford, D. Harman and J. Miquel), Alan R. Liss, New York, pp. 77–79

103. Bandy, B. and Davidson, A.J. (1990) Mitochondrial mutations may increase oxidative stress: implications for carcinogenesis and aging. *Free Radicals in Biology and Medicine*, **8**, 523–539

104. Clark, J.M. and Beardsley, G.P. (1986) Thymine glycol lesions terminate chain elongation by DNA polymerase I *in vitro*. *Nucleic Acids Research*, **14**, 737–749

105. Richter, C., Park, J.-W. and Ames, B.N. (1988) Normal oxidative damage to mitochondrial and nuclear DNA is extensive. *Proceedings of the National Academy of Sciences of the USA*, **85**, 6465–6467

106. Breimer, L.H. (1990) Molecular mechanisms of oxygen radical carcinogenesis and mutagenesis: the role of DNA base damage. *Molecular Carcinogenesis*, **3**, 188–197

107. Lutz, W.K. (1990) Endogenous genotoxic agents and processes as a basis of spontaneous carcinogenesis. *Mutation Research*, **238**, 287–295

108. Lochner, A., Niekerk, I.V. and Kotze, J.C.N. (1981) Mitochondrial acyl-CoA, adenine nucleotide translocase activity and oxidative phosphorylation in myocardial ischaemia. *The Journal of Molecular and Cellular Cardiology*, **13**, 991–997

109. Kotaka, K., Miyazaki, Y., Ogawa, K. *et al.* (1982) Reversal of ischemia-induced mitochondrial dysfunction after coronary reperfusion. *The Journal of Molecular and Cellular Cardiology*, **14**, 223–231

110. Rouslin, W. (1983) Mitochondrial complexes I, II, III, IV and V in myocardial ischemia and autolysis. *American Journal of Physiology*, **244**, H743–H748

111. Piper, H.M., Sezer, O., Schleyer, M. *et al.* (1985) Development of ischemia-induced damage in defined mitochondrial subpopulations. *The Journal of Molecular and Cellular Cardiology*, **17**, 885–896

112. Asimakis, G.K. and Sordahl, L.A. (1981) Intramitochondrial adenine nucleotides and energy-linked functions of heart mitochondria. *American Journal of Physiology*, **241**, H672–H681

113. LaNoue, K.F., Watts, J.A. and Koch, C.D. (1981) Adenine nucleotide transport during cardiac ischemia. *American Journal of Physiology*, **241**, H663–H671

114. Asimakis, G.K. and Conti, V.R. (1984) Myocardial ischemia: correlation of mitochondrial adenine nucleotide and respiratory function. *The Journal of Molecular and Cellular Cardiology*, **16**, 439–448

115. Asimakis, G.K., Wilson, D.E. and Conti, V.R. (1985) Release of AMP and adenosine from rat heart mitochondria. *Life Sciences*, **37**, 2373–2380

116. Shug, A., Lerner, E., Elson, C. and Shrago, E. (1971) The inhibition of adenine nucleotide translocase by oleoyl CoA and its reversal in rat liver mitochondria. *Biochemical and Biophysical Research Communications*, **43**, 557–563

117. Shug, A.L., Shrago, E., Bittar, N., Folts, J.D. and Koke, J.R. (1975) Acyl CoA inhibition of adenine nucleotide translocation in ischemic myocardium. *American Journal of Physiology*, **228**, 689–692

118. Katz, A.M. and Messineo, F.C. (1981) Lipid-membrane interactions and the pathogenesis of ischemic damage in the myocardium. *Circulation Research*, **48**, 1–16

119. Jacobus, W.E., Tiozzo, R., Lugli, G., Lehninger, A.L. and Carafoli, E. (1975) Aspects of energy-linked calcium accumulation by rat heart mitochondria. *The Journal of Biological Chemistry*, **250**, 7863–7870

120. Nicholls, D.G. and Crompton, M. (1980) Mitochondrial calcium transport. *FEBS Letters*, **111**, 261–268

121. Carafoli, E. (1985) The homeostasis of calcium in heart cells. *The Journal of Molecular and Cellular Cardiology*, **17**, 203–212

122. Haworth, R.A., Hunter, D.R. and Berkoff, H.A. (1981) Contracture in isolated adult rat heart cells: role of Ca^{2+}, ATP, and compartmentation. *Circulation Research*, **49**, 1119–1128

123. Rouslin, W., Erickson, J.L. and Solaro, R.J. (1986) Effects of oligomycin and acidosis on rates of ATP depletion in ischemic heart muscle. *American Journal of Physiology*, **250**, H503–H508

124. Otani, H., Tanaka, H., Inoue, T. *et al.* (1984) In vitro study on contribution of oxidative metabolism of isolated rabbit heart mitochondria to myocardial reperfusion injury. *Circulation Research*, **55**, 168–175

125. Arroyo, C.M., Kramer, J.H., Leiboff, R.H. *et al.* (1987) Spin trapping of oxygen and carbon-centered free radicals in ischemic canine myocardium. *Free Radicals in Biology and Medicine*, **3**, 313–316

126. McCord, J.M. (1988) Free radicals and myocardial ischemia: overview and outlook. *Free Radicals in Biology and Medicine*, **4**, 9–14

127. Guarnieri, C., Flamigni, F. and Caldarera, C.M. (1980) Role of oxygen in the cellular damage induced by re-oxygenation of hypoxic heart. *The Journal of Molecular and Cellular Cardiology*, **12**, 797–808

128. Meerson, F.Z., Kagan, V.E., Kozlov, Y.P., Belkina, L.M. and Arkhipenko, Y.V. (1982) The role of lipid perioxidation in pathogenesis of ischemic damage and the antioxidant protection of the heart. *Basic Research in Cardiology*, **77**, 465–485

129. Ferrari, R., Ceconi, C., Curello, S. *et al.* (1985) Oxygen mediated myocardial damage during ischemia and reperfusion: role of the cellular defences against oxygen toxicity. *The Journal of Molecular and Cellular Cardiology*, **17**, 937–945

130. Shoubridge, E.A., Karpati, G. and Hastings, K.E.M. (1990) Deletion mutants are functionally dominant over wild-type mitochondrial genomes in skeletal muscle fiber segments in mitochondrial disease. *Cell*, **62**, 43–49

131. Torroni, A., Stepien, G., Hodge, J.A. and Wallace, D.C. (1990) Neoplastic transformation is associated with coordinate induction of nuclear and cytoplasmic oxidative phosphorylation genes. *The Journal of Biological Chemistry*, **265**, 20589–20593

132. Schon, E.A., Rizzuto, R., Moraes, C.T. *et al.* (1989) A direct repeat is a hotpsot for large-scale deletion of human mitochondrial DNA. *Science*, **244**, 346–349

133. Ikebe, S.-i., Tanaka, M., Ohno, K. *et al.* (1990) Increase of deleted mitochondrial DNA in the striatum in Parkinson's disease and senescence. *Biochemical and Biophysical Research Communications*, **170**, 1044–1048

134. Ozawa, T., Tanaka, M., Sugiyama, S. *et al.* (1990) Multiple mitochondrial DNA deletions exist in cardiomyocytes of patients with hypertrophic or dilated cardiomyopathy. *Biochemical and Biophysical Research Communications*, **170**, 830–836

135. Corbisier, P. and Remacle, J. (1990) Involvement of mitochondria in cell degeneration. *European Journal of Cell Biology*, **51**, 173–182

136. Langston, J.W., Ballard, P., Tetrud, J.W. and Irwin, I. (1983) Chronic parkinsonism in humans due to a product of meperidine-analog synthesis. *Science*, **219**, 979–980

137. Nicklas, W.J., Vyas, I. and Heikkila, R.E. (1985) Inhibition of NADH-linked oxidation in brain mitochondria by 1-methyl-4-phenly-pyridine, a metabolite of the neurotoxin, 1-methyl-4-phenyl-1,2,3,6-tetrahydropyridine. *Life Sciences*, **36**, 2503–2508

138. Singer, T.P., Castagnoli, N.Jr., Ramsay, R.P. and Trevor, A.J. (1987) Biochemical events in the development of Parkinsonism induced by 1-methyl-4-phenyl-1,2,3,6-tetrahydropyridine. *The Journal of Neurochemistry*, **49**, 1–8

139. Ginsberg, M.D. (1980) Carbon monoxide. In *Experimental and Clinical Neurotoxicology* (eds P.S. Spencer and H.H. Schaumberg), Williams & Wilkins, Baltimore, pp. 374–394

140. Uitti, R.J., Rajput, A.H., Ashenhurst, E.M. and Rozdilsky, B. (1985) Cyanide-induced parkinsonism: a clinicopathologic report. *Neurology*, **35**, 921–925

141. Schapira, A.H.V., Cooper, J.M., Dexter, D. *et al.* (1990) Mitochondrial complex I deficiency in Parkinson's disease. *The Journal of Neurochemistry*, **54**, 823–827

142. Parker, W.D., Jr., Boyson, S.J. and Parks, J.K. (1989) Abnormalities of the electron transport chain in idiopathic Parkinson's disease. *Annals of Neurology*, **26**, 719–723

143. Bindoff, L.A., Birch-Machin, M., Cartlidge, N.E.F., Parker, W.D., Jr. and Turnbull, D.M. (1989) Mitochondrial function in Parkinson disease. *The Lancet*, **i**, 49

144. Shoffner, J.M., Watts, R.L., Juncos, J.L., Torrino, A. and Wallace, D.C. (1991) Mitochondrial oxidative phosphorylation defects in Parkinson's disease. *Annals of Neurology*, **30**, 332–339

145. Whitehouse, P.J., Price, D.L., Struble, R.G., Clark, A.W., Coyle, J.T. and DeLong, M.R. (1982) Alzheimer's disease and senile dementia: loss of neurons in the basal forebrain. *Science*, **215**, 1237–1239

146. Parker, W.D., Filley, C.M. and Parks, J.K. (1990) Cytochrome *c* oxidase deficiency in Alzheimer's disease. *Neurology*, **40**, 1302–1303

147. Sims, N.R., Finegan, J.M., Blass, J.P., Bowen, D.M. and Neary, D. (1987) Mitochondrial function in brain tissue in primary degenerative dementia. *Brain Research*, **436**, 30–38

148. Peterson, C. and Goldman, J.E. (1986) Alterations in calcium content and biochemical processes

in cultured skin fibroblasts from aged and Alzheimer's donors. *Proceedings of the National Academy of Science of the USA*, **83**, 2758–2762

149. Blass, J.P., Baker, A.C., Ko, L. and Black, R.S. (1990) Induction of Alzheimer's antigens by an uncoupler of oxidative phosphorylation. *Archives of Neurology*, **47**, 864–869

150. Hansen, L.A., Masliah, E., Terry, R.D. and Mirra, S.S. (1989) A neuropathological subset of Alzheimer's disease with concomitant Lewy body disease and spongiform change. *Acta Neuropathologica*, **78**, 194–201

151. Saraiva, A.A., Borges, M.M., Madeira, M.D., Tavares, M.A. and Paula-Barbosa, M.M. (1985) Mitochondrial abnormalities in cortical dendrites from patients with Alzheimer's disease. *The Journal of Submicroscopic Cytology*, **17**, 459–464

12
Mitochondrial dysfunction in neurodegenerative disorders and aging

A.H.V. Schapira

This chapter will seek to review the evidence that a decline in mitochondrial function may contribute to certain neurodegenerative disorders, to the aging process and to programmed cell death.

PARKINSON'S DISEASE

Parkinson's disease (PD) is characterized clinically by bradykinesia, rigidity and tremor and pathologically by the death of dopaminergic neurons in the substantia nigra with Lewy bodies in some surviving neurons. Several aetiological mechanisms have been invoked to explain this selective cell death including heredity, infection, autoimmunity and environmental toxins. Specific biochemical abnormalities in the substantia nigra are now being defined in PD. The relationship of these to neuronal cell death and their connection to the processes outlined above will improve our understanding of the cause of PD.

The MPTP model of PD

The 1-methyl-4-phenyl-1,2,3,6-tetrahydropyridine (MPTP) model of PD has provided many valuable insights into the possible causes of the idiopathic disease and the mechanisms by which neuronal cell death may be induced [1,2]. The structural resemblance of MPTP to the herbicide paraquat initially led to the suggestion that MPTP caused cell damage through the generation of free radicals [3,4]. However, it became clear that MPTP and its metabolite 1-methyl-4-phenylpyridine (MPP$^+$) could enter into redox cycling only with difficulty because of the high potential of MPTP [5]. Nevertheless, there is some evidence that oxidative damage may indeed be involved in MPTP toxicity, at least under certain conditions [6–8].

MPTP must be converted to MPP$^+$ by glial monoamine oxidase B (MAO-B) in order to cause dopamine containing cell death [9]. Inhibition of MAO-B by, for instance, pargyline, prevents the development of parkinsonism in MPTP-treated primates. MPP$^+$ is a substrate for the dopamine reuptake pathway [10] and so is preferentially concentrated into the neurons of the substantia nigra. MPP$^+$ is also a substrate for an energy-dependent mechanism to concentrate lipophilic cations into mitochondria [11]. This results in millimolar levels of intramitochondrial MPP$^+$ by concentrating cytoplasmic levels several hundredfold. Once inside mitochondria, MPP$^+$ has direct access to NADH-CoQ$_1$ reductase (Complex I), of which it is a specific inhibitor both *in vitro* [12] and *in vivo* [13]. The resulting fall in ATP levels is thought to account for the neuronal death induced by MPTP. This is supported in culture studies by the direct correlation between increasing levels of MPP$^+$, falling levels of ATP and increasing rates of cell death [14].

The mechanism by which MPP$^+$ inhibits Complex I is not yet fully understood. It is clear, however, that MPP$^+$ binds only loosely to Complex I and all inhibition (up to 40% at 10 mM) within 15 min of incubation can be reversed by washing [15]. Competitive binding experiments with rotenone and piericidin suggest that MPP$^+$ interacts with the same site of Complex I as these other inhibitors [16], namely the ND1 subunit [17]. Recent experiments have demonstrated that prolonged incubation (> 15 min) of MPP$^+$ with inverted mitochondria can, under certain conditions, produce progressive (up to 78%), irreversible and specific inhibition of Complex I [15]. This irreversible inhibition can be prevented by co-incubation with free-radical scavengers. These results imply that MPP$^+$ can inter-act with Complex I to increase the production of free radicals and that these in turn damage Complex I to further inhibit its activity irreversibly. The selectivity of the free-radical damage for Complex I suggests either that the radicals were generated in the vicinity of Complex I and were short-lived, or that the initial interaction of MPP$^+$ with Complex I made the latter more vulnerable to oxidative damage. These observations serve to emphasize the close and reciprocal relationship between free radicals and Complex I and may have implications for the mechanisms of cell damage active in PD.

Mitochondrial function in PD

The Complex I inhibition detected in the MPTP model was the clue to examine mitochondrial function in idiopathic PD. The identification of a specific Complex I defect in PD substantia nigra provided a direct biochemical link between the idiopathic disease and the neurotoxic model, and stimulated the search for a similar defect in more accessible tissues. Thus far, mitochondrial respiratory chain activity in PD has been investigated in brain, skeletal muscle and blood.

BRAIN

A significant decrease in Complex I activity has now been identified in a large study of post-mortem PD substantia nigra (n = 17) when compared with age-matched controls (n = 22) [18]. The decline in Complex I activity was 38% when expressed per mg of total protein, or 30% when corrected for any variation in citrate synthase activity. The value of the latter expression is that it corrects for any difference in mitochondrial mass between samples. Although these results are statistically significant on a group to group comparison, there is nevertheless overlap between the two. The activities of other respiratory chain enzymes (Complexes II–IV) did not show any statistically significant differences between the two groups.

Other studies have examined the anatomical specificity of the Complex I defect within the PD brain. Normal activities of all respiratory chain complexes have been found in cerebral cortex, cerebellum, tegmentum, caudate nucleus and globus pallidum [18,19]. However, one group did find a decrease in Complex III activity in striatum but normal function in cerebral cortex [20].

The disease specificity of the Complex I defect has been investigated by examining respiratory chain function in multiple system atrophy (MSA). MSA is characterized by neuronal loss in the substantia nigra which is often more severe than in PD. If the Complex I deficiency in PD were simply due to neuronal degeneration, then a similar defect should be apparent in MSA. In fact, mitochondrial function was found to be normal in MSA substantia nigra [19]. The fact that these patients had also been on L-Dopa for similar durations and total doses as the PD patients also suggested that L-Dopa did not cause the Complex I deficiency seen in PD. This is supported by the finding that oral loading of rats with L-Dopa does not affect Complex I activity as judged by assay of skeletal muscle respiratory chain enzymes [21].

Enzyme spectrophotometric assay of post-mortem frozen substantia nigra is limited to the use of homogenates because of the small amounts of tissue available. Intact mitochondria cannot be isolated from frozen tissue although a fraction enriched in mitochondrial membranes can be prepared; this usually requires more tissue than is available from such samples and prevents the use of any form of correction for variation in mitochondrial mass. Thus, Complex I deficiency identified in PD is a measure of both glial and neuronal components. The latter represents only a very small proportion (1–5%) of the total cell count, although its proportion per mg of total protein or, more importantly, its contribution to mitochondrial mass may be greater. Nevertheless, a defect of Complex I activity of the extent identified (30–38%) is almost certain to represent a decline in glial as well as neuronal Complex I. Hattori *et al.* [22] have sought to evaluate these separate components by the use of immunohistochemistry. Their results showed that there was a higher proportion of substantia nigral neurons staining poorly with Complex I antibodies in PD than in controls. However, the loss of neuronal Complex I staining did not always correlate with cell degeneration, suggesting that either the loss of Complex I precedes neuronal death or it is unrelated to it. It is difficult to imagine that a neuron, which is highly dependent on oxidative phosphorylation, could survive without any Complex I. The presence of a small proportion of control neurons with no Complex I staining raises similar problems with interpretation. This study also found a decrease in Complex II staining in some PD neurons compared with controls. Complex III and IV staining was no different in the neurons from either group. The Complex I staining of the glial cells in both groups was light and no distinction could be made between the two in this study.

Several studies have tried to address the question of the possible molecular mechanisms that may underlie the Complex I deficiency in PD. At the protein level, two groups have reported the results of immunoblotting studies. In one, deficiencies in certain low-molecular mass Complex I subunits were found in corpus striatum [23]. Polypeptide profiles of Complexes II, III and IV were normal. The other study analysed substantia nigra Complex I polypeptide profiles using holoenzyme and subunit-specific antisera and found no difference in cross-reacting subunits [24].

As seven Complex I subunits are encoded by mitochondrial DNA (mtDNA), this genome became an obvious target of interest in PD. Southern blot analysis of mtDNA from substantia nigra did not demonstrate the presence of any significant proportion of deleted mtDNA [25,26]. Polymerase chain reaction (PCR) amplification was then used to detect small amounts of mtDNA bearing the 5 kb 'common' deletion that might not have been apparent on Southern blotting. The first report using this technique suggested a sixteenfold increase in this deletion in striatum from PD brain [27]. However, subsequent reports using this same technique demonstrated no increase in this deletion in PD substantia nigra when samples were correctly age-matched [28]. In addition, it was estimated that even at age 80 years, mtDNA from control or PD substantia nigra contained only one in five thousand to ten thousand molecules with this 5 kb deletion. However, these results do not exclude the possibility that other deletions or point mutations of mtDNA may contribute to the Complex I defect. Indeed, an increase in polymorphisms affecting ND subunits has been suggested in PD [29]. However, it must be emphasized that the natural mutation rate of mtDNA is high, pathogenic mutations of mtDNA tend to be heteroplasmic and significant mutations should run true to the disease.

SKELETAL MUSCLE

Reports of respiratory chain defects in PD skeletal muscle have been conflicting. The first report by Bindoff *et al.* [30] documented a decline in activities of Complexes I, II and IV to 60, 51 and 60% of age-matched control means respectively. Wallace *et al.* [31] found a decrease in Complex I activity in three of six PD patients (mean deficiencies 92%, 90% and 86%), Complex IV deficiency in one and normal activities in two. Analysis of frozen biopsied muscle from 26 PD patients compared with age-matched controls showed multiple respiratory chain defects [32]. A selective deficiency of Complex I was found in five untreated PD patients [33]. Only one study has analysed skeletal muscle mitochondrial function *post mortem* and found a 49% decrease in Complex I activity in four PD samples [34]. However, polarographic and enzymic assay of fresh mitochondria isolated from nine PD patients did not show any defect when compared with similar age- and mobility-matched controls [18]. Similarly the study of 16 PD patients by similar techniques has failed to demonstrate any respiratory chain defect in skeletal muscle (M. Zeviani, personal communication). The recent analysis of seven PD patients by magnetic resonance spectroscopy also failed to detect any respiratory chain defect (D. Taylor, personal communication).

Limb weakness is not a recognized clinical feature in PD. It is therefore surprising when levels of Complex I deficiency that are normally associated with severe myopathy in patients with mitochondrial myopathy are found in skeletal muscle from patients with PD. In addition, it would be expected that the severity of respiratory chain dysfunction as is found in some PD patients would be associated with a lactic acidosis. It has recently been shown that PD patients do not have a lactic acidosis nor become acidotic on exercise [34–36].

It may be that some of the variation in results of mitochondrial function in PD can be explained by differences in analytical method, the effects of storage and an inherent heterogeneity amongst the PD population.

PLATELETS

Parker *et al.* [37] described a mean 55% decrease in Complex I activity in platelet mitochondria from ten PD patients compared with eight controls. This severity of defect, however, has not been reproduced by others using platelet mitochondria isolated by similar methods [36], or by using platelet homogenates [18].

A recent study has analysed platelet mitochondria from 25 PD patients and compared respiratory function with 15 matched controls [38]. A group to group analysis showed a mild (16%) but specific and significant (P = 0.005) decrease in citrate synthase-corrected Complex I activity in the PD patients. The overlap between the PD and control groups, however, was such that no distinction could be made between the two on an individual Complex I activity result. Platelet Complex I activity, therefore, cannot be used as a diagnostic test for PD.

The presence of a mild Complex I defect in a proportion of PD patients may seem at odds with the apparent specificity of the defect in the CNS. However, platelets have certain pharmacological characteristics which render them particularly susceptible to, for instance, a circulating toxin [39]. These include a high concentration of MAO-B and specific amine-uptake mechanisms capable of concentrating MPP^+. Platelets are often used as a tool for the study of neurological disorders. Their record in accurately reflecting primary pathogenetic biochemical changes in other tissues, however, is not good. It is therefore important to have a critical approach to the interpretation of the results of platelet mitochondrial function in neurodegenerative disorders. Any possible role that mitochondrial dysfunction may play in the pathogenesis of PD must be considered in the context of other biochemical changes that have been identified in this disorder. The most relevant of these are the increase in nigral iron and the evidence for oxidative damage.

Iron

Several studies have now described an increase in iron content in PD substantia nigra [40,41], with levels no different from controls in other brain areas [42]. However, elevated iron levels are also found in the brain in other neurodegenerative disorders [43]. In contrast with Huntington's disease and MSA, where elevated iron levels are accompanied by an increase in ferritin, ferritin levels are decreased in PD [44]. This suggests that the high nigral iron level may be unbound and so potentially highly reactive. Neuromelanin, an autoxidation product of L-Dopa, binds iron [45] and high concentrations of iron are therefore found in the melanized neurons of the substantia nigra [46]. Also, there seems to be a direct relationship between a neuron's melanin content and its susceptibility to degenerate [47]. In cultures of rat mesencephalic dopaminergic cells, however, elevated iron levels were non-specifically toxic to both neurons and glia [48]. This suggests that additional factors must be responsible for the selective neuronal death in PD.

Iron has the ability to enhance free-radical damage through the generation of highly reactive hydroxyl radicals via the Haber–Weiss reaction. This potential for iron-loaded cells to accentuate free radical production prompted the search for evidence of increased oxidative stress in PD.

Oxidative stress in PD

A decrease in the content of polyunsaturated fatty acids together with an increase in malondialdehyde levels in the PD substantia nigra provided substantial evidence for oxidative damage in this tissue at the time of death [49]. Previous studies had shown a decrease in the activity of catalase [50] and glutathione peroxidase [51], and a decrease in reduced glutathione levels [41,52]. The subsequent demonstration of increased levels of mitochondrial superoxide dismutase in PD [53] provided further evidence of a response to free-radical generation in the PD substantia nigra. However, this is in contrast with the finding of normal levels of vitamin E [54] and ascorbic acid [41] in PD brain, two compounds that would be expected to be increased in the presence of active oxidative stress.

Mitochondria and free radicals

The mitochondrial respiratory chain is one of the most important sources of reactive oxygen and hydrogen peroxide in the aerobically respiring cell [55,56]. Under normal circumstances, electrons remain tightly bound to the respiratory chain to limit superoxide formation. The major source of superoxide ions in the normally functioning respiratory chain is thought to be cytochrome b_{566} [57] or reduced ubiquinone [58]. Inhibition of Complex I or Complex III can lead to enhanced free-radical production [59,60]. Mitochondrial respiratory chain function can also be impaired by direct oxidative damage. The complex most susceptible to free-radical damage is uncertain. *In vitro* studies using rat brain or beef heart mitochondria suggest that Complex I [61] or Complexes I–III [62] are impaired. *In vivo* experiments using 2-cyclohexane-1-one to decrease levels of cerebral reduced glutathione showed that Complex IV activity was most sensitive to inactivation by free radicals and was the first of the respiratory chain enzymes to be affected with Complex I activity only being affected after Complexes II and III [63].

These results are directly relevant to the findings of mitochondrial dysfunction and oxidative damage in PD. For instance, it could be argued that the Complex I deficiency in PD substantia nigra is secondary to the co-existing oxidative damage. However, the *in vivo* and some of the *in vitro* studies suggest that specific Complex I deficiency is not the pattern of respiratory chain dysfunction induced by free radicals. Alternatively, the defective Complex I activity could itself produce the oxidative damage seen in PD. Two lines of evidence support the latter proposition.

(1) Under certain conditions, MPP+ leads to severe and specific inhibition of Complex I via the generation of free radicals [15]. The selectivity of the radical species for Complex I in this situation might be related to an increased susceptibility for damage determined by MPP+ binding. This defective Complex I function might not only lead to increased leakage of superoxide radicals but might also render the protein particularly sensitive to their damaging and inhibitory effects upon enzyme activity. If the Complex I defect in PD were primary, increased free-radical generation would ensue which, in the presence of other promoters, e.g. iron, could result in enhanced Complex I deficiency, ATP depletion and cell death.

(2) In an attempt to produce an *in vivo* model of the biochemical events in PD substantia nigra, Hartley and colleagues iron-loaded dopaminergic PC 12 cells in culture. This resulted in an increase in malondialdehyde levels and a decrease in reduced glutathione levels indicating active free-radical damage. Mitochondrial function was also impaired with a 20% decrease in Complex IV and an 11% decrease in Complex I activity. The results of this

'*in vivo*' model also suggest that oxidative damage does not induce the specific Complex I defect seen in PD. In addition, even with the high intracellular iron levels achieved in this culture model (which were of the same order of magnitude as found in PD nigra), only a mild Complex I defect was induced – less than one-third of the severity seen in PD substantia nigra homogenates.

Presymptomatic PD

Approximately 8–10% of 'control' brains studied at *post mortem* are found to have evidence of nigra cell death and Lewy bodies in substantia nigra neurons [64]. The area of nigra affected by the neuronal loss in these brains is the same as that affected in PD and is distinct from the pattern of nerve cell death seen with age [65]. These observations have led to the concept of these brains being from presymptomatic PD patients. Such brains are thus very valuable in identifying the earliest biochemical changes as distinct from those that might be secondary to degeneration. Preliminary investigation of a group of presymptomatic PD brains has provided some very interesting results (D.T. Dexter *et al.*, unpublished work). Surprisingly iron and ferritin levels were no different from controls. The levels of reduced glutathione were, however, decreased. Complex I activities were intermediate between controls and PD patients, but were not statistically different at *n*=6.

These results suggest that oxidative stress, as determined by glutathione levels, may precede Complex I deficiency. However, the numbers used were small and only a partial decrease in Complex I activity would require more samples to show significance from control. In addition, it is intriguing to note that a decrease in reduced glutathione is one of the first events following inhibition of Complex I and is probably a general phenomenon associated with mitochondrial dysfunction [66].

A working hypothesis

It would seem reasonable to assume that despite which is the forerunner, both oxidative stress and mitochondrial dysfunction contribute to dopaminergic cell death in PD. The high concentration of iron in the nigra will enhance both these processes directly or indirectly. Either or both of these phenomena could be related to genetic or environmental factors.

GENETIC
There is increasing evidence for a genetic component to PD. The recent description of a large American–Italian kindred with autosomal dominant PD [67], together with other family studies suggesting autosomal dominant PD with variable penetrance [68], has prompted a reappraisal of previously negative twin data [69–71]. Such studies have been limited by the ability to determine early or subclinical PD in the twin pair. To some extent this has been overcome with striatal fluoro-Dopa-uptake studies with positron emission tomography. Assessment by this technique suggests that PD may be more prevalent amongst identical twins [72]. The theoretical possibility that mtDNA may contribute to PD

has not been supported by family studies, which have not shown any maternal pattern of inheritance [68].

There may, of course, be more than one gene which is associated with PD. A particular gene defect, such as may occur in the American–Italian family, may have high penetrance; other genes may have variable penetrance. Thus there may be a 'gene pool' of other potential mutations, two or more of which may be required to determine PD susceptibility. Candidates for this susceptibility pool include Complex I genes, genes for iron and free-radical metabolism, and genes that determine xenobiotic metabolism. In this respect, the recent description of the increased incidence of the cytochrome *P*-450 2D6 allele amongst PD patients is of particular interest [73,74]. The relevance of these findings is supported by the abnormal xenobiotic metabolism of sulphur compounds [75] and the enhanced production of *N*-methylated derivatives in PD [76]. This potential genetic heterogeneity may in part explain the variation in results seen in studies of PD skeletal muscle mitochondrial function. Similarly, those patients with the lowest platelet mitochondrial Complex I defect might be those expressing a genotype directly or indirectly decreasing Complex I activity. The difference between these individuals with PD and those controls with similar Complex I activities (and genotypes) may lie in their exposure to specific toxins.

ENVIRONMENTAL

Certain environmental agents have been associated with small groupings of PD, e.g. well-water drinking [77–79] and use of agrochemicals [80]. However, epidemiological data on the influence of the environment on PD have often been circumstantial and frequently contradictory. Nevertheless, the example of MPTP toxicity has shown that certain chemicals may induce changes that closely resemble idiopathic PD clinically, pharmacologically and biochemically. The role of such toxins in the production of the majority of PD cases must depend upon their distribution and availability, and the susceptibility of individuals to them. It is at this level that a genetic susceptibility might determine an individual's sensitivity to a given toxin, e.g. through the cytochrome *P*-450 system. Thus a common toxin could superimpose upon an uncommon genetic susceptibility to produce the observed prevalence of PD. Alternatively, an uncommon toxin could interact with a relatively common genetic predisposition. In this respect it is interesting to speculate why only a small proportion of those individuals exposed to MPTP actually developed parkinsonism. The most likely explanation of this would be variation in batch contamination and MPTP exposure, but another possibility might be that certain metabolic phenotypes rendered a small group of individuals particularly sensitive to the effects of this toxin.

There seems to be increasing evidence for a genetic susceptibility to PD upon which an environmental factor or factors might act to initiate the cascade of biochemical events that result in dopaminergic cell death. Complex I deficiency and oxidative damage may be the initiators of the cascade or simply be the terminal events that cause cell death. Either or both of these may be the targets for toxins, environmentally derived or endogenously produced. A clearer picture of the sequence of events associated with neuronal cell death and the molecular basis of the Complex I defect will provide important new insights into the cause of PD. Attention is now focusing on candidates for the PD susceptibility genes and environmental toxins capable of following metabolic pathways that might lead to selective dopaminergic cell death.

ALZHEIMER'S DISEASE

Defects of energy metabolism have long been sought as an explanation for neuronal death in Alzheimer's disease (AD). Abnormalities on positron emission tomography have been found consistently [81,82] and a defect of pyruvate dehydrogenase has been described in AD brain [83,84]. Changes in enzyme activities as measured in frozen post-mortem brain tissue are, of course, subject to the same difficulties of interpretation as in PD. A recent study has analysed respiratory chain activities in AD brain and found a small (17–26%) decrease in Complex IV activity in frontal and temporal cortices [85]. However, these changes were paralleled by lower citrate synthase levels in these areas, presumably representing decreased mitochondrial mass in AD – probably a reflection of neuronal degeneration. Whether Complex IV activities in AD would have fallen in the control range if corrected for the citrate synthase is not known. Similar decreases in Complex IV activity have been shown in MSA but have subsequently fallen into the control range when corrected for any variations of mitochondrial mass. A further study analysing the entire mitochondrial respiratory chain in AD temporal cortex could not identify any defect after citrate synthase correction (J.M. Cooper *et al.*, unpublished work).

An increase in platelet fluidity [86] together with a decrease in endoplasmic reticulum NADH–cytochrome *c* reductase activity [87] have been described in AD. In addition a severe 50% decrease in platelet Complex IV activity has also been described in a small group of AD patients [88]. The relationship of such a severe defect in platelet mitochondrial Complex IV to a normal or mildly decreased level of the same enzyme in AD brain is uncertain.

HUNTINGTON'S DISEASE

Huntington's disease (HD) is dominantly inherited via a highly penetrant variable length trinucleotide repeat on the short arm of chromosome 4. The biochemical correlate of this molecular genetic mutation is unknown. PET studies have demonstrated reduced glucose metabolism in HD caudate nucleus [89]. The possibility that a mitochondrial respiratory chain defect might contribute to such an abnormality was enhanced by the parental sex effect known to operate in HD. In this, the offspring of affected mothers develop the clinical features of HD much later than those of affected fathers. This raised the possibility that there may be a maternal 'protective' effect transmitted via mtDNA.

One study found a significant decrease in respiratory activity in HD brain with deficiencies of Complex IV and cytochrome aa_3 [90]. Another study showed a severe deficiency of citrate synthase-corrected Complex II/III activity but normal Complex I and IV activities in HD caudate [91]. As the first study used succinate as substrate for polarographic analysis, the results of these two papers are compatible with a deficiency of Complexes II/III. However, cytochrome *b* is the only mtDNA-encoded subunit of these two complexes, and levels of this cytochrome have been found to be normal in HD [90]. In addition, analysis of mtDNA by restriction length polymorphism did not reveal any significant mutation in a large HD family [92]. *In vitro* translation of mtDNA-encoded products from lymphoblast cell lines of members of this family also failed to show any significant abnormality. There is, therefore, little evidence to incriminate mtDNA as playing a contributory role in HD. The parental sex effect may be related to genomic imprinting [93] and may be seen in other diseases with trinucleotide repeats.

A study using platelet mitochondria from HD patients showed a 72% decrease in Complex I activity [94] but this contrasts with the normal Complex I activity found in HD caudate nucleus (see above).

MITOCHONDRIAL FUNCTION IN AGING

Aging is associated with a gradual decline in normal tissue function. This effect is most apparent in brain and muscle, two organs highly dependent on oxidative phosphorylation for energy requirements. It is not surprising therefore that mitochondrial function has become a target for those interested in the mechanisms involved in aging. Changes in the structure and function of mitochondria as well as mutations of mtDNA have now been identified in senescence.

Structure

Mitochondrial vacuolization, cristal rupture and the accumulation of intramito-chondrial paracrystalline inclusions have all been associated with aging tissue [95,96]. Changes in the size and number of mitochondria with age are dependent upon the tissue and species studied. Appearances similar to ragged red fibres [97], as well as an increase in cytochrome oxidase-negative fibres [98], have been seen in aging human skeletal muscle. However, these histochemical changes are uncommon and even cytochrome oxidase-negative fibres, when taken as a proportion of total fibre count, are rare in aged muscle. Cytochrome oxidase-negative fibres accumulate with age in limb, heart, diaphragm and extraocular muscles [99]. These fibres are randomly distributed at low density ($55/cm^2$ and $370/cm^2$ in limb and eye muscles respectively) even in patients aged 71–97. Densities of cytochrome oxidase-negative fibres much greater than this may be seen in limb muscles of patients with mitochondrial myopathy and no clinical muscle weakness or biochemical defect.

The lipid composition of mitochondrial membranes changes with age. Cholesterol content and phospholipid (cardiolipin) levels are both decreased [100]. The activity of the phosphate carrier also declined with age but could be restored by treatment with acetyl-L-carnitine [101].

Function

Several reports have now documented a decline of mitochondrial function with age. Trounce *et al.* [102] found that the most significant decrease in activity was associated with Complex I, but also found reduced activities of Complexes II–IV, in skeletal muscle of patients up to age of 90 years. Cooper *et al.* [103] demonstrated similar effects in limb muscle, with decreased activities of 59% for Complex I and 47% for Complex IV in patients aged around 70 years. A third report again found a decline in mitochondrial function with age in skeletal muscle but on this occasion the decrease was confined to Complex I [104].

Human liver mitochondria also show a marked decrease in oxygen utilization with age which apparently involves NAD^+- and succinate-linked substrates [105].

There was an increase in mitochondrial uncoupling with age, suggesting that the mitochondrial membranes became leaky.

Mitochondrial DNA

The idea that mutations of mtDNA [106,107] or mitochondrial oxidative damage [108] might accumulate with age and oxidative stress and contribute to the aging process has been expressed for some time. As discussed above, the respiratory chain is an important source of superoxide ions, the generation of which might increase if the activity of individual complexes is decreased. There is evidence from the flight muscle mitochondria of the housefly [109], and liver, heart and brain mitochondria of the rat [110], that superoxide and hydrogen peroxide generation increase with age. MtDNA is particularly susceptible to the mutagenic effects of free radicals because of its localization within the matrix, its lack of a histone coat [111] and inadequate repair mechanisms [112]. mtDNA has a high rate of spontaneous mutations and is particularly susceptible to the effect of genotoxins [113,114]. The accumulation of mtDNA mutations might result in a decline in the activity of those complexes with mtDNA-encoded subunits. This would produce a self-perpetuating and amplifying cycle of respiratory chain dysfunction and free-radical production.

There is now ample evidence that mtDNA deletions increase with age. Piko *et al.* [115] used thermal denaturation to determine mismatch in nucleotide sequences in mtDNA from young and senescent rats. No change in denaturation temperatures was observed, suggesting that mutation load was below 0.2%, the resolution limit of the technique. Using electron micrographs of mismatched segments in liver mtDNA, approximately 5% of senescent mtDNA contained abnormalities suggestive of deletions or insertions. The level of oxidative damage to mtDNA as determined by 8-hydroxydeoxyguanosine levels is approximately 1/8000 bases as compared with 1/130 000 bases in nuclear DNA [116].

There is now good evidence that the 5 kb mtDNA 'common' deletion accumulates with age in human skeletal muscle [103,117], liver [118] and other tissues [119]. The proportion of deleted molecules varied between tissues in these studies: diaphragm gave the highest amplification signal, heart and brain were intermediate, and liver gave the lowest signal [119]. Quantification of this deletion in senescent human muscle, however, suggested that less than 1 in 5000 mtDNA molecules were affected [103]. Such a low level of this mutation alone is unlikely to result in the decline in respiratory function observed in this tissue. However, this does not exclude the possibility that other mtDNA deletions or point mutations may develop and contribute to the defects. To support this proposition are recent observations that a 7.4 kb mtDNA deletion increases in human heart with age [120], a 3.6 kb deletion increases in skeletal muscle [121] and multiple mtDNA deletions are found in aged human diaphragm [122].

The corollary of an increasing mtDNA mutation load must be a qualitative or quantitative change in encoded products. Concentrations of mitochondrial RNA (ribosomal and messenger) have been shown to decline with age in rat brain and heart [123] and this seems to be due to a decrease of approximately 50% in transcription rate rather than a fall in mtDNA levels [124]. Indeed it has been suggested that mtDNA levels may rise with age [125]. It was proposed that this decrease in transcription may be related to falling ATP levels which in turn were caused by oxidative stress-induced increased membrane permeability. It is of interest that the reduction in mtDNA transcription rate has been reported to be reversed by pretreatment with acetyl-L-carnitine [123].

APOPTOSIS

Cell death occurs by necrosis or apoptosis. The latter refers to programmed cell death and has been observed in a wide range of multicellular organisms including several human tissues. Apoptosis plays an important role in development. This is particularly so in the nervous system [126] where the clearance of excess neurons is regulated by synaptic connections, and the supply of nerve growth factor. Whether this reliance upon external factors of cell survival extends beyond development is unclear.

Apoptosis requires protein synthesis and expression of endonucleases that cut the cell's DNA into fragments of 185 kDa. Multiples of this fragment produce a ladder-like appearance on agarose gel electrophoresis which is characteristic of apoptosis. Whether the signal for cell suicide is a positive phenomenon or the result of withdrawal of survival factors is not known. Studies in the nematode *Caenorhabditis elegans* have identified two genes, *ced-3* and *ced-4*, the expression of which are necessary for cell death [127]. A third gene, *ced-9*, blocks the action of these genes and prevents apoptosis [128]. The human counterpart of *ced-9* is the Bcl-2 proto-oncogene, overexpression of which prevents or delays cell death in lymphocytes [129,130] and cultured sympathetic neurons [131]. Bcl-2 is a 25 kDa protein with a unique sequence that has no substantial homology to other proto-oncogene products and has properties of an integral membrane protein. Reports on the intracellular localization of Bcl-2 are conflicting; some have suggested that it is an inner mitochondrial membrane protein [132,133] whilst others have suggested the protein resides on the outer mitochondrial and perinuclear membranes [134]. This discrepancy may reflect the different cells studied, the wider expression being found in lymphoma and breast carcinoma cells. The mechanisms by which Bcl-2 may prevent apoptosis are not known.

The recent description of MPP$^+$-induced apoptosis in cerebellar granule cells [135] may, if reproduced, have important implications for our understanding not only of the mechanism of action of MPP$^+$ but also for the role that programmed cell death may have in neurodegenerative disorders such as PD. It is tempting to speculate on some relationship between Bcl-2 action and respiratory chain function that might be affected by MPP$^+$. Present evidence suggests that mtDNA does not undergo apoptosis [136] and so would not play a role in this process. This is supported by the recent observation that Bcl-2 can protect ρ^o cells from apoptosis [137]. The fact that these cells do not have a functioning respiratory chain also suggests that Bcl-2 action is not mediated via an effect upon mitochondrial respiration.

REFERENCES

1. Davis, G.C., Williams, A.C., Markey, S.P., Ebert, M.H., Caine, E.D., Reichert, C.M. and Kopin, I.J. (1979) Chronic parkinsonism secondary to intravenous injection of mepiridine analogues. *Psychiatry Research*, **1**, 649–654

2. Langston, J.W., Ballard, P., Tetrud, J.W. and Irwin, I. (1983) Chronic parkinsonism in humans due to a product of mepiridine analog synthesis. *Science*, **219**, 979–980

3. Johannessen, J.N., Adams, J.D., Schuller, H.M., Bacon, J.P. and Markey, S.P. (1986) 1-Methyl-4-phenylpyridine induces oxidative stress in the rodent. *Life Sciences*, **38**, 743–749

4. Chacón, J.N., Chedekel, M.R., Land, E.J. and Truscott, T.G. (1989) Chemically induced Parkinson's disease II: Intermediates in the oxidation, and reduction reactions of the 1-methyl-4-phenyl-2,3-dihydropyridine ion, and its deprotonated form. *Biochemical and Biophysical Research Communications*, **158**, 63–71

5. Frank, D.M., Arora, P.K., Bloomer, J.L. and Sayre, L.M. (1987) Model study on the bioreduction of paraquat, MPP+, and analogues. Evidence against a redox cycling mechanism in MPTP neurotoxicity. *Biochemical and Biophysical Research Communications*, **147**, 1095–1104

6. Poirier, J., Donaldson, J. and Barbeau, A. (1985) The specific vulnerability of the substantia nigra to MPTP is related to the presence of transition metals. *Biochemical and Biophysical Research Communications*, **128**, 25–33

7. Sinha, B.K., Singh, Y. and Krishna, G. (1986) Formation of superoxide, and hydroxyl radicals from 1-methyl-4-phenylpyridinium ion (MPP+): reductive activation by NADPH cytochrome P-450 reductase. *Biochemical and Biophysical Research Communications*, **135**, 583–588

8. Chacón, J.N., Chedekel, M.R., Land, E.J. and Truscott, T.G. (1987) Chemically induced Parkinson's disease: intermediates in the oxidation of 1-methyl-4-phenyl-1,2,3,6-tetrahydropyridine to the 1-methyl-4-phenylpyridinium ion. *Biochemical and Biophysical Research Communications*, **144**, 957–964

9. Salach, J.I., Singer, T.P., Castagnoli, N. and Trevor, A. (1984) Oxidation of the neurotoxic amine MPTP by monoamine oxidases A and B and suicide inactivation of the enzymes by MPTP. *Biochemical and Biophysical Research Communications*, **125**, 831–835

10. Javitch, J.A., D'Amato, R.J., Strittmatter, S.M. and Snyder, S.H. (1985) Parkinsonism-inducing MPTP: uptake of the metabolite MPP+ by dopamine neurones explains selective toxicity. *Proceedings of the National Academy of Sciences of the USA*, **82**, 2173–2177

11. Ramsay, R.R. and Singer, T.P. (1986) Energy-dependent uptake of N-methyl-4-phenylpyridinium, the neurotoxic metabolite of 1-methyl-4-phenyl-1,2,3,6-tetrahydropyridine, by mitochondria. *Journal of Biological Chemistry*, **261**, 7585–7587

12. Nicklas, W.J., Vyas, I. and Heikkila, R.E. (1985) Inhibition of NADH-linked oxidation in brain mitochondria by MPP+, a metabolite of the neurotoxin MPTP. *LIfe Sciences*, **36**, 2503–2508

13. Mizuno, Y., Suzuki, K., Sone, N. and Saitoh, T. (1988) Inhibition of mitochondrial respiration by MPTP in mouse brain *in vivo*. *Neuroscience Letters*, **91**, 349–353

14. Di Monte, D. and Smith, T.M. (1988) Free radicals, lipid peroxidation and 1-methyl-4-phenyl-1,2,3,6-tetrahydropyridine (MPTP)-induced parkinsonism. *Review of Neurosciences*, **2**, 67–81

15. Cleeter, M.J.W., Cooper, J.M. and Schapira, A.H.V. (1992) Irreversible inhibition of mitochondrial complex I by 1-methyl-4-phenylpyridinium: evidence for free radical involvement. *Journal of Neurochemistry*, **58**, 786–789

16. Ramsay, R.R., Krueger, M.J., Youngster, S.K., Gluck, M.R., Casida, J.E. and Singer, T.P. (1991) Interaction of MPP+ and its analogs with the rotenone/piericidin binding site of NADH dehydrogenase. *Journal of Neurochemistry*, **51**, 1184–1190

17. Earley, F.G.P., Patel, S.D., Ragan, C.I. and Attardi, G. (1987) Photolabelling of a mitochondrially encoded subunit of NADH dehydrogenase with [3H]dihydrorotenone. *FEBS Letters*, **219**, 108–113

18. Mann, V.M., Cooper, J.M., Krige, D., Daniel, S.E., Schapira, A.H.V. and Marsden, C.D. (1992) Brain, skeletal muscle and platelet homogenate mitochondrial function in Parkinson's disease. *Brain*, **115**, 333–342

19. Schapira, A.H.V., Mann, V.M., Cooper, J.M., Dexter, D., Daniel, S.E., Jenner, P., Clark, J.B. and Marsden, C.D. (1990) Anatomic and disease specificity of NADH CoQ reductase (complex I) deficiency in Parkinson's disease. *Journal of Neurochemistry*, **55**, 2142–2145

20. Mizuno, Y., Suzuki, K. and Ohta, S. (1990) Postmortem changes in mitochondrial respiratory enzymes in brain and a preliminary observation in Parkinson's disease. *Journal of the Neurological Sciences*, **96**, 49–57

21. Dagani, F., Ferrari, R., Anderson, J.J. and Chase, T.N. (1991) L-Dopa does not affect electron transfer chain enzymes and respiration of rat muscle mitochondria. *Movement Disorders*, **6**, 315–319

22. Hattori, N.B., Tanaka, M., Ozawa, T. and Mizuno, Y. (1991) Immunohistochemical studies on complexes I, II, III and IV of mitochondria in Parkinson's disease. *Annals of Neurology*, **30**, 563–571

23. Mizuno, Y., Ohta, S., Tanaka, M., Takamiya, S., Suziki, K. and Sato, T. (1989) Deficiencies in complex I subunits of the respiratory chain in Parkinson's disease. *Biochemical and Biophysical Research Communications*, **163**, 1450–1455

24. Schapira, A.H.V., Cooper, J.M., Dexter, D., Clark, J.B., Jenner, P. and Marsden, C.D. (1990) Mitochondrial complex I deficiency in Parkinson's disease. *Journal of Neurochemistry*, **54**, 823–827
25. Schapira, A.H.V., Holt, I.J., Sweeney, M., Harding, A.E., Jenner, P. and Marsden, C.D. (1990) Mitochondrial DNA analysis in Parkinson's disease. *Movement Disorders*, **5**, 294–297
26. Lestienne, P., Nelson, J., Riederer, P., Jellinger, K. and Reichmann, H. (1990) Normal mitochondrial genome in brain from patients with Parkinson's disease and complex I defect. *Journal of Neurochemistry*, **55**, 1810–1812
27. Ozawa, T., Tanaka, M., Ikebe, S., Ohno, K., Kondo, T. and Mizuno, Y. (1990) Quantitative determination of deleted mitochondrial DNA relative to normal DNA in parkinsonian striatum by a kinetic PCR analysis. *Biochemical Biophysical Research Communications*, **172**, 483–489
28. Mann, V.M., Cooper, J.M. and Schapira, A.H.V. (1992) Quantitation of a mitochondrial DNA deletion in Parkinson's disease. *FEBS Letters*, **299**, 218–222
29. Ozawa, T., Tanaka, M., Ino, H., Ohno, K., Sano, T., Wada, Y., Yoneda, M., Tanno, Y., Miyatake, T., Tanaka, T., Itoyama, S., Ikebe, S., Hattori, N. and Mizuno, Y. (1991) Distinct clustering of point mutations in mitochondrial DNA among patients with mitochondrial encephalomyopathies and with Parkinson's disease. *Biochemical and Biophysical Research Communications*, **176**, 938–946
30. Bindoff, L.A., Birch-Machin, M., Cartlidge, N.E., Parker Jr, W.D. and Turnbull, D.M. (1989) Mitochondrial function in Parkinson's disease. *Lancet*, **i**, 49
31. Wallace, D.C., Shoffner, J.M., Watts, R.L., Juncos, J.L. and Torrino, A. (1992) Mitochondrial oxidative phosphorylation defects in Parkinson's disease. *Annals of Neurology*, **32**, 113–114
32. Blin, O., Desnuelle, C., Rascol, O., Borg, M., Peyro, H., Azulay, J.P., Billé, F., Figarella, D., Pontier, F., Pellissier, J.F., Montractruc, J.L. and Serratrice, G. (1992) Mitochondrial respiratory failure in skeletal muscle from 26 patients with PD. *Movement Disorders*, **7** (Suppl. 1), 72
33. Borg, M., Desnuelle, C. and Ohatel, M. (1992) Respiratory chain defects in skeletal muscle from de novo patients with PD. *Journal of Neurology*, **239** (Suppl.), S87
34. Hattori, Y.N., Yoshino, H., Kondo, T., Mizuno, Y. and Horai, S. (1992) Is Parkinsons's disease a mitochondrial disorder? *Journal of the Neurological Sciences*, **10**, 29–33
35. Di Monte, D., Tetrud, J.W. and Langston, J.W. (1991) Blood lactate in Parkinson's disease. *Annals of Neurology*, **29**, 342–343
36. Bravi, D., Anderson, J.J., Dagani, F.D., Davis, T.L., Ferrari, R., Gillespie, M. and Chase, T.N. (1992) Effect of aging and dopaminomimetic therapy on mitochondrial respiratory function in Parkinson's disease. *Movement Disorders*, **3**, 228–231
37. Parker, W.D., Boyson, S.J. and Parks, J.K. (1989) Abnormalities of the electron transport chain in idiopathic Parkinson's disease. *Annals of Neurology*, **26**, 719–723
38. Krige, D., Carroll, M.T., Cooper, J.M., Marsden, C.D. and Schapira, A.H.V. (1992) Platelet mitochondrial function in Parkinson's disease. *Annals of Neurology*, **32**, 782–788
39. Da Prada, M., Cesura, A.M., Launay, J.M. and Richards, J.G. (1988) Platelets as a model for neurones? *Experientia*, **44**, 115–126
40. Dexter, D.T., Wells, F.R., Agid, Y., Lees, A.J., Jenner, P. and Marsden, C.D. (1987) Increased nigral iron content in post-mortem parkinsonian brain. *Lancet*, **ii**, 1219–1220
41. Reiderer, P., Sofic, E., Rausch, W.D., Schmidt, B., Reynolds, G.P., Jellinger, K. and Youdim, M.B.H. (1989) Transition metals, ferritin, glutathione and ascorbic acid in parkinsonian brains. *Journal of Neurochemistry*, **52**, 515–520
42. Sofic, E., Paulus, W., Jellinger, K., Riederer, P. and Youdim, M.B.H. (1991) Selective increase of iron in substantia nigra of patients with Parkinson's disease. *Journal of Neurochemistry*, **56**, 978–982
43. Dexter, D.T., Jenner, P., Schapira, A.H.V. and Marsden, C.D. (1992) Alterations in levels of iron, ferritin, and other trace metals in neurodegenerative diseases affecting the basal ganglia. *Annals of Neurology*, **32**, S94–S100
44. Dexter, D.T., Carayon, A., Vidailhet, M., Ruberg, M., Agid, F., Agid, Y., Lees, A.J., Wells, F.R., Jenner, P. and Marsden, C.D. (1990) Decreased ferritin levels in brain in Parkinson's disease. *Journal of Neurochemistry*, **55**, 16–20
45. Jellinger, K., Kienzl, E., Rumpelmair, G., Riederer, P., Stachelberger, H., Ben-Shacher, D. and Youdim, M.B.H. (1992) Iron–melanin complex in substantia nigra of parkinsonian brain: an x-ray microanalysis. *Journal of Neurochemistry*, **59**, 1168–1171
46. Hirsch, E.C., Brandel, J.-P., Galle, P., Javoy-Agid, F. and Agid, Y. (1991) Iron and aluminium increase in the substantia nigra of patients with Parkinson's disease: an x-ray microanalysis. *Journal of Neurochemistry*, **56**, 446–451
47. Kastner, A., Hirsch, E.C., Lejeune, O., Javoy-Agid, F., Rascol, O. and Agid, Y. (1992) Is the vulnerability of neurons in the substantia nigra of patients with Parkinson's disease related to their neuromelanin content? *Journal of Neurochemistry*, **59**, 1080–1089

48. Michel, P.M., Vyas, S. and Agid, Y. (1992) Toxic effects of iron for cultured mesencephalic dopaminergic neurons derived from rat embryonic brains. *Journal of Neurochemistry*, **59**, 118–127

49. Dexter, D.T., Carter, C.J., Wells, F.R., Javoy-Agid, F., Agid, Y., Lees, A., Jenner, P. and Marsden, C.D. (1989) Basal lipid peroxidation in substantia nigra is increased in Parkinson's disease. *Journal of Neurochemistry*, **52**, 381–389

50. Ambani, L.M., Van Woert, M.H. and Murphy, S. (1975) Brain peroxidase and catalase in Parkinson's disease. *Archives of Neurology*, **32**, 114–118

51. Kish, S.J., Morito, C. and Hornykiewicz, O. (1985) Glutathione peroxidase activity in Parkinson's disease. *Neuroscience Letters*, **58**, 343–346

52. Perry, T.L., Godin, D.V. and Hansen, S. (1982) Parkinson's disease: a disorder due to nigral glutathione deficiency. *Neuroscience Letters*, **33**, 305–310

53. Saggu, H., Cooksey, J., Dexter, D., Wells, F.R., Lees, A., Jenner, P. and Marsden, C.D. (1989) A selective increase in particulate superoxide dismutase activity in Parkinson's substantia nigra. *Journal of Neurochemistry*, **53**, 692–697

54. Dexter, D.T., Ward, R.J., Wells, F.R., Daniel, S.E., Lees, A.J., Peters, T.J., Jenner, P. and Marsden, C.D. (1992) α-Tocopherol levels in brain are not altered in Parkinson's disease. *Annals of Neurology*, **32**, 591–593

55. Boveris, A., Oshino, N. and Chance, B. (1972) The cellular production of hydrogen peroxide. *Biochemical Journal*, **128**, 617–630

56. Boveris, A. and Chance, B. (1973) The mitochondrial generation of hydrogen peroxide. *Biochemical Journal*, **134**, 707–716

57. Nohl, H. and Jordan, W. (1986) The mitochondrial site of superoxide formation. *Biochemical and Biophysical Research Communications*, **138**, 533–539

58. Cadenas, E., Boveris, A., Ragan, C.I. and Stoppani, A.O.M. (1977) Production of superoxide radicals and hydrogen peroxide by NADH–ubiquinone reductase and ubiquinol–cytochrome *c* reductase from beef-heart mitochondria. *Archives of Biochemistry and Biophysics*, **180**, 248–257

59. Turrens, J.F. and Boveris, A. (1980) Generation of superoxide anion by the NADH dehydrogenase of bovine heart mitochondria. *Biochemical Journal*, **191**, 421–427

60. Hasegawa, E., Takeshige, K., Oishi, T., Murai, Y. and Minikami, S. (1990) 1-Methyl-4-phenylpyridinium (MPP$^+$) induces NADH dependent superoxide formation, and enhances NADH-dependent lipid peroxidation in bovine heart submitochondrial particles. *Biochemical and Biophysical Research Communications*, **170**, 1049–1055

61. Hillered, L. and Ernster, L. (1983) Respiratory activity of isolated rat brain mitochondria following in vitro exposure to oxygen radicals. *Journal of Cerebral Blood Flow and Metabolism*, **3**, 207–214

62. Zhang, Y., Marcillat, O., Giulivi, C., Ernster, I. and Davies, K.J. (1990) The oxidative inactivation of mitochondrial electron transport chain components and ATP. *Journal of Biological Chemistry*, **265**, 16330–16336

63. Benzi, G., Curti, D., Pastoris, O., Marzatico, F., Villa, R.F. and Dagani, F. (1991) Sequential damage in mitochondrial complexes by peroxidative stress. *Neurochemistry Research*, **16**, 1295–1302

64. Gibb, W.R.G. and Lees, A.J. (1988) The relevance of the Lewy body in the pathogenesis of idiopathic Parkinson's disease. *Journal of Neurology, Neurosurgery and Psychiatry*, **51**, 745–752

65. Fearnley, J.M. and Lees, A.J. (1991) Ageing and Parkinson's disease: substantia nigra regional selectivity. *Brain*, **114**, 2283–2301

66. Mithöfer, K., Sandy, M.S., Smith, M.T. and Di Monte, D. (1992) Mitochondrial poisons cause depletion of reduced glutathione in isolated hepatocytes. *Archives of Biochemistry and Biophysics*, **295**, 132–136

67. Golbe, L.I., Di Iorio, G., Bonavita, V., Miller, D.C. and Duvoisin, R.C. (1990) A large kindred with autosomal dominant Parkinson's disease. *Annals of Neurology*, **27**, 276–282

68. Maraganore, D.M., Harding, A.E. and Marsden, C.D. (1991) A clinical and genetic study of familial Parkinson's disease. *Movement Disorders*, **6**, 205–211

69. Ward, C.D., Duvoisin, R.C., Ince, S.E., Nutt, J.D., Eldridge, R. and Calne, D.B. (1983) Parkinson's disease in 65 pairs of twins and in a set of quadruplets. *Neurology*, **33**, 815–824

70. Marsden, C.D. (1987) Parkinson's disease in twins. *Journal of Neurology, Neurosurgery and Psychiatry*, **50**, 105–106

71. Martilla, R.J., Kaprio, J., Koskenvuo, M.D. and Rinne, U.K. (1988) Parkinson's disease in a nationwide twin cohort. *Neurology*, **38**, 1217–1219

72. Burn, D.J., Mark, M.H., Playford, E.D. *et al.* (1992) Parkinson's disease in twins studied with ^{18}F-dopa and positron emission tomography. *Neurology*, **42**, 1894–1900

73. Smith, C.A.D., Gough, A.C., Leigh, P.N. *et al.* (1992) Debrisoquine hydroxylase gene polymorphism and susceptibility to Parkinson's disease. *Lancet*, **339**, 1375–1377

74. Armstrong, M., Daly, A.K., Cholerton, S., Bateman, D.N. and Idle, J.R. (1992) Mutant debriso-quine hydroxylation genes in Parkinson's disease. *Lancet*, **339**, 1017–1018
75. Steventon, G.B., Heafield, M.T.E., Waring, R.H. and Williams, A.C. (1989) Xenobiotic metabolism in Parkinson's disease. *Neurology*, **39**, 883–887
76. Green, S., Buttrum, S., Molloy, H. *et al.* (1991) *N*-Methylation of pyridines in Parkinson's disease. *Lancet*, **338**, 120–121
77. Rajput, A.H., Uitti, R.J., Stern, W. and Laverty, W. (1986) Early onset Parkinson's disease and childhood environment. *Advances in Neurology*, **45**, 295–297
78. Tanner, C. (1985) Well-water drinking and Parkinson's disease. *Proceedings of the VIII International Symposium on Parkinson's Disease*, New York, 11
79. Jiménez-Jiménez, F.J., Mateo, D. and Giménez-Roldán, S. (1992) Exposure to well water and pesticides in Parkinson's disease: a case-control study in the Madrid areas. *Movement Disorders*, **7**, 149–152
80. Semahuk, K.M., Love, E.J. and Lee, R.G. (1992) Parkinson's disease and exposure to agricultural work and pesticide chemicals. *Neurology*, **42**, 1328–1335
81. Duara, R., Grady, C., Haxby, J.V. *et al.* (1986) Positron emission atrophy in Alzheimer's disease. *Neurology*, **36**, 879–887
82. Haxby, J.V., Grady, C.L., Duara, R., Schlageter, N., Berg, G. and Rapoport, S.I. (1986) Neocortical metabolic abnormalities precede non-memory cognitive deficits in early Alzheimer-type dementia. *Archives of Neurology*, **43**, 882–885
83. Perry, E.K., Perry, R.H. and Tomlinson, B.E. (1980) Coenzyme-A acetylating enzymes in Alzheimer disease: possible cholinergic 'compartment' of pyruvate dehydrogenase. *Neuroscience Letters*, **18**, 105–110
84. Sorbi, S., Bird, E.D. and Blass, J.P. (1983) Decreased pyruvate dehydrogenase complex activity in Huntington and Alzheimer brain. *Annals of Neurology*, **13**, 72–78
85. Kish, S.B., Bergeron, C., Rajput, A. *et al.* (1992) Brain cytochrome oxidase in Alzheimer's disease. *Journal of Neurochemistry*, **59**, 776–779
86. Zubenko, G.S., Cohen, B.M., Reynolds, C.F., Boller, F., Malinakova, I. and Keefe, N. (1987) Platelet membrane fluidity in Alzheimer's disease and major depression. *American Journal of Psychiatry*, **144**, 860–868
87. Zubenko, G.S. (1989) Endoplasmic reticulum abnormality in Alzheimer's disease: selective alteration in platelet NADH–cytochrome *c* reductase activity. *Journal of Geriatric Psychiatry and Neurology*, **2**, 3–10
88. Parker, W.D., Filley, C.M. and Parks, J.K. (1990) Cytochrome oxidase deficiency in Alzheimer's disease. *Neurology*, **40**, 1302–1303
89. Kuhl, D.E., Phelps, M.E., Markham, C.H., Metter, E.J., Riege, W.H. and Winter, J. (1982) Cerebral metabolism and atrophy in Huntington's disease determined by [18]FDG and computed tomographic scan. *Annals of Neurology*, **12**, 425–434
90. Brennan, W.A., Bird, E.D. and Aprille, J.R. (1985) Regional mitochondrial respiratory activity in Huntington's disease brain. *Journal of Neurochemistry*, **44**, 1948–1950
91. Mann, V.M., Cooper, M.J., Javoy-Agid, Y., Jenner, P. and Schapira, A.H.V. (1990) Mitochondrial function and parental sex effect in Huntington's disease. *Lancet*, **336**, 749
92. Irwin, C.C., Wexler, N.S., Young, A.B. *et al.* (1989) The role of mitochondrial DNA in Huntington's disease. *Journal of Molecular Neuroscience*, **1**, 129–136
93. Reik, W. (1988) Genomic imprinting: a possible mechanism for the parental origin effect in Huntington's disease. *Journal of Medical Genetics*, **25**, 805–808
94. Parker, W.D., Boyson, S.J., Luder, A.S. and Parks, J.K. (1990) Evidence for a defect in NADH: ubiquinone oxidoreductase (complex I) in Huntington's disease. *Neurology*, **40**, 1231–1234
95. Feldman, M.L. and Navaratnam, V. (1981) Ultrastructural changes in atrial myocardium of the ageing rat. *Journal of Anatomy*, **133**, 7–17
96. Frenzel, H. and Fiemann, J. (1984) Age dependent structural changes in the myocardium of rats. A quantitive light and electron microscopic study on the right and left chamber wall. *Mechanisms of Ageing and Development*, **27**, 24–41
97. Jennekens, F.G.I., Tomlinson, B.E. and Walton, J.N. (1971) Histochemical aspects of five limb muscles in old age. An autopsy. *Journal of the Neurological Sciences*, **14**, 259–276
98. Müller-Höcker, J. (1989) Cytochrome *c* oxidase deficient cardiomyocytes in the human heart – an age-related phenomenon. *American Journal of Pathology*, **134**, 1167–1173
99. Müller-Höcker, J. (1992) Mitochondria and ageing. *Brain Pathology*, **2**, 149–158
100. Ruggiero, F.M., Cafagna, F., Petruzella, V., Gadaleta, M.N. and Quagliariello, E. (1992) Lipid composition in synaptic and nonsynaptic mitochondria from rat brains and effect of aging. *Journal of Neurochemistry*, **59**, 487–491

101. Paradies, G., Ruggiero, F.M., Gadaleta, M.N. and Quagliariello, E. (1992) The effect of aging and acetyl-L-carnitine on the activity of the phosphate carrier and on the phospholipid composition in rat heart mitochondria. *Biochimica et Biophysica Acta*, **1103**, 324–326
102. Trounce, I., Byrne, E. and Marzuki, S. (1989) Decline in skeletal muscle mitochondrial respiratory chain function, possible factor in ageing. *Lancet*, **i**, 637–639
103. Cooper, J.M., Mann, V.M. and Schapira, A.H.V. (1992) Analyses of mitochondrial respiratory chain function and mitochondrial DNA deletion in human skeletal muscle: effect of ageing. *Journal of the Neurological Sciences*, **113**, 91–98
104. Cardellach, F., Galofre, J., Cusso, R. and Urbano-Marquez, A. (1989) Decline in skeletal muscle mitochondrial respiratory chain function with ageing. *Lancet*, **ii**, 44–45
105. Yen, T.C., Chen, K.L., King, S.H., Yeh, S.H. and Wei, Y.H. (1989) Liver mitochondrial respiratory functions decline with age. *Biochemical and Biophysical Research Communications*, **165**, 994–1003
106. Miquel, J., Economos, A.C., Fleming, J. and Johnson, Jr, J.E. (1980) Mitochondrial role in cell ageing. *Experimental Gerontology*, **15**, 575–591
107. Linnane, A., Marzuki, S., Ozawa, T. and Tanaka, M. (1989) Mitochondrial DNA mutations as an important contributor to ageing and degenerative diseases. *Lancet*, **i**, 642–645
108. Harman, D. (1972) The biologic clock: the mitochondria. *Journal of the American Geriatric Society*, **20**, 145–147
109. Farmer, K.J. and Sohal, R.S. (1989) Relationship between superoxide anion radical generation and aging in the housefly, *Musca domestica*. *Free Radical Biology of Medicine*, **7**, 23–29
110. Sohal, R.S. and Allen, R.G. (1990) Oxidative stress as a causal factor in differentiation and ageing: a unifying hypothesis. *Experimental Gerontology*, **25**, 499–522
111. Caron, F., Jacq, C. and Rouviere-Yaniv, J. (1979) Characterisation of a histone-like protein extracted from yeast mitochondria. *Proceedings of the National Academy of Sciences of the USA*, **76**, 4265–4269
112. Clayton, D.A. (1982) Replication of animal mitochondrial DNA. *Cell*, **28**, 693–705
113. Backer, J.M. and Weistein, I.B. (1980) Mitochondrial DNA is a major cellular target for a dihydrodiol-epoxide derivative of benzo(a)pyrene. *Science*, **209**, 297–299
114. Niranjan, B.G., Bhat, N.K. and Avadhani, N.G. (1981) Preferential attack of mitochondrial DNA by aflotoxin B1 during hepatocarcinogenesis. *Science*, **215**, 73–75
115. Piko, L., Hougham, A.J. and Bulpitt, K.J. (1988) Studies of sequence heterogeneity of mitochondrial DNA from rat and mouse tissue: evidence for an increased frequency of deletion/additions with age. *Mechanisms of Age and Development*, **43**, 279–293
116. Ames, B.N. (1989) Endogenous DNA damage as related to cancer and aging. *Mutation Research*, **214**, 41–46
117. Cortopassi, G.A. and Arnheim, N. (1990) Detection of a specific mtDNA deletion in tissues of older humans. *Nucleic Acid Research*, **8**, 6927–6933
118. Yen, T.C., Su, J.H., King, K.L. and Wei, Y.H. (1991) Age associated 5kb deletion in human liver. *Biochemical and Biophysical Research Communications*, **178**, 124–131
119. Cortopassi, G.A., Shibata, D., Soong, N.-W. and Arnheim, N. (1992) A pattern of accumulation of a somatic deletion of mitochondrial DNA in aging human tissues. *Proceedings of the National Academy of Sciences of the USA*, **89**, 7370–7374
120. Hattori, K., Tanaka, M., Sugiyama, S. *et al.* (1991) Age-dependent increase in deleted mitochondrial DNA in the human heart: possible contributory factor to presbycardia. *American Heart Journal*, **121**, 1735–1742
121. Katayama, M., Tanaka, M., Yamamoto, H., Ohbayashi, T., Nimura, Y. and Ozawa, T. (1991) Deleted mitochondrial DNA in the skeletal muscle of aged individuals. *Biochemistry International*, **25**, 47–56
122. Torii, K., Sugiyama, S., Tanaka, M. *et al.* (1992) Aging-associated deletions of human diaphragmatic mitochondrial DNA. *American Journal of Respiratory Cell Molecular Biology*, **6**, 543–549
123. Gadaleta, M.N., Petruzzella, V., Renis, M., Fracasso, F. and Cantatore, P. (1990) Reduced transcription of mitochondrial DNA in the senescent rat. Tissue dependence and effect of acetyl-L-carnitine. *Euopean Journal of Biochemistry*, **187**, 501–506
124. Fernandez-Silva, P., Petruzzella, V., Fracasso, F., Gadaleta, M.N. and Cantatore, P. (1991) Reduced synthesis of mtRNA in isolated mitochondria of senescent rat brain. *Biochemical and Biophysical Research Communications*, **176**, 645–653
125. Gadaleta, M.N., Rainaldi, G., Lezza, A.M.S., Milella, F., Fracasso, F. and Cantatore, P. (1992) Mitochondrial DNA copy number and mitochondrial DNA deletion in adult and senescent rats. *Mutation Research*, **275**, 181–193
126. Oppenheim, R.W. (1991) Cell death during development of the nervous system. *Annual Review of Neuroscience*, **14**, 453–501

127. Yuan, J.Y. and Horvitz, H.R. (1990) The *Caenorhabditis elegans* genes ced-3 and ced-4 act cell autonomously to cause programmed cell death. *Developmental Biology*, **138**, 33–41
128. Hengartner, M.O., Ellis, R.E. and Horvitz, H.R. (1992) *Caenorhabditis elegans* gene ced-9 protects cells from programmed cell death. *Nature*, **356**, 494–499
129. Vaux, D.L., Cory, S. and Adams, J.M. (1988) Bcl-2 gene promotes haemopoietic cell survival and cooperates with c-*myc* to immortalize pre-B cells. *Nature*, **335**, 440–442
130. Sentman, C.L., Shutter, J.R., Hockenbery, D., Kanagawa, O. and Korsmeyer, S.J. (1991) Bcl-2 inhibits multiple forms of apoptosis but not negative selection of thymocytes. *Cell*, **67**, 879–888
131. Garcia, I., Martinou, I., Tsujimoto, Y. and Martinou, J.-C. (1992) Prevention of programmed cell death of sympathetic neurons by the bcl-2 proto-oncogene. *Science*, **258**, 302–304
132. Liu, Y.-J., Mason, D.Y., Johnson, G.D. *et al.* (1991) Germinal center cells express bcl-2 protein after activation by signals which prevent protein that blocks apoptosis. *European Journal of Immunology*, **21**, 1905–1910
133. Hockenbery, D., Nuñez, G., Milliman, C., Schreiber, R.D. and Korsmeyer, S.J. Bcl-2 is an inner mitochondrial membrane protein that blocks programmed cell death. *Nature*, **348**, 334–336
134. Monaghan, P., Robertson, D., Amos, T.A.S., Dyer, M.J.S., Mason, D.Y. and Greaves, M.F. Ultrastructural localization of Bcl-2 protein. *Journal of Histochemistry and Cytochemistry*, **40**, 1819–1825
135. Dipasquale, B., Marini, A.M. and Youie, R.J. (1991) Apoptosis and DNA degradation induced by 1-methyl-4-phenylpyridinium in neurons. *Biochemical and Biophysical Research Communications*, **181**, 1442–1448
136. Murgia, M., Pizzo, P., Sandona, D., Zanovello, P., Rizzuo, R. and Di Virgilio, F. (1992) Mitochondrial DNA is not fragmented during apoptosis. *Journal of Biological Chemistry*, **267**, 10939–10941
137. Jacobson, M.D., Burne, J.F., King, M.P., Miyashita, T., Reed, J.C. and Raff, M.C. (1993) Bcl-2 blocks apoptosis in cells lacking mitochondrial DNA. *Nature*, **361**, 365–369

Index